Wilson Disease

Wilson Disease
Pathogenesis, Molecular Mechanisms, Diagnosis, Treatment and Monitoring

Edited By

Karl Heinz Weiss

Michael Schilsky

Academic Press is an imprint of Elsevier
125 London Wall, London EC2Y 5AS, United Kingdom
525 B Street, Suite 1650, San Diego, CA 92101, United States
50 Hampshire Street, 5th Floor, Cambridge, MA 02139, United States
The Boulevard, Langford Lane, Kidlington, Oxford OX5 1GB, United Kingdom

© 2019 Elsevier Inc. All rights reserved.

No part of this publication may be reproduced or transmitted in any form or by any means, electronic or mechanical, including photocopying, recording, or any information storage and retrieval system, without permission in writing from the publisher. Details on how to seek permission, further information about the Publisher's permissions policies and our arrangements with organizations such as the Copyright Clearance Center and the Copyright Licensing Agency, can be found at our website: www.elsevier.com/permissions.

This book and the individual contributions contained in it are protected under copyright by the Publisher (other than as may be noted herein).

Notices
Knowledge and best practice in this field are constantly changing. As new research and experience broaden our understanding, changes in research methods, professional practices, or medical treatment may become necessary.

Practitioners and researchers must always rely on their own experience and knowledge in evaluating and using any information, methods, compounds, or experiments described herein. In using such information or methods they should be mindful of their own safety and the safety of others, including parties for whom they have a professional responsibility.

To the fullest extent of the law, neither the Publisher nor the authors, contributors, or editors, assume any liability for any injury and/or damage to persons or property as a matter of products liability, negligence or otherwise, or from any use or operation of any methods, products, instructions, or ideas contained in the material herein.

Library of Congress Cataloging-in-Publication Data
A catalog record for this book is available from the Library of Congress

British Library Cataloguing-in-Publication Data
A catalogue record for this book is available from the British Library

ISBN 978-0-12-811077-5

For information on all Academic Press publications
visit our website at https://www.elsevier.com/books-and-journals

Publisher: Stacy Masucci
Acquisition Editor: Stacy Masucci
Editorial Project Manager: Samuel Young
Production Project Manager: Maria Bernard
Designer: Greg Harris

Typeset by SPi Global, India

Contents

Contributors ... xi
Foreword ... xiii

Part I
History of Wilson Disease

1. History of Wilson Disease

James S. Dooley and Rupert Purchase

Introduction ... 3
The Definition of Wilson Disease ... 3
Description of Kayser-Fleischer Rings; Unity of Neurological Definition; Resolving Pseudosclerosis; Liver Disease ... 5
  Corneal Ring ... 5
  Pseudosclerosis ... 5
  Liver Disease ... 6
Toxin ... Copper ... 6
Evolution of Diagnostic Tests for Wilson Disease ... 6
  Kayser-Fleischer Rings ... 6
  Tissue Copper ... 7
  Serum Ceruloplasmin ... 7
Copper as a Therapeutic Target: The Discovery of Drugs for the Treatment of Wilson Disease ... 7
  British Anti-Lewisite ... 8
  D-Penicillamine ... 8
  Trientine (Triethylenetetramine) ... 8
  Zinc Salts ... 8
  Ammonium Tetrathiomolybdate ... 9
Liver Transplantation ... 9
Molecular Genetics ... 9
  Cloning of the Wilson Disease Gene *ATP7B* ... 10
  Function of ATP7B ... 10
  Mutation Analysis ... 10
Future Directions ... 10
Conclusion ... 10
Further Reading on the History of Wilson Disease ... 11
References ... 11

Part II
Molecular Mechanisms

2. Normal Human Copper Metabolism

Cynthia Abou Zeid and Stephen G. Kaler

Introduction ... 17
Copper in Human Health ... 17
Dietary Copper ... 18
Copper Absorption ... 18
Copper Transport and Excretion ... 20
Conclusion ... 20
References ... 21

3. ATP7B Function

Hannah Pierson and Svetlana Lutsenko

The Biochemical Function of ATP7B ... 23
Cellular Factors That Affect ATP7B Function ... 25
  Intracellular Trafficking Allows ATP7B to Switch Between the Biosynthetic and Homeostatic Functions ... 25
  Glutaredoxin 1 Interacts With ATP7B and Regulates ATP7B Trafficking ... 25
  Atox1 ... 26
  Kinase-Mediated Phosphorylation ... 27
Tissue Specific Functions of ATP7B ... 27
  ATP7B Role in the Liver ... 27
  ATP7B in Intestine ... 28
  ATP7B in the Brain and Other Tissues ... 29
Conclusions ... 29
Acknowledgment ... 29
References ... 29

4. Biochemical and Cellular Properties of ATP7B Variants

Samuel Jayakanthan, Courtney McCann and Svetlana Lutsenko

Introduction	33
Molecular Features of ATP7B and Catalytic Cycle	34
Physical Properties of ATP7B Variants	35
Wilson Disease Mutation in Various Domains	44
Mutations in the N-Terminal Domain Are Rare But Have Significant Consequences	44
Actuator (A-Domain) and ATP Binding Domains (ATPBD) Are Major Targets of WD Mutations	45
The H1069Q Mutant	46
The Mutations Within the TM Domain Revealed Long-Range Interactions Between Different Parts of ATP7B	46
Single Nucleotide Polymorphisms May Play a Larger Role in Disease Than Previously Thought	47
Conclusions	48
References	48

5. Animal Models of Wilson Disease

Dominik Huster

Copper Toxicity and Liver Disease in Rodent Models of Wilson Disease	51
Long-Evans Cinnamon Rat	51
LPP $Atp7b^{-/-}$ Rat	53
Toxic Milk Mouse	53
Toxic Milk Mouse From the Jackson Laboratory (tx-j Mouse)	54
$Atp7b^{-/-}$ Mouse	55
Other Animal Models for Copper Storage Disorders	56
Dogs With Copper-Associated Hepatitis	56
North Ronaldsay Sheep	57
Conclusion and Research Perspectives	57
Acknowledgment	59
References	59

6. Cellular Copper Toxicity: A Critical Appraisal of Fenton-Chemistry-Based Oxidative Stress in Wilson Disease

Hans Zischka and Josef Lichtmannegger

Introduction	66
Oxidative Stress in WD Patients and Related Animal Models	67
Treating Copper Toxicity in WD Rats	71
Conclusions	77
Acknowledgments	78
References	78

Part III
Epidemiology

7. Epidemiology of Wilson Disease

Thomas Damgaard Sandahl and Peter Ott

Introduction	85
Incidence and Prevalence of Wilson Disease	85
The Sardinian Case	85
Prevalence in Other Populations	86
Biochemical Screening for Wilson Disease	87
Population Based Genetic Methods	87
What to Learn From Different Studies of Prevalence?	89
Clinical Presentation	89
Age at Presentation	89
Presenting Phenotype	90
Gender Differences	92
Diagnostic Delay	92
Prognosis of Wilson Disease	92
Acknowledgment	92
References	92

Part IV
Diagnosis

8. The Diagnostic Approach to Wilson Disease

Michelle Angela Camarata and Aftab Ala

Establishing a Diagnosis of Wilson Disease	97
The Leipzig Score and Its Utility in Diagnostic Work Up of a Wilson Diagnosis	97
Differential Diagnosis	97
Clinical Features Suggestive of Wilson Disease (WD) Diagnosis	99
Liver Disease	100
Neurological/Neuropsychiatric Disease	100
Kayser-Fleischer Rings	100
Other Manifestations	100
Acute Hepatitis and Wilson Disease	100
Investigations	101
Ceruloplasmin	101
Nonceruloplasmin Bound Copper (NCC)	101
Urinary Excretion of Copper	101
Concentration of Copper in the Liver	102
Liver Histology	102

Neuroradiology	102
Molecular Genetic Studies	103
Family Screening	103
References	103

9. The Genetics of Wilson Disease

Michelle Angela Camarata and Si Houn Hahn

ATP7B Gene and ATPase	105
Molecular Structure of *ATP7B*	105
ATP7B (P-Type ATPase) Protein Structure	106
Variants in the *ATP7B* Gene	106
Regional Gene Frequency	107
Genotype-Phenotype Correlation	107
Clinical Molecular Diagnosis	109
Population Screening	110
Conclusion	110
References	111

10. Biochemical Markers

Aurélia Poujois and France Woimant

Traditional Biochemical Markers of Wilson Disease	115
Ceruloplasmin	115
Serum Copper and Serum Nonceruloplasmin-Bound Copper (NCC)	117
Urinary Excretion of Copper	118
Liver Biopsy With Copper Determination	118
Exchangeable Copper and REC: New Specific and Sensible Biochemical Markers of Wilson Disease	119
Conclusion	122
References	122

11. Diagnosis of Hepatic Wilson Disease

Palittiya Sintusek, Eirini Kyrana and Anil Dhawan

Introduction	125
Parameters in the Diagnostic Scoring System	125
Ceruloplasmin	125
Serum Copper	129
Urinary Copper	129
Liver Copper	130
Eye Signs	133
Coombs-Negative Hemolytic Anemia	133
Neurological and Psychological Symptoms and Signs	133
Genetic ATP7B Mutation Analysis	134
Proposed Parameters That Facilitate WD Diagnosis	134
AST to ALT Ratio	134
Alkaline Phosphatase (ALP) to Total Bilirubin (TB) Ratio	135
Alkaline Phosphatase (ALP)	136
Others	136
References	136

12. Liver Pathology of Wilson Disease

Hans Peter Dienes and Peter Schirmacher

Pathology of Wilson Disease	139
References	144

13. Neurological Wilson Disease

Tomasz Litwin, Petr Dusek and Anna Członkowska

Introduction	145
Pathophysiology of Neurological WD Symptoms	145
Neurologic Symptoms of WD	147
Tremor	147
Dystonia	148
Parkinsonism	149
Ataxia	149
Chorea	150
Dysarthria	150
Dysphagia	150
Drooling	150
Gait and Posture Disturbances	150
Other, Selected Neurological Symptoms in WD	151
Epilepsy	151
Neuropathies	151
Restless Leg Syndrome	151
Sleep Disturbances	151
Neurologic Symptoms of Hepatic Encephalopathy (HE) in WD	152
Ophthalmologic Signs of WD	152
Neuroimaging in WD	153
Conclusions	154
References	154

14. Psychiatric Symptoms in WD

Paula C. Zimbrean

Introduction	159
Prevalence of Psychiatric Symptoms in WD	159
Mood and Anxiety Disorders	159
Psychotic Disorders	162
Sleep Disturbances	162
Other Psychiatric Symptoms	162
Cognitive Deficits	163
The Diagnosis of Psychiatric Presentations in WD	163

Laboratory Data and Neuroimaging 164
Comment on Genetic Studies 165
Significance of Psychiatric Symptoms in WD 165
Treatment of Psychiatric Symptoms in WD 166
Conclusion 166
References 167

Part V
Treatment Decisions

15. General Considerations and the Need for Liver Transplantation

Michelle Angela Camarata, Karl Heinz Weiss and Michael L. Schilsky

Introduction 173
Liver Transplant in Patients Presenting With ALF as Their Initial Presentation 173
Therapies Used to Bridge Liver Transplant 174
Liver Transplant in Patients With Established WD 175
Liver Transplantation for Neuropsychiatric Manifestations of WD 176
Options for Liver Transplantation in Wilson Disease 176
  Auxiliary Liver Grafts 178
  Living Donor Liver Transplant 178
  Outcomes Following LT 178
  Immunosuppression 179
Conclusion 180
References 180

16. Chelation Therapy: D-Penicillamine

Peter Ferenci

Pharmacology 183
  Drug-Drug Interactions 183
  Mode of Action 183
Use of D-Penicillamine in Wilson Disease 184
Side Effects 184
Efficacy 184
References 185

17. Trientine for Wilson Disease: Contemporary Issues

Eve A. Roberts

Chemistry 187
Pharmacology 188
Therapeutics: Efficacy/Safety 189
Adverse Effects 190
Emerging Applications 190

Social Issues: Real or Effective Nonavailability of Trientine 190
Conclusion 192
References 193

18. Tetrathiomolybdate (TTM)

Christian Rupp and Karl Heinz Weiss

Introduction 197
Development and Early Clinical Studies 197
Bis-Choline TTM (WTX-101) 198
Pharmacokinetics 198
Mode of Action 199
Efficacy of Bis-Choline TTM 199
Safety and Tolerability of Bis-Choline TTM 200
Conclusion 200
References 200

19. Zinc Therapy of Wilson Disease

Roderick H.J. Houwen

Introduction 203
Mechanism 203
Dosage 203
Efficacy 203
Pregnancy 205
Side Effects 205
Overtreatment 205
Combination Therapy 205
Where Do We Go From Here 206
References 206

20. Symptomatic Treatment of Residual Neurological or Psychiatric Disease

Ana Vives-Rodriguez and Daphne Robakis

Neurological Manifestations 209
Parkinsonism 209
Dystonia 210
Treatment Options in Dystonia 210
  Botulinum Toxin 211
  Oral Medications 211
  Surgical Treatment for Dystonia 211
Tremor 211
  Propranolol 212
  Primidone 212
Cerebellar Ataxia 213
Cognitive Impairment 213
Psychiatric Manifestations 213
Conclusion 214
References 214

21. Other Treatment Regimens and Emerging Therapies

Christian Rupp and Karl Heinz Weiss

Response Guided Therapy	217
Other Drugs	217
Acid Sphingomyelinase	217
Curcurmin	218
Chinese Herbal Medicines	218
Plant Decapeptide OSIP108	218
Methanobactin	218
Therapeutic Plasma Exchange (TPE)	218
Mutation Specific Therapy	218
Gene Transfer	219
Hepatocyte/Tissue Transfer	219
References	219

Part VI
Monitoring

22. Monitoring of Medical Therapy and Copper End Points

France Woimant and Aurélia Poujois

The Initial Phase	223
Frequency of the Follow Up	224
Clinical Follow Up	224
Biological Monitoring (Except Cupric Assessment)	225
Copper Monitoring	225
The Maintenance Phase	226
Frequency of Follow Up During the Maintenance Phase	227
Clinical Follow Up	227
Biological Monitoring (Except Cupric Assessment)	228
Copper Monitoring	228
Other Noninvasive Assessment of Wilson Disease	228
Red Flags	229
Conclusions	230
References	230

Part VII
Main Challenges in Diagnosis and Treatment

23. Special Treatment Considerations for Wilson Disease

Michelle Angela Camarata, Karl Heinz Weiss and Michael L. Schilsky

Treatment Choices and Their Influence on the Monitoring of Therapy for Outcome and Safety	235
Hepatic	235
Neurologic	237
Monitoring Copper Parameters on Treatment	239
What Constitutes Treatment Failure?	240
Alternative Reasons for Changing Therapy	241
Earliest Age for Initiating Treatment	242
Treatment in Pregnancy	243
References	243

Index	247

Contributors

Numbers in parenthesis indicate the pages on which the authors' contributions begin.

Cynthia Abou Zeid (17), Molecular Medicine Branch, *Eunice Kennedy Shriver* National Institute of Child Health and Human Development, National Institutes of Health, Bethesda, MD, United States

Aftab Ala (97), Department of Clinical and Experimental Medicine, University of Surrey; Department of Gastroenterology and Hepatology, Royal Surrey County Hospital, Guilford, United Kingdom

Michelle Angela Camarata (97,105,173,235), Department of Gastroenterology and Hepatology, Royal Surrey County Hospital; Department of Clinical and Experimental Medicine, University of Surrey, Guilford, United Kingdom; Department of Surgery, Section of Transplant and Immunology, Yale School of Medicine, New Haven, CT, United States

Anna Członkowska (145), 2nd Department of Neurology, Institute Psychiatry and Neurology, Warsaw, Poland

Anil Dhawan (125), Paediatric Liver, GI and Nutrition Centre, King's College Hospital, London, United Kingdom

Hans Peter Dienes (139), Medical University of Vienna, Institute of Clinical Pathology, Vienna, Austria

James S. Dooley (3), UCL Institute for Liver and Digestive Health, University College London, London, United Kingdom

Petr Dusek (145), Department of Neurology and Center of Clinical Neuroscience, First Faculty of Medicine and General University Hospital, Charles University in Prague, Prague, Czech Republic

Peter Ferenci (183), Internal Medicine 3, Gastroenterology and Hepatology, Medical University of Vienna, Vienna, Austria

Si Houn Hahn (105), Division of Genetic Medicine, Department of Pediatrics, University of Washington School of Medicine, Seattle Children's Hospital, Seattle, WA, United States

Roderick H.J. Houwen (203), Department of Pediatric Gastroenterology, University Medical Center Utrecht, Utrecht, The Netherlands

Dominik Huster (51), Department of Gastroenterology and Oncology, Deaconess Hospital Leipzig, Academic Teaching Hospital, University of Leipzig, Leipzig, Germany

Samuel Jayakanthan (33), Department of Physiology, Johns Hopkins University School of Medicine, Baltimore, MD, United States

Stephen G. Kaler (17), Molecular Medicine Branch, *Eunice Kennedy Shriver* National Institute of Child Health and Human Development, National Institutes of Health, Bethesda, MD, United States

Eirini Kyrana (125), Paediatric Liver, GI and Nutrition Centre, King's College Hospital, London, United Kingdom

Josef Lichtmannegger (65), Institute of Molecular Toxicology and Pharmacology, Helmholtz Center Munich, German Research Center for Environmental Health, Neuherberg, Germany

Tomasz Litwin (145), 2nd Department of Neurology, Institute Psychiatry and Neurology, Warsaw, Poland

Svetlana Lutsenko (23,33), Department of Physiology, Johns Hopkins University School of Medicine, Baltimore, MD, United States

Courtney McCann (33), Department of Physiology, Johns Hopkins University School of Medicine, Baltimore, MD, United States

Peter Ott (85), Department of Hepatology and Gastroenterology, Aarhus University Hospital, Aarhus, Denmark

Hannah Pierson (23), Department of Physiology, Johns Hopkins University School of Medicine, Baltimore, MD, United States

Aurélia Poujois (115,223), French National Reference Centre for Wilson Disease, Neurology Department, Lariboisière Hospital, Assistance Publique-Hôpitaux de Paris, Paris, France

Rupert Purchase (3), Department of Chemistry, School of Life Sciences, University of Sussex, Falmer, Brighton, United Kingdom

Daphne Robakis (209), Department of Neurology, Yale School of Medicine, New Haven, CT, United States

Eve A. Roberts (187), Department of Pharmacology & Toxicology, University of Toronto, Toronto; History of Science and Technology Programme, University of King's College, Halifax; Department of Paediatrics; Department of Medicine, University of Toronto, Toronto, Canada

Christian Rupp (197,217), Internal Medicine IV, University Hospital Heidelberg, Heidelberg, Germany

Thomas Damgaard Sandahl (85), Department of Hepatology and Gastroenterology, Aarhus University Hospital, Aarhus, Denmark

Michael L. Schilsky (173,235), Division of Digestive Diseases and Transplant and Immunology, Department of Medicine and Surgery, Yale University School of Medicine, New Haven, CT, United States

Peter Schirmacher (139), Institute of Pathology, University Hospital Heidelberg, Heidelberg, Germany

Palittiya Sintusek (125), Division of Gastroenterology and Hepatology, King Chulalongkorn Memorial Hospital, Chulalongkorn University, Bangkok, Thailand; Paediatric Liver, GI and Nutrition Centre, King's College Hospital, London, United Kingdom

Ana Vives-Rodriguez (209), Department of Neurology, Yale School of Medicine, New Haven, CT, United States

Karl Heinz Weiss (173,197,217,235), Internal Medicine IV, University Hospital Heidelberg, Heidelberg, Germany

France Woimant (115,223), French National Reference Centre for Wilson Disease, Neurology Department, Lariboisière Hospital, Assistance Publique-Hôpitaux de Paris, Paris, France

Paula C. Zimbrean (159), Department of Psychiatry and Surgery (Transplant), Yale University, New Haven, CT, United States

Hans Zischka (65), Institute of Toxicology and Environmental Hygiene, Technical University of Munich, Munich; Institute of Molecular Toxicology and Pharmacology, Helmholtz Center Munich, German Research Center for Environmental Health, Neuherberg, Germany

Foreword

Wilson disease is a rare disorder in which there is an increasing understanding of the underlying pathophysiology of the disease and advances in diagnosis and treatment are being made. Though the molecular basis of Wilson disease was identified as mutation of the copper transporting ATPase, ATP7B, much remains to be learned given the wide range of phenotypic expression of this disease and enhancing the multidisciplinary approach needed for the care of afflicted patients. Current advances in diagnosis and treatment of Wilson disease not only utilize new information derived from our knowledge of the underlying pathophysiology of this disorder but also draw from the well of decades of treatment experience and clinical expertise. Being a rare disease with few clinical trials, this expertise in science and practice is critical to share so that we may advance basic and clinical research, improve on current clinical care, and help fill in our knowledge gaps that hopefully will aid future generations of patients with Wilson disease.

Therefore, we are grateful that so many international experts unselfishly shared their expertise, knowledge, and views on Wilson disease. As editors, we hope we have tried to capture the breadth and depth of the science and clinical areas of import on this disorder. We personally thank our colleagues for their invaluable contributions to this edition and sincerely hope that together, we have created a useful work for those wishing to learn more about Wilson disease.

We want to thank Stacy Manucci, senior editor from Elsevier, for championing this project and our managing editor at Elsevier, Sam Young, for all his help and ready availability to respond to our requests.

But finally, none of this would have any meaning if not for the support and courage of our Wilson disease patients and their families. Their willingness to participate in many of our research efforts and clinical studies will continue to generate invaluable contributions to future generations of patients.

Michael L. Schilsky
Karl Heinz Weiss

Part I

History of Wilson Disease

Chapter 1

History of Wilson Disease

James S. Dooley* and Rupert Purchase[†]
*UCL Institute for Liver and Digestive Health, University College London, London, United Kingdom, [†]Department of Chemistry, School of Life Sciences, University of Sussex, Falmer, Brighton, United Kingdom

INTRODUCTION

There can be few diseases seen in the practice of medicine of such surpassing interest as Wilson's disease (hepatolenticular degeneration, progressive lenticular degeneration, pseudosclerosis of Westphal-Strümpell). So protean are the signs and symptoms of this disease that it must be considered in the differential diagnosis of many syndromes. Thus, nervous system involvement is but one facet of an illness involving many organs … But the interest of Wilson's disease spreads further than this. The very process of unravelling the biochemical lesion has led to a much wider understanding of the handling of copper in the body and hence to the formulation of logically designed therapy, so making Wilson's disease not only one of the very few diseases of the nervous system but also of the liver for which there is a specific and effective treatment.

Walshe (1976) [1]

By the end of the 20th century, five notable developments had led to an understanding of Wilson disease (McKusick 277900 [2]) and its treatment:

1. Samuel Alexander Kinnier Wilson's seminal publication in 1912 [3] describing the symptoms, signs, and pathology of this condition that now bears his name.
2. The recognition that Wilson disease is inherited as an autosomal recessive disorder [4, 5].
3. The role of copper in the pathogenesis of Wilson disease [6, 7].
4. The successful therapeutic use of agents that reduce copper levels in Wilson disease patients [8–13].
5. The identification in 1993 of the gene (*ATP7B*) mutated in Wilson disease [14–16].

Despite progress, much remains to be achieved. Wilson disease remains a challenging condition to recognize, to diagnose, and often to treat. Future developments in cellular and molecular biology should lead to more successful outcomes for all Wilson disease patients.

THE DEFINITION OF WILSON DISEASE

Kinnier Wilson's MD Thesis, for which he was awarded the Gold Medal at Edinburgh University in 1911 and which was published in *Brain* in 1912 [3], is a remarkable achievement for such a young man (Fig. 1). He had qualified only 9 years before in 1902 from Edinburgh, Bachelor of Medicine (MB), and completed a BSc with first class honors the following year. He first worked at the National Hospital for the Paralysed and Epileptic (later the National Hospital for Nervous Diseases), Queen Square, London, in 1904 after completing a fellowship in Paris with the neurologist Pierre Marie at Bicêtre Hospital. After taking his membership of the Royal College of Physicians of London in 1907, he produced an acclaimed translation of the book by Meige and Feindel on tics and their treatment and in 1908 published work on apraxia. He then took up a British Medical Association Research Fellowship between 1909 and 1911, during which he did the work for his MD Thesis at Queen Square.

Several accounts from others are worth repeating here. Wilson was the first specialist neurologist appointed in the UK at King's College Hospital in around 1918—they had all been general physicians with an interest in neurology until then. According to Macdonald Critchley who had been his House Physician, his registrar, and then his junior colleague, Wilson was interested in examination and detail but with the goal of addressing "why?" that is, why had such and such happened. He was always "probing, questioning, and speculating"; "he was the Marco Polo of the extrapyramidal system—but much

FIG. 1 Samuel Alexander Kinnier Wilson (1878–1937). A photograph from the early 1920s. *(Reprinted by kind permission of his son, James Kinnier Wilson).*

more than that" [17]. This is clearly reflected in his thesis. He had a broad interest in neurology, and he was three-quarters of the way through writing his textbook of neurology [18] when he died aged 58 in 1937.

Within 10 years of starting his clinical career, Wilson had uncovered a new neurological entity, having collected 12 patients with what he recognized was a shared pattern of disease. He himself saw four of these patients and in three of these recorded their features and their pathology—particularly neuropathology—in minute detail. He also reviewed two cases seen at Queen Square previously and six other cases from the literature. His 213-page publication in the journal *Brain* is a lesson in observation, analysis, and synthesis of this new and until then confusing constellation of features. The detailed description of each patient's neurological symptoms, signs, and, when done, neuropathology from postmortem is stunning—especially for someone only 10 years into his career.

On page 436 of his paper in *Brain* [3], Wilson wrote the following:

Progressive lenticular degeneration may be defined as a disease which occurs in young people, which is often familial but not congenital or hereditary; it is essentially and chiefly a disease of the extrapyramidal motor system, and is characterised by involuntary movements, usually of the nature of tremor, dysarthria, dysphagia, muscular weakness, spasticity, and contractures with progressive emaciation; with these may be associated emotionalism and certain symptoms of a mental nature. It is progressive, and, after a longer or shorter period, fatal. Pathologically it is characterised predominantly by bilateral degeneration of the lenticular nucleus, and in addition cirrhosis of the liver is constantly found, the latter morbid condition rarely, if ever, giving rise to symptoms during the life of the patient.

His observations led to a new concept of neurological wiring, with the basal ganglia, in which the pathological findings were seen, being recognized for their importance in extrapyramidal function—a new departure in neurology. This distinguished symptoms, signs, and pathology from the pyramidal system and corticospinal tract.

Broussolle et al. [19] state the following in their detailed review of Wilson's work and contributions: "To summarize, Wilson's description and analysis of the motor disorder suggest a combination of dystonia, akinesia, and extrapyramidal rigidity. Indeed, in order to emphasize his astute observations, Wilson introduced in 1912 for the first time in the literature the terms "extrapyramidal system" to refer to the basal ganglia and "extrapyramidal signs" or "extrapyramidal syndrome" to discuss the symptoms and signs associated with basal ganglia pathology. These concepts were later formally expanded and expounded in his Croonian lectures (published in the *Lancet* in 1925). This terminology was subsequently largely used in the neurological literature during the 20th century."

Wilson recognized the presence of cirrhosis of the liver as common to all his and previous cases, and although it has been written that he considered this was without clinical sequelae, in his paper, he clearly recognizes hepatological features (such as ascites) as being present in one patient. One of the patients died of massive hematemesis, but it is not surprising that at that time, a link between cirrhosis and esophageal varices was not made—and indeed, Preble [20] in 1900 had only just published a paper describing a series of patients with cirrhosis and bleeding varices. So, it was early days in hepatology. Interestingly, two of the patients reviewed by Wilson had episodes of jaundice some years before their neurological presentation, but whether these were viral hepatitis or hemolysis related to Wilson disease can never be known.

Wilson's description resonates with the range of neurological features still seen today in such patients. He hypothesized as to the possible cause (familial, yes; a toxin, perhaps; and alcohol and syphilis, no). Some have discussed in the literature why Wilson did not call the syndrome inherited. Human genetics was in its very early days, and Broussolle et al. [19] have questioned whether Wilson was simply saying that he did not think the condition dominantly inherited.

In his 1912 *Brain* paper, Wilson recognized features of liver disease in a minority of his patients, but he did not record the presence of Kayser-Fleischer rings. These had been reported by others in the early 1900s (see "Corneal Ring" and "Kayser-Fleischer Rings" sections). Some years later, in the 1930s, at a meeting of the Section of Ophthalmology of the Royal Society of Medicine, Wilson described two children with Kayser-Fleischer rings in the cornea, and in the ensuing discussion, the possibility was raised that these rings were the result of metallic deposits, including copper, and that Wilson disease was a consequence of altered copper metabolism [21].

There were no treatments available to Wilson, and he saw patients within families who began neurologically normal and developed the features later. The mean survival from neurological onset in those who died was around 3 years [3, 19].

Wilson's paper was a fundamentally important landmark in neurology. His incredible contribution brought together his own observations with those of others and, subsequently, but not for some years, recognition of the cause. This was followed by the development of diagnostic tests and effective treatment around 40 years later. In the 1990s, the disease-related gene was cloned, and its function was dissected, and more recently, gene therapy for Wilson disease in an experimental model has been reported [22].

DESCRIPTION OF KAYSER-FLEISCHER RINGS; UNITY OF NEUROLOGICAL DEFINITION; RESOLVING PSEUDOSCLEROSIS; LIVER DISEASE

Corneal Ring [23, 24]

The greenish-brown ring in the cornea was first described by Kayser in 1902 in a patient thought to have multiple sclerosis, and Fleischer in 1903 described the same appearance in a case of "pseudosclerosis" and another with multiple sclerosis. In 1912, Fleischer drew together the link between corneal pigmentation, tremor, mental problems, and cirrhosis (cited by Hoogenraad [25]). Kayser-Fleischer (KF) rings are now recognized as pathognomonic of Wilson disease, though they have been reported in cholestasis due to primary biliary cirrhosis, but this is exceptionally unusual [24].

Pseudosclerosis [26, 27]

In his paper [3], Wilson included previous reports of patients with neurological syndromes and cirrhosis, which were unexplained, from Gowers, in 1888 and 1906, and Ormerod, in 1890. He recognized the description of what now seems the first report of a patient with Wilson disease by F.T. von Frerichs in his textbook *A Clinical Treatise on Diseases of the Liver* from 1860 (Murchison's English translation). In 1883, Westphal reported two cases with a neurological picture with tremor where multiple sclerosis, well known at that time, was considered but not supported by necropsy data. He used "pseudosclerosis" to describe this new condition. Strümpell in 1898 and 1899 reported three other cases of "pseudosclerosis," the third of whom had cirrhosis. Gowers had termed the condition he described "tetanoid chorea" and had described the association with cirrhosis in both patients (1888, 1906).

It took some years for "pseudosclerosis" to be recognized as being Wilson disease. Wilson [3] made the point in his paper that this "unsatisfactory and makeshift expression…….. coined for certain cases which were said to resemble disseminated sclerosis clinically, but not pathologically, should be abandoned, and that re-examination of the subject should be undertaken." In 1920, Spielmeyer reported the neuropathology of pseudosclerosis and Wilson disease and came to the conclusion that they were the same condition [26, 27]. In 1921, Hans Christian Hall in Paris reviewed 64 cases in the literature and four of their own cases and concluded that pseudosclerosis and Wilson disease were identical [4]. Hall gave the disorder the name "hepatolenticular degeneration." A review of the complexities of the difference in opinion that existed has been published [27].

Liver Disease

In Wilson's paper, as already pointed out, cirrhosis was recognized as part of the disorder, but was not considered to produce clinical problems. In 1916, Byrom Bramwell, whose house physician Wilson had been and who had inspired Wilson to study neurology [17, 19], reported a family in which four siblings had died of "acute fatal cirrhosis" [28]. Early evidence that liver disease might be the only clinical feature of the disease was published between 1925 and 1929 by Barnes and Hurst [29–31], who described a family where three of eight children had typical Wilson disease but hepatic symptoms came before neurological changes; a fourth child died of severe liver disease without neurological symptoms.

In 1953, Franklin and Bauman [32] reported a series of 11 patients, of whom five had hepatic signs and symptoms preceding those due to neurological involvement. They pointed out the frequency and severity of the hepatic features and emphasized that in their series, the diagnosis of Wilson disease was not made before the appearance of neurological disease. Papers from Chalmers et al. [33] and Walshe [34] in 1957 and 1962, respectively, further documented the importance of thinking of Wilson disease in cases of idiopathic cirrhosis, jaundice, or other evidence of liver disease in children and young adults.

In 1976, [1] Walshe reported that the presentation of 112 patients with Wilson disease could be considered as hepatic in 51 (45%), neurological in 43 (38%), and "presymptomatic" in 14 (12%). Better recognition of Wilson disease in patients with hepatic disease reflected not only improvement in diagnostic approaches but also the impact of publications and education of specialists and the development of hepatology as a subspeciality, in its own right, of gastroenterology.

TOXIN … COPPER

Interestingly, Rumpel in 1913 described increased hepatic copper and silver in one of Fleischer's patients with "pseudosclerosis" (cited by Hoogenraad [35]). However, the Kayser-Fleischer rings were attributed to silver. In 1922, Siemerling and Oloff (cited by Walshe [24]) reported Kayser-Fleischer rings and a sunflower cataract—the latter similar to the cataract associated with foreign bodies containing copper—in a patient with pseudosclerosis. They postulated that this disease could be due to the accumulation of copper in viscera, eyes, and the nervous system.

The finding of "marked excess" of hepatic and brain copper in a Wilson disease patient was reported by Glazebrook in 1945 [6], though de Meyenbourg (1929), Haurowitz (1930), and Lüthy (1931) have also been credited with this finding (cited by Cumings [36]).

Mandelbrote et al. [37] in 1948 while studying mobilization of copper by British Anti-Lewisite (BAL) ("British Anti-Lewisite" section) in patients with multiple sclerosis found a high urinary copper in one of the "control" patients who had Wilson disease, and this increased after administration of BAL.

Later in 1948, Cumings [7] in the first controlled study reported high copper levels in the liver and brain of Wilson disease patients. In Wilson disease, hepatic copper was 39.4–156.5 mg/100 g dry weight, compared with normal values of 3.7–17.2 mg. Basal ganglia copper in Wilson disease was also considerably higher than in normal tissue. Both Glazebrook [6] and Cumings [7] recognized the deleterious consequences of an excess of copper on the liver and other organs of Wilson disease patients. In addition, Glazebrook was aware of the enzymatic inhibitory properties of copper (II). Also, Cumings [7] had cited a study by Mallory published in *The Journal of Medical Research* in 1921 reporting liver necrosis and cirrhosis in rabbits chronically poisoned with oral copper acetate—as part of studies on the cause of hemochromatosis in which it was hypothesized that copper might play a part. The first treatment for Wilson disease followed from the findings of raised copper levels in the basal ganglia, liver, and urine, even though debate around the role of copper continued through the 1950s into the mid-1960s. Sternlieb and Scheinberg clearly defined copper as cause and effect in 1968 [38] when they showed benefit of treating asymptomatic patients with D-penicillamine.

EVOLUTION OF DIAGNOSTIC TESTS FOR WILSON DISEASE

There was progress both in diagnosis and treatment in the 1950s and 1960s; that for treatment will be covered in "Copper as a Therapeutic Target: The Discovery of Drugs for the Treatment of Wilson Disease" section. Early diagnostic approaches are considered here, including observations on serum ceruloplasmin in Wilson disease from the early 1950s.

Kayser-Fleischer Rings

Kayser-Fleischer (KF) rings had been recognized as characteristic of some patients with pseudosclerosis in the early years of the century (see "Corneal Ring" section), and pseudosclerosis became accepted as a form of Wilson disease, as discussed previously. It has subsequently become acknowledged that KF rings are pathognomonic of Wilson disease [23], but they may not be present, particularly in presymptomatic patients and those presenting with hepatic disease [24, 39].

Tissue Copper

Although liver, brain, and urinary copper had been shown to be high in Wilson disease, only urine samples were straightforward to obtain—though needing care to insure a complete collection without contamination.

Serum Ceruloplasmin

Between 1947 and 1951, Holmberg and Laurell [40] had reported the purification, characterization, and crystallization of what they named ceruloplasmin, a copper-binding protein, in serum. Around 1948, Laurell had tested a sample from a patient that he had been told had Wilson disease, and the result was normal. It was not until later that he was informed that the patient did *not* have Wilson disease (cited in Ref. [23], p. 5). In 1952, Scheinberg and Gitlin [41] showed decreased ceruloplasmin levels in patients with Wilson disease. Cartwright et al. [42] reported in 1954 that the serum "direct" copper (i.e., not bound to ceruloplasmin) was increased in Wilson disease.

Interestingly, in *Clinical Neurology* by Elliott, Hughes, and Turner published in 1952 [43], not only the value of Kayser-Fleischer rings is noted for diagnosis, but also urine may show amino acids, a finding reported by Uzman and Denny-Brown in 1948 [44]. Liver biopsy could be valuable in showing cirrhosis.

These findings provided the bedrock for potential diagnosis that could be straightforward with typical neurology and positive Kayser-Fleischer rings—but it is clear that other phenotypes are common, and the full panel of tests may be necessary with serum ceruloplasmin and copper, urinary copper, and if required liver copper. Genetic testing came much later (see "Molecular Genetics" section).

COPPER AS A THERAPEUTIC TARGET: THE DISCOVERY OF DRUGS FOR THE TREATMENT OF WILSON DISEASE

With the identification of copper overload in Wilson disease patients, treatment of the disorder based on the sequestration of copper emerged from the 1940s onward. Advances in chemistry in the 19th and 20th centuries, particularly in the chemistry of coordination compounds (chemical substances in which a central metal atom is surrounded by nonmetal atoms or groups of atoms, called ligands, joined to it by chemical bonds [45]), underpin the rationale for treating Wilson disease. Five agents used for treating Wilson disease are described here (see "British Anti-Lewisite," "D-Penicillamine," "Trientine (Triethylenetetramine)," "Zinc Salts," and "Ammonium Tetrathiomolybdate" sections), and their chemical structures are shown in Fig. 2.

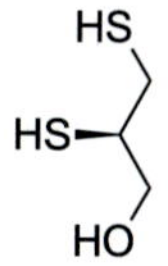

British Anti-Lewisite (BAL; 2,3-dimercapto-1-propanol; dimercaprol). The (*R*)-isomer is shown here. Clinically, the racemate is administered (intramuscularly). A bidentate ligand. The two sulfhydryl groups chelate (soft) metals.

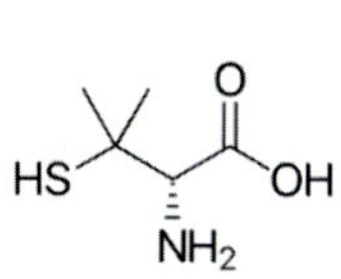

D-Penicillamine (3-mercapto-D-valine). Clinically, only the (*S*)-isomer (shown here) is administered (orally). Nominally, a bidentate or a tridentate ligand, but can reductively chelate metals, and forms mixed Cu(I)/Cu(II) cluster compounds with copper. The toxicity of the (*R*)-isomer, L-penicillamine, precludes its use as a chelating agent. The stereochemical configuration of D-penicillamine originates from the stereochemistry of the penicillin nucleus, which can undergo chemical degradation to D-penicillamine without producing a racemate.

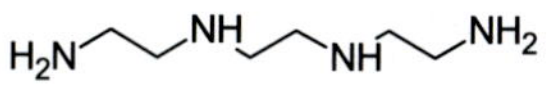

Trientine (triethylenetetramine). Clinically, the dihydrochloride is administered (orally). The four amino groups participate in the chelation of a metal ion.

Ammonium tetrathiomolybdate $(NH_4)_2MoS_4$. Administered orally. Forms copper—molybdenum—sulfur cluster compounds with food (preventing absorption of dietary copper), and with albumin.

Zinc sulfate heptahydrate $ZnSO_4 \cdot 7H_2O$ (and **zinc acetate dihydrate** $Zn(OAc)_2 \cdot 2H_2O$). Administered orally, zinc(II) induces intestinal metallothionein, which can preferentially bind copper thereby decreasing dietary copper absorption. Also induces hepatic metallothionein.

FIG. 2 Five drugs used to treat Wilson disease.

British Anti-Lewisite

British Anti-Lewisite (BAL), 2,3-dimercapto-1-propanol, was developed in Oxford during the Second World War by Sir Rudolf Peters and his team as an antidote against arsenical warfare poisoning [46, 47]. In 1946, BAL O-glucoside was reported to increase urinary copper in normal man [48], and as mentioned in the "Toxin … Copper" section, in 1948, Mandelbrote et al. reported one patient with Wilson disease in their control group with a baseline high urinary copper that increased after BAL [37].

This led Cumings to suggest that BAL might be used therapeutically to treat patients with Wilson disease [7]. In 1951, Cumings and Denny-Brown independently reported the use of BAL in Wilson disease [49, 50]. The agent had to be given by repeated intramuscular injection, with a range of attendant unpleasant complications and adverse effects. Clearly, an alternative and effective oral agent was needed.

Two water-soluble analogues of BAL have been developed—sodium 2,3-dimercaptopropane-1-sulfonate (Unithiol; Dimaval) and 2,3-dimercaptosuccinic acid. These two drugs can be given orally and have been investigated for the treatment of heavy metal poisoning [51]. However, experience of their use for Wilson disease is limited [52].

D-Penicillamine

In the early 1950s, John Walshe was using paper chromatography to identify urinary amino acids in patients with liver disease and detected a novel band in the urine of one patient [53]. After initial enthusiasm that this was a new amino acid, it was pointed out that the patient was being treated with penicillin and that this could be the explanation. Indeed, this was β,β-dimethylcysteine, that is, penicillamine, and D-penicillamine (Fig. 2) had been characterized as a degradation product of penicillin in 1943—and it had been synthesized [54]. (Paper chromatography would not have revealed the stereochemistry of Walshe's eluted amino acids).

Walshe took up a Fulbright fellowship in the Harvard Medical School at Boston City Hospital in the United States in 1954–55 with the hepatologist, Dr. Charles Davidson. In the spring of 1955, Dr. Davidson was asked by Professor Denny-Brown to see a patient with neurological and hepatic Wilson disease who had not responded well to BAL. Walshe suggested that they try penicillamine, surmising from its chemical structure that it might bind copper [55]. Penicillamine was obtained and given to the patient who had a cupruresis. Walshe published the first seminal articles on the use of this oral agent for treating Wilson disease in 1956 [8, 56]. For this preliminary work, Walshe had been supplied with both DL-penicillamine and D-penicillamine [8]. A literature search revealed marked differences in toxicity between the D- and L-isomers [57], and subsequent investigations in Wilson disease with this drug have used D-penicillamine (Fig. 2). D-Penicillamine does have adverse effects in a significant proportion of patients [58], and its use [59] and mechanism of action [60–63] have given rise to debate, but it has remained in the forefront of treatment of patients with Wilson disease in the 60 years since its introduction [64].

Trientine (Triethylenetetramine)

A patient with Wilson disease under the care of Dr. Walshe could not receive D-penicillamine, because of adverse effects, and an alternative oral treatment was needed. In discussion with the Cambridge University biochemist, Dr. H. B. F. (Hal) Dixon, the known copper-chelating agent triethylenetetramine (trientine) [65–67] was considered a likely option, and Walshe pursued this. It has a different chemical structure compared with D-penicillamine (Fig. 2), and therefore, cross-reactions would not be expected. The first report of its use (administered as the dihydrochloride [68, 69]) for a Wilson disease patient intolerant of D-penicillamine was in 1969 [9]. Walshe has noted that trientine and D-penicillamine act on different pools of copper in the body [70]. Trientine is now accepted as a valuable chelator in Wilson disease, with fewer adverse effects than D-penicillamine [71].

Zinc Salts

Zinc is an important alternative to chelating agents for the treatment of Wilson disease, although not without debate. Aware that zinc sulfate reduced hepatic copper levels in sheep [72], Gerrit Schouwink in Holland began to use zinc sulfate for the treatment for Wilson disease in the late 1950s [73]. He carried out copper balance studies in patients and showed that there was negative copper balance when zinc sulfate was given (cited by Hoogenraad [74]).

In 1978, Tjaard Hoogenraad and his colleagues studied Schouwink's patients and reported that zinc treatment had been effective [75]. Hoogenraad reviewed a longer-term follow-up and a larger group of zinc-treated patients with Wilson disease, in his monograph [76]. Overall, he concluded that zinc sulfate is safe and effective for the treatment of Wilson disease, though with a caveat regarding patients with portal hypertension "doing well" on penicillamine.

Dr. George Brewer in Ann Arbor, Michigan, the United States, had considered zinc for the treatment of patients with Wilson disease, following his observation of copper deficiency in patients with sickle cell anemia treated with zinc acetate [77, 78]. He has subsequently published a considerable series of papers and abstracts on copper balances, ^{64}Cu uptake, and the treatment of Wilson disease patients with zinc [13, 79, 80], contributing in a major way to the management of patients using zinc salts.

Both Hoogenraad [81] and Brewer [82] published on the mechanism of action of zinc in the gut confirming previous findings [83] that zinc induces the synthesis of intestinal metallothionein. This protein preferentially sequesters copper within the duodenal enterocyte, thus limiting the absorption of dietary copper.

Ammonium Tetrathiomolybdate

Observations by veterinary scientists also led to the introduction of a fifth recognized treatment for Wilson disease, ammonium tetrathiomolybdate. Molybdate-induced copper deficiencies in cattle and sheep, potentiated by the addition of sulfate dietary supplements [72, 84], were explained by the production of thiomolybdates in the animals' rumen, which with copper(I) or copper(II) formed insoluble copper thiomolybdates [85, 86].

The first tentative use of tetrathiomolybdate (as the easily synthesized ammonium salt [87]) to treat a patient with Wilson disease was reported by Walshe in 1984 [10, 88]. In a series of papers [89, 90], 1991–2009, George Brewer in the United States has published a methodical study of the use of tetrathiomolybdate for treating the neurological form of Wilson disease. Brewer has also contributed the only randomized study (ammonium tetrathiomolybdate and zinc vs trientine dihydrochloride and zinc) [91], commendable in this rare condition.

Recently, clinical trials have begun on the treatment of Wilson disease with the choline salt of tetrathiomolybdate (WTX-101) [92, 93].

The use of ammonium tetrathiomolybdate in Michigan has prompted investigations into its mechanisms of action. For example, the structure of the copper-molybdenum cluster bound in the rat liver and kidneys has been examined by X-ray absorption spectroscopy [94, 95]. Also, a stable purple complex from the reaction of ammonium tetrathiomolybdate with the yeast copper(I) trafficking protein Atx1 has been isolated and characterized by X-ray crystallography [96]—leading to speculation about the involvement of the drug in copper trafficking pathways.

LIVER TRANSPLANTATION

In discussing treatment, it is important to include liver transplantation, first reported in this condition in 1971 [97], as a modality that may be the only lifesaving procedure for some patients: those with fulminant liver failure and those with decompensated chronic liver disease unresponsive to medical treatment. Replacement of the liver cures the metabolic defect. The survival of patients after transplant mirrors that of other etiologies.

MOLECULAR GENETICS

The pathophysiology of Wilson disease was the subject of numerous studies over many years. Was there increased copper absorption or reduced loss? Then, using different approaches, Frommer in 1974 [98] and Gibbs and Walshe in 1980 [99] reported defective biliary excretion in patients with Wilson disease. Animal models, in particular the Long-Evans Cinnamon rat [100] and the toxic milk mouse [101], did play an important part in the study of derangement of copper metabolism (see Chapter 5). However, it awaited the cloning of the gene responsible before the cellular defect in Wilson disease, and these animal models, could be pieced together.

Kinnier Wilson had considered that the condition he had identified was familial [3]. In 1921, Hall had suggested that it might be autosomal recessive [4]; this was strongly supported by data published by Bearn in 1953 and 1960 [5, 102].

The chromosomal location of the gene responsible for Wilson disease was provided by linkage analysis by Frydman et al. in 1985 [103], and this showed that the gene locus was on chromosome 13 linked to that of esterase D. Later in 1990, Houwen et al. [104], using restriction fragment length polymorphisms (RFLPs), gave a more specific position between the markers D13S31 and D13S55 of the q14-q21 band.

Cloning of the Wilson Disease Gene *ATP7B*

In 1993, three groups [14, 15, 16] independently reported the localization and identification of the Wilson disease gene. This achievement depended to a major part on knowledge of the molecular genetics of Menkes disease. This is a condition where there is high intestinal and renal/urinary copper, but defective absorption of copper from the intestine results in severe copper deficiency and thence a variety of features. The gene for Menkes disease, which encodes a copper-transporting ATP7Aase protein and is situated on the X chromosome, had been cloned early in 1993 [105]. Bull et al. and Yamaguchi et al. used probes based on the copper-binding domain of ATP7A, to identify cDNAs, which led them to identify the Wilson disease gene on chromosome 13 [14, 15, 105]. Tanzi et al. took a different approach starting with a degenerate oligonucleotide based on a novel heavy metal binding site in the amyloid β-protein precursor [16, 105]. The three groups established that the Wilson disease gene encodes a copper-transporting ATP7Base protein homologous to the Menkes disease gene product. Mutations or deletions in the gene encoding the copper-transporting ATPase ATP7A are associated with Menkes disease, while mutations in the gene encoding the copper-transporting ATPase ATP7B result in Wilson disease [106].

The Wilson disease gene has a similar structure to that of the Menkes disease gene but with clearly a different function. That so much progress was made with regard to these two diseases within such a short period reflects tremendous effort and ingenuity, but also highlights the rapidly developing field of molecular biology.

Function of ATP7B

Analysis of the gene product has led to a model whereby, under normal intracellular copper concentrations, ATP7B transports copper into the Golgi (for association with apoceruloplasmin and secretion into the circulation) (see Chapter 3). With high hepatocyte copper concentrations, ATP7B transports copper into vesicles that carry the copper to the bile canaliculus of the hepatocyte, leading to biliary excretion of copper. These findings explain the low biliary copper found in patients with Wilson disease where ATP7B is not functioning normally.

Mutation Analysis

The Wilson Disease Mutation Database, compiled at the University of Alberta and updated until 2010 [107], records more than 500 disease-related mutations of the *ATP7B* gene. This database owes much to the work and commitment to Wilson disease of Professor Diane Wilson Cox—and indeed, she was the senior author on the paper by Bull et al. reporting the cloning of *ATP7B*.

Most patients are compound heterozygotes; there are no clear-cut phenotype/genotype associations (see Chapter 9), except that homozygosity for one of the more common mutations, H1069D (in some populations), appears to associate with late, neurological presentation [108].

Laboratory techniques are advancing so rapidly that whereas only a few laboratories could do genetic analysis as part of the diagnostic path in the past, this is now becoming more widely available and affordable.

FUTURE DIRECTIONS

The debate continues on best practice in the treatment of Wilson disease, and this is covered in later chapters. There is a large literature on treatment of Wilson disease, but virtually all reports are retrospective analyses of patients. Although more studies, ideally randomized, are needed, they would be challenging to do. Studies of new agents would be informative and one hopes therapeutically positive.

Other issues needing to be addressed will be covered in later chapters but particularly the pathophysiology of neurological deterioration in some patients after the start of treatment (Why? How best to manage?) and the recognition and management (of patient and family) of patients with a psychiatric/psychological presentation. The latter was recognized in Wilson's publication of 1912 and still remains a major challenge clinically.

CONCLUSION

The centennial journey from Wilson's initial detailed description of progressive lenticular degeneration in 1912 has seen developments in recognition, diagnosis, understanding of the pathogenic mechanism(s), and treatment of this disease. As

John Walshe noted [1], this has been accompanied by a deeper awareness of copper biochemistry [109], whose complexity we are perhaps only now beginning to appreciate [110].

Wilson disease is a rare orphan condition, and foremost among the goals for the future are educational initiatives, particularly among neurologists and hepatologists, so that this condition does not go unrecognized. Early diagnosis and treatment are paramount. Treatment is not without challenges, but optimal monitoring lifelong is central to a successful outcome.

However, perhaps, the most useful advance would be the recognition by health systems and hospitals of the value of multidisciplinary teams in the management of these patients. Rare disease initiatives will lead the way. Meanwhile, the paramount contribution of patient support organizations to support individuals with this rare condition and to work with clinical specialists is fundamental to progress in the future.

FURTHER READING ON THE HISTORY OF WILSON DISEASE

In addition to the literature cited in the preceding sections, other historical perspectives of Wilson disease are available in reviews by Walshe [111–115], Brewer [116], Westermark [117], Trocello et al. [118], Schilsky [119], Lanska [120], Tanner [121], and Purchase [122, 123]. Alastair Compston has written commentaries on the *Brain* papers by Wilson and Mandelbrote et al. from 1912 [3] and 1948 [37], respectively [124, 125]. Photographic and cinematographic records from the mid-1920s of some of Kinnier Wilson's patients, including Wilson disease patients, have been made public [126]. The monographs by Cumings [36] and Owen [127] review the literature of Wilson disease up to the mid-1950s and early 1980s, respectively. Solioz has chronicled the discovery in the 1990s of copper-transporting proteins, including ATP7A and ATP7B [128].

REFERENCES

[1] Walshe JM. Wilson's disease (hepatolenticular degeneration). In: Vinken PJ, Bruyn GW, editors. Handbook of clinical neurology. Metabolic and deficiency diseases of the nervous system. Part 1, vol. 27. Amsterdam: North-Holland Publishing Company; 1976. p. 379–414.

[2] Online Mendelian Inheritance in Man (OMIM). Entry #277900 Wilson disease. Available from: https://omim.org/entry/277900. [Accessed 22 December 2018].

[3] Wilson SAK. Progressive lenticular degeneration: a familial nervous disease associated with cirrhosis of the liver. Brain 1912;34:295–509.

[4] Hall HC. La dégénérescence hépato-lenticulaire. Maladie de Wilson—Pseudo-sclérose. Paris: Masson et Cie; 1921.

[5] Bearn AG. Genetic and biochemical aspects of Wilson's disease. Am J Med 1953;15:442–9.

[6] Glazebrook AJ. Wilson's disease. Edinb Med J 1945;52:83–7.

[7] Cumings JN. The copper and iron content of brain and liver in the normal and hepato-lenticular degeneration. Brain 1948;71:410–5.

[8] Walshe JM. Penicillamine. A new oral therapy for Wilson's disease. Am J Med 1956;21:487–95.

[9] Walshe JM. Management of penicillamine nephropathy in Wilson's disease: a new chelating agent. Lancet 1969;294:1401–2.

[10] Walshe JM. Copper: its role in the pathogenesis of liver disease. Semin Liver Dis 1984;4:252–63.

[11] Brewer GJ, Dick RD, Johnson V, Wang Y, Yuzbasiyan-Gurkin V, Kluin KJ, Fink JK, Aisen A. Treatment of Wilson's disease with ammonium tetrathiomolybdate. I. Initial therapy in 17 neurologically affected patients. Arch Neurol 1994;51:545–54.

[12] Schouwink G. De hepatocerebrale degeneratie, met een onderzoek naar de zinkstofwisseling [MD thesis]. University of Amsterdam; 1961 [Cited by Hoogenraad T. Wilson's disease. London: W. B. Saunders; 1996. p. 10].

[13] Brewer GJ, Hill GM, Prasad AS, Cossack ZT, Rabbani P. Oral zinc therapy for Wilson's disease. Ann Intern Med 1983;99:314–20.

[14] Bull PC, Thomas GR, Rommens JM, Forbes JR, Cox DW. The Wilson disease gene is a putative copper transporting P-type ATPase similar to the Menkes gene. Nat Genet 1993;5:327–37.

[15] Yamaguchi Y, Heiny ME, Gitlin JD. Isolation and characterization of a human liver cDNA as a candidate gene for Wilson's disease. Biochem Biophys Res Commun 1993;197:271–7.

[16] Tanzi RE, Petrukhin K, Chernov I, et al. The Wilson disease gene is a copper transporting ATPase with homology to the Menkes disease gene. Nat Genet 1993;5:344–50.

[17] Critchley M. Remembering Kinnier Wilson. Mov Disord 1988;3:2–6.

[18] Wilson SAK. Neurology. London: Edward Arnold; 1940.

[19] Broussolle E, Trocello J-M, Woimant F, Lachaux A, Quinn N. Samuel Alexander Kinnier Wilson. Wilson's disease, Queen Square and neurology. Rev Neurol (Paris) 2013;169:927–35.

[20] Preble RB. Conclusions based on sixty cases of fatal gastro-intestinal hemorrhage due to cirrhosis of the liver. Am J Med Sci 1900;119:263–80.

[21] Kinnier Wilson SA. Kayser–Fleischer ring in cornea in two cases of Wilson's disease (progressive lenticular degeneration). Proc R Soc Med 1934;27:297–8.

[22] Murillo O, Luqui DM, Gazquez C, et al. Long-term metabolic correction of Wilson's disease in a murine model by gene therapy. J Hepatol 2016;64:419–26.

[23] Scheinberg IH, Sternlieb I. Wilson's disease. Philadelphia: W. B. Saunders Company; 1984. p. 93–8.
[24] Walshe JM. The eye in Wilson disease. Q J Med 2011;104:451–3.
[25] Hoogenraad TU. Wilson's disease. London: W. B. Saunders Company Ltd.; 1996. p. 5.
[26] Spielmeyer W. Die histopathologische zusammengehörigkeit der Wilsonschen krankheit und der pseudosklerose. Z Ges Neurol Psychiatr 1920;57:312–51.
[27] Boide P, Wagner A, Steinberg H. Wilson-Krankheit und Westphal-Strümpell-Pseudosklerose. Nervenarzt 2011;82:1335–42.
[28] Bramwell B. Familial cirrhosis of the liver: four cases of acute fatal cirrhosis in the same family, the patients being respectively nine, ten, fourteen and fourteen years of age; suggested relationship to Wilson's progressive degeneration of the lenticular nucleus. Edinb Med J 1916;17:90–9.
[29] Barnes S, Hurst EW. Hepato-lenticular degeneration. Brain 1925;48:279–333.
[30] Barnes S, Hurst EW. A further note on hepato-lenticular degeneration. Brain 1926;49:36–60.
[31] Barnes S, Hurst EW. Hepato-lenticular degeneration: a final note. Brain 1929;52:1–5.
[32] Franklin EC, Bauman A. Liver dysfunction in hepatolenticular degeneration; a review of eleven cases. Am J Med 1953;15:450–8.
[33] Chalmers TC, Iber FL, Uzman LL. Hepatolenticular degeneration (Wilson's disease) as a form of idiopathic cirrhosis. New Engl J Med 1957;256:235–42.
[34] Walshe JM. Wilson's disease. The presenting symptoms. Arch Dis Child 1962;37:253–6.
[35] Hoogenraad TU. Wilson's disease. London: W. B. Saunders Company Ltd.; 1996. p. 8
[36] Cumings JN. Heavy metals and the brain. Oxford: Blackwell Scientific Publications; 1959. p. 50–1.
[37] Mandelbrote BM, Stanier MW, Thompson RHS, Thruston MN. Studies on copper metabolism in demyelinating diseases of the central nervous system. Brain 1948;71:212–28.
[38] Sternlieb I, Scheinberg IH. Prevention of Wilson's disease in asymptomatic patients. New Engl J Med 1968;278:352–9.
[39] Hoogenraad TU. Wilson's disease. London: W. B. Saunders Company Ltd.; 1996. p. 115–6.
[40] Holmberg CG, Laurell C-B. Investigations in serum copper. III: caeruloplasmin as an enzyme. Acta Chem Scand 1951;5:476–80.
[41] Scheinberg IH, Gitlin D. Deficiency of ceruloplasmin in patients with hepatolenticular degeneration (Wilson's disease). Science 1952;116:484–5.
[42] Cartwright GE, Hodges RE, Gubler CJ, Mahoney JP, Daum K, Wintrobe MM, Bean WB. Studies on copper metabolism. XIII. Hepatolenticular degeneration. J Clin Invest 1954;33:1487–501.
[43] Elliott FA, Hughes B, Turner JWA. Clinical neurology. London: Cassell and Company; 1952. p. 259–60.
[44] Uzman L, Denny-Brown D. Amino-aciduria in hepato-lenticular degeneration (Wilson's disease). Am J Med Sci 1948;215:599–611.
[45] Kauffman GB, editor. Coordination chemistry: a century of progress. ACS symposium series no. 565. Washington, DC: American Chemical Society; 1994.
[46] Peters RA, Stocken LA, Thompson RHS. British Anti-Lewisite (BAL). Nature 1945;156:616–9.
[47] Stocken LA, Thompson RHS. Reactions of British Anti-Lewisite (BAL) with arsenic and other metals in living systems. Physiol Rev 1949;29:168–94.
[48] McCance RA, Widdowson EM. Observations on the administration of BAL-Intrav to man. Nature 1946;157:837.
[49] Cumings JN. The effects of B.A.L. in hepatolenticular degeneration. Brain 1951;74:10–22.
[50] Denny-Brown D, Porter H. The effect of BAL (2,3-dimercaptopropanol) on hepatolenticular degeneration (Wilson's disease). New Engl J Med 1951;245:917–25.
[51] Chisolm Jr. JJ, Thomas DJ. The role of DMPS and other chelating agents in the management of childhood lead poisoning. In: Scheinberg IH, Walshe JM, editors. Orphan diseases and orphan drugs. Manchester: Manchester University Press; 1986. p. 86–97.
[52] Walshe JM. Unithiol in Wilson's disease. Br Med J 1985;290:673–4.
[53] Walshe JM. Disturbances of aminoacid metabolism following liver injury: a study by means of paper chromatography. Q J Med 1953;22:483–506.
[54] Clarke HT, Johnson JR, Robinson SR, editors. The chemistry of penicillin. Princeton University Press; 1949. p. 15–6, 455–72.
[55] Walshe JM. Copper: quest for a cure—Bentham eBooks. 2009 eISBN: 978-1-60805-060-4.
[56] Walshe JM. Wilson's disease, new oral therapy. Lancet 1956;270:25–6.
[57] Wilson JE, du Vigneaud V. Inhibition of the growth of the rat by L-penicillamine and its prevention by aminoethanol and related compounds. J Biol Chem 1950;184:63–70.
[58] Scheinberg IH, Sternlieb I. Wilson's disease. Philadelphia: W. B. Saunders Company; 1984. p. 134–45.
[59] Członkowska A, Litwin T. Treatment of Wilson's disease—another point of view. Expert Opin Orphan Drugs 2015;3:239–43.
[60] Birker PJMWL, Freeman HC. Structure, properties, and function of a copper(I)-copper(II) complex of D-penicillamine: pentathallium(I) μ_8-chlorododeca(D-penicillaminato)-octacuprate(I)hexacuprate(II) n-hydrate. J Am Chem Soc 1977;99:6890–9.
[61] Laurie SH, Prime DM. The formation and nature of the mixed valence copper–D-penicillamine–chloride cluster in aqueous solution and its relevance to the treatment of Wilson's disease. J Inorg Biochem 1979;11:229–39.
[62] Tran-Ho L-C, May PM, Hefter GT. Complexation of copper(I) by thioamino acids. Implications for copper speciation in blood plasma. J Inorg Biochem 1997;68:225–31.
[63] Königsberger L-C, Königsberger E, Hefter G, May PM. Formation constants of copper(I) complexes with cysteine, penicillamine and glutathione: implications for copper speciation in the human eye. Dalton Trans 2015;44:20413–25.
[64] Teive HAG, Barbosa ER, Lees AJ. Wilson's disease: the 60th anniversary of Walshe's article on treatment with penicillamine. Arq Neuropsiquiatr 2017;75:69–71.
[65] Jonassen HB, Meibohm AW. Inorganic complex compounds containing polydentate groups. V. Formation constants of the triethylenetetramine–copper(II) and nickel(II) complex ions. J Phys Chem 1951;55:726–33.

[66] Dixon HBF. The chemistry of trientine. In: Scheinberg IH, Walshe JM, editors. Orphan diseases and orphan drugs. Manchester: Manchester University Press; 1986. p. 23–32.

[67] Laurie SH, Sarkar B. Potentiometric and spectroscopic study of the equilibria in the aqueous copper(II)–3,6-diazaoctane-1,8-diamine system and an equilibrium-dialysis examination of the ternary system of human serum albumin-copper(II)–3,6-diazaoctane-1,8-diamine. J Chem Soc Dalton Trans 1977;1822–7.

[68] Dixon HBF, Gibbs K, Walshe JM. Preparation of triethylenetetramine dihydrochloride for the treatment of Wilson's disease. Lancet 1972;299:853.

[69] Purchase R. The purification of triethylenetetramine and its dihydrochloride for the treatment of Wilson's disease. J Chem Res 2005;233–5.

[70] Walshe JM. Copper chelation in patients with Wilson's disease: a comparison of penicillamine and triethylene tetramine dihydrochloride. Q J Med 1973;42:441–52.

[71] European Association for the Study of the Liver. EASL clinical practice guidelines: Wilson's disease. J Hepatol 2012;56:671–85.

[72] Dick AT. Studies on the assimilation and storage of copper in crossbred sheep. Aust J Agr Res 1954;5:511–44.

[73] Schouwink G. The continuing story of copper and zinc. In: Scheinberg IH, Walshe JM, editors. Orphan diseases and orphan drugs. Manchester: Manchester University Press; 1986. p. 56–61.

[74] Hoogenraad TU. Wilson's disease. London: W. B. Saunders Company Ltd.; 1996. p. 156.

[75] Hoogenraad TU, van Hattum J, Van den Hamer CJ. Management of Wilson's disease with zinc sulphate. Experience in a series of 27 patients. J Neurol Sci 1987;77:137–46.

[76] Hoogenraad TU. Wilson's disease. London: W. B. Saunders Company Ltd.; 1996. p. 164–9, 177.

[77] Brewer GJ, Shoomaker EB, Leichtman DA, Kruckeberg WC, Brewer LF, Meyers N. The uses of pharmacological doses of zinc in the treatment of sickle cell anemia. In: Brewer GJ, Prasad AS, editors. Zinc metabolism: current aspects in health and disease. New York: Alan R. Liss; 1977. p. 241–58.

[78] Prasad AS, Brewer GJ, Schoomaker EB, Rabbini P. Hypocupremia induced by zinc therapy in adults. JAMA, J Am Med Assoc 1978;240:2166–8.

[79] Brewer GJ. Raulin Award Lecture: Wilson's disease therapy with zinc and tetrathiomolybdate. J Trace Elem Exp Med 2000;13:51–61.

[80] Brewer GJ. Zinc and tetrathiomolybdate for the treatment of Wilson's disease and the potential efficacy of anticopper therapy in a wide variety of diseases. Metallomics 2009;1:199–206.

[81] Hoogenraad TU. Wilson's disease. London: W. B. Saunders Company Ltd.; 1996. p. 156–7.

[82] Yuzbasiyan-Gurkan V, Grider A, Nostrant T, Cousins RJ, Brewer GJ. Treatment of Wilson's disease with zinc: X. Intestinal metallothionein induction. J Lab Clin Med 1992;120:380–6.

[83] Hall AC, Young BW, Bremner I. Intestinal metallothionein and the mutual antagonism between copper and zinc in the rat. J Inorg Biochem 1979;11:57–66.

[84] Dick AT. Influence of inorganic sulphate on the copper–molybdenum interrelationship in sheep. Nature 1953;172:637–8.

[85] Suttle NF. Recent studies of the copper–molybdenum antagonism. Proc Nutr Soc 1974;33:299–305.

[86] Dick AT, Dewey DW, Gawthorne JM. Thiomolybdates and the copper–molybdenum–sulfur interaction in ruminant nutrition. J Agric Sci 1975;85:567–8.

[87] Laurie SH. Thiomolybdates—simple but very versatile reagents. Eur J Inorg Chem 2000;2443–50.

[88] Walshe JM. Tetrathiomolybdate (MoS4) as an 'anti-copper' agent in man. In: Scheinberg IH, Walshe JM, editors. Orphan diseases and orphan drugs. Manchester: Manchester University Press; 1986. p. 76–85.

[89] Brewer GJ, Dick RD, Yuzbasiyan-Gurkin V, Tankanow R, Young AB, Kluin KJ. Initial therapy of patients with Wilson's disease with tetrathiomolybdate. Arch Neurol 1991;48:42–7.

[90] Brewer GJ, Askari F, Dick RB, Sitterly J, Fink JK, Carlson M, Kluin KJ, Lorincz MT. Treatment of Wilson's disease with tetrathiomolybdate: V. Control of free copper by tetrathiomolybdate and a comparison with trientine. Transl Res 2009;154:70–7.

[91] Brewer GJ, Askari F, Lorincz MT, Carlson M, Schilsky M, Kluin KJ, Hedera P, Moretti P, Fink JK, Tankanow R, Dick RB, Sitterly J. Treatment of Wilson disease with ammonium tetrathiomolybdate. IV. Comparison of tetrathiomolybdate and trientine in a double-blind study of treatment of the neurologic presentation of Wilson disease. Arch Neurol 2006;63:521–7.

[92] Lee VE, Schulman JM, Stiefel EI, Lee CC. Reversible precipitation of bovine serum albumin by metal ions and synthesis, structure and reactivity of new tetrathiometallate chelating agents. J Inorg Biochem 2007;101:1707–18.

[93] Weiss KH, Askari FK, Czlonkowska A, Ferenci P, Bronstein JM, Bega D, Ala A, Nicholl D, Flint S, Olsson L, Plitz T, Bjartmar C, Schilsky ML. Bischoline tetrathiomolybdate in patients with Wilson's disease: an open-label, multicentre, phase 2 study. Lancet Gastroenterol Hepatol 2017;2:869–76.

[94] George GN, Pickering IJ, Harris HH, Gailer J, Klein D, Lichtmannegger J, Summer K-H. Tetrathiomolybdate causes formation of hepatic copper–molybdenum clusters in an animal model of Wilson's disease. J Am Chem Soc 2003;125:1704–5.

[95] Zhang L, Lichtmannegger J, Summer KH, Webb S, Pickering IJ, George GN. Tracing copper–thiomolybdate complexes in a prospective treatment for Wilson's disease. Biochemistry 2009;48:891–7.

[96] Alvarez HM, Xue Y, Robinson CD, Canalizo-Hernández MA, Marvin RG, Kelly RA, Mondragón A, Penner-Hahn JE, O'Halloran TV. Tetrathiomolybdate inhibits copper trafficking proteins through metal cluster formation. Science 2010;327:331–4.

[97] DuBois RS, Rodgerson DO, Martineau G, Shroter G, Giles G, Lilly J, Halgrimson CG, Starzl TE, Sternlieb I, Scheinberg IH. Orthotopic liver transplantation for Wilson's disease. Lancet 1971;297:505–8.

[98] Frommer DJ. Defective biliary excretion of copper in Wilson's disease. Gut 1974;15:125–9.

[99] Gibbs K, Walshe JM. Biliary excretion of copper in Wilson's disease. Lancet 1980;316:538–9.
[100] Li Y, Togashi Y, Sato S, Emoto T, Kang J-H, Takeichi N, Kobayashi H, Kojima Y, Une Y, Uchino J. Spontaneous hepatic copper accumulation in Long-Evans Cinnamon rats with hereditary hepatitis: a model of Wilson's disease. J Clin Invest 1991;87:1858–61.
[101] Rauch H. Toxic milk, a new mutation affecting copper metabolism in the mouse. J Hered 1983;74:141–4.
[102] Bearn AG. A genetical analysis of thirty families with Wilson's disease (hepatolenticular degeneration). Ann Hum Genet Lond 1960;24:33–43.
[103] Frydman M, Bonné-Tamir B, Farrer LA, Conneally PM, Magazanik A, Ashbel S, Goldwitch Z. Assignment of the gene for Wilson disease to chromosome 13: linkage to the esterase D locus. Proc Natl Acad Sci U S A 1985;82:1819–21.
[104] Houwen RH, Scheffer H, te Meerman GJ, van der Vlies P, Buys CHCM. Close linkage of the Wilson's disease locus to D13S12 in the chromosomal region 13q21 and not to ESD in 13q14. Hum Genet 1990;85:560–2.
[105] Chelly J, Monaco AP. Cloning the Wilson disease gene. Nat Genet 1993;5:317–8.
[106] Lutsenko S, Barnes NL, Bartee MY, Dmitriev OY. Function and regulation of human copper-transporting ATPases. Physiol Rev 2007;87:1011–46.
[107] Wilson Disease Mutation Database. http://www.wilsondisease.med.ualberta.ca/database.asp. [Accessed 22 December 2018].
[108] Stapelbroek JM, Bollen CW, Ploos van Amstel JK, van Erpecum KJ, van Hattum J, van den Berg LH, Klomp LWJ, Houwen RHJ. The H1069Q mutation in *ATP7B* is associated with late and neurologic presentation in Wilson disease: results of a meta-analysis. J Hepatol 2004;41:758–63.
[109] Lutsenko S. Copper trafficking to the secretory pathway. Metallomics 2016;8:840–52.
[110] Dodani SC, Firl A, Chan J, Nam CI, Aron AT, Onak CS, Ramos-Torres KM, Paek J, Webster CM, Feller MB, Chang CJ. Copper is an endogenous modulator of neural circuit spontaneous activity. Proc Natl Acad Sci U S A 2014;111:16280–5.
[111] Walshe JM. Wilson's disease: yesterday, today, and tomorrow. Mov Disord 1988;3:10–29.
[112] Walshe JM. Treatment of Wilson's disease: the historical background. Q J Med 1996;89:553–5.
[113] Walshe JM. History of Wilson's disease: 1912 to 2000. Mov Disord 2006;21:142–7.
[114] Walshe JM. The conquest of Wilson's disease. Brain 2009;132:2289–95.
[115] Walshe JM. History of Wilson disease: a personal account. In: Czlonkowska A, Schilsky ML, editors. Handbook of clinical neurology. Wilson disease, vol. 142 (3rd series). Amsterdam: Elsevier; 2017. p. 1–5.
[116] Brewer GJ. Commentary: landmark articles on copper in the field of human health. J Trace Elem Exp Med 2001;14:191–4.
[117] Westermark K. The man behind the syndrome: S. A. Kinnier Wilson. J Hist Neurosci 1993;2:143–50.
[118] Trocello J-M, Broussolle E, Girardot-Tinant N, Pelosse M, Lachaux A, Lloyd C, Woimant F. Wilson's disease, 100 years later…. Rev Neurol (Paris) 2013;169:936–43.
[119] Schilsky ML. A century for progress in the diagnosis of Wilson disease. J Trace Elem Med Biol 2014;28:492–4.
[120] Lanska DJ. History of Wilson's disease. In: Encyclopedia of the neurological sciences. 2nd ed. San Diego: Academic Press; 2014. p. 775–6.
[121] Tanner S. A history of Wilson disease. In: Kerkar N, Roberts EA, editors. Clinical and translational perspectives on Wilson disease. San Diego: Academic Press; 2018. p. 1–11.
[122] Purchase R. The treatment of Wilson's disease, a rare genetic disorder of copper metabolism. Sci Prog 2013;96:19–32.
[123] Purchase R. The link between copper and Wilson's disease. Sci Prog 2013;96:213–23.
[124] Compston A. From the archives. Brain 2009;132:1997–2001.
[125] Compston A. From the archives. Brain 2013;136:688–91.
[126] Reynolds EH, Healy DG, Lees AJ. A film of patients with movement disorders made in Queen Square, London in the mid-1920s by Samuel Alexander Kinnier Wilson. Mov Disord 2011;26:2453–9.
[127] Owen Jr. CA. Wilson's disease. Park Ridge, NJ: Noyes Publications; 1981.
[128] Solioz M. The copper rush of the nineties. Metallomics 2016;8:824–30.

Part II

Molecular Mechanisms

Chapter 2

Normal Human Copper Metabolism

Cynthia Abou Zeid and Stephen G. Kaler
Molecular Medicine Branch, Eunice Kennedy Shriver *National Institute of Child Health and Human Development, National Institutes of Health, Bethesda, MD, United States*

INTRODUCTION

Copper is a trace metal classified as an essential micronutrient. It is only needed in small amounts but is nonetheless crucial for proper human development. Copper is involved in major metabolic pathways, primarily as a cofactor for various metalloenzymes. This trace metal can readily gain or donate electrons, a property that renders it valuable for assisting oxidases in the reduction of molecular oxygen [1]. Nevertheless, this also heightens the risk of accumulation of reactive oxygen species and toxic oxidative stress. Consequently, copper metabolism is tightly regulated both on the general systemic level and on the cellular level.

Total copper levels in the blood are a result of balance from copper absorption via the small intestine and copper excretion in the bile. Regulation occurs at both levels, but biliary excretion is the primary determinant of copper homeostasis. Disruption of this mechanism by inherited mutations in copper transport genes causes copper accumulation in various organs such as the liver, basal ganglia, and eyes, leading to toxicity and organ dysfunction, as seen in Wilson disease. On the cellular level, physiological processes are maintained by a more elaborate regulation mechanism, which involves several copper transporters, chaperones, and binding proteins. This chapter will review the physiological and cellular mechanisms of normal copper metabolism.

COPPER IN HUMAN HEALTH

Much can be learned about the role of copper as an essential component for proper growth and neurodevelopment by examining conditions in which this element is present either in excess or at much lower levels than normal. In states of copper deficiency resulting from genetic mutations, e.g., Menkes disease (ATP7A), intestinal resections, or improper dietary intake, growth, neurodevelopment, the immune system, and connective tissue structures can be severely affected [2]. Conversely, copper presence in amounts that exceed the body's regulatory capacity causes a buildup of this metal in various organs such as the liver and the basal ganglia of the brain (e.g., Wilson disease), and may lead to major organ dysfunction and death [3]. This has led to the discovery of copper involvement in various metabolic pathways, as an essential component for proper development and function of different systems. Copper exerts its physiological role primarily by aiding the function of cuproenzymes throughout the body. In the nervous system, examples of such reactions are those catalyzed by dopamine β-hydroxylase (DBH). DBH is a copper-dependent enzyme responsible for catecholamine production. It catalyzes the reaction that yields norepinephrine from dopamine and is thus essential for various dopaminergic neuronal signaling pathways. Furthermore, copper is thought to have a protective role against oxidative stress in multiple organs including the central nervous system [4]. Copper is also a cofactor for cytochrome *c* oxidase, the last enzyme in the electron transport chain of the mitochondria, where electrons provided by copper are used to convert molecular oxygen into water molecules. Copper also serves as a cofactor for Cu/Zn superoxide dismutase (SOD1), which is responsible for the detoxification of reactive oxygen species [5]. Low copper levels have been associated with myeloneuropathies and neurodegenerative diseases such as Alzheimer's disease, dementia, and amyotrophic lateral sclerosis, suggesting a role for this element in proper myelination [6, 7] and protection from neurodegeneration [4, 8].

Connective tissue and blood vessel development are physiological processes relying on copper-dependent enzymes. Lysyl oxidase is responsible for collagen and elastin cross-linking, which increases the resistance of connective tissue fibers. Superoxide dismutase 3, an extracellular isoform of SOD, acts as an antioxidant and radical scavenger and plays a role in the prevention or suppression of skin inflammation [9]. Copper also functions as a cofactor for tyrosinase, the enzyme responsible for the synthesis of melanin pigment in the skin and retinal epithelial pigment. Additionally, copper plays an important role in the metabolism of iron, a metal present in much higher concentrations in the body. Copper is a cofactor of ceruloplasmin, a ferroxidase responsible of the oxidation of ferrous iron (Fe^{2+}) into ferric iron (Fe^{3+}). Ferric iron is then exported out of the cells and is able to bind with its transporter transferrin in the circulation. In addition, hephaestin, the enzyme responsible for the transport of dietary iron from the intestinal lumen into the blood, also relies on copper for proper functioning [10]. Modulation of iron availability by copper can be deleterious and cause iron-deficient anemia but may also be used as an antimicrobial strategy. Several transition metals like iron and copper are needed for bacterial and fungal growth, and restriction of their availabilities to microbes can help to fight infections by these pathogens [11]. Conversely, because of its capacity to induce oxidative stress and cell death, copper toxicity could be exploited as a host defense against bacterium and fungi [12]. Copper is therefore essential to diverse physiological processes like nervous system development, connective tissue metabolism, metal homeostasis, and protection against pathogens and oxidative stress.

DIETARY COPPER

Copper is readily available in the diet and present in a variety of foods [13]. Whole grains, beans, nuts, chocolate, potatoes, shrimp, and organ meats are some examples of copper-rich nutrients. In adults, the recommended dietary allowance (RDA) for copper is 900 μg/day, with different requirements depending on age and sex. Copper needs are largely met by a balanced diet, and copper deficiency rarely manifests as a result of dietary deficiency alone. The average adult has a median intake of 1.0–1.6 mg of copper per day [1].

COPPER ABSORPTION

Copper absorption was initially studied by following the uptake of radiolabeled copper. These experiments showed that the stomach and duodenum are the two main sites of copper absorption, which occurs early on in the digestion [14–16]. Interestingly, copper uptake seems to depend on dietary availability of the nutrient: absorption is higher in low-copper diets and lower in high-copper diets [16], suggesting that the intestinal barrier is the first regulator of total body copper levels [17]. After the digestion of copper-containing nutrients, dietary copper has to undergo reduction to yield the cuprous ion form (Cu^+), prior to its absorption [18]. This is thought to be accomplished by cytochrome *b* reductase 1 (CYBR1) or six-transmembrane epithelial antigen of the prostate (STEAP) reductase [19]. To get from the intestinal lumen into the blood via the intestinal epithelial cells, Cu^+ has to be absorbed at the apical membrane of the enterocyte, pass through the cytoplasm while avoiding oxidative damage to intracellular components, and then traverse the basolateral membrane to enter the bloodstream. To do so, in physiological states, copper absorption in the small intestines occurs primarily via carrier-mediated transport [18]. In the intestinal brush border, two copper transporters are thought to play a role in copper uptake from the intestinal lumen to the enterocyte: copper transporter 1 (CTR1) and divalent metal transporter 1 (DMT1) (Fig. 1). On the apical aspect of the enterocyte, the reduced cuprous ion binds with high affinity to copper transporter 1 (CTR1), a protein encoded by the *SLC31A1* gene. This high-affinity copper transporter takes up Cu^+ from the intestinal lumen to the epithelial cell, in an energy-independent manner [2, 3]. CTR1 is the major copper importer present on the apical site of enterocytes, but conflicting reports have also shown a possible role of DMT1 in copper transfer into intestinal cells. DMT1 is a well-known iron transporter responsible for Fe^{2+} uptake into enterocytes. Copper ions are thought to bind nonspecifically to this transporter, which may represent an alternative mechanism of Cu^+ entry into the cell, perhaps only during iron deficiency [19–21]. In addition, when luminal concentrations of the trace metal increase, transporter-independent paracellular copper diffusion could be an alternative route of copper absorption, as reported in weanling and suckling rats [22]. Older animals develop a saturable copper uptake system, and paracellular transport does not seem to exist under normal conditions [23].

Once in the cytoplasm, Cu^+ is directed to the intracellular compartments where it is needed and to other copper transport proteins. This process is possible because of small proteins called copper chaperones that keep free Cu^+ ions from causing intracellular damage by binding and transporting them. In particular, the CTR1 carboxyl terminus and intracellular loops interact with the metallochaperone, ATOX1 [24]. Antioxidant 1 copper chaperone (ATOX1) is responsible for copper transport in the cytosol and delivery to two major energy-dependent copper pumps: ATP7A and ATP7B. ATP7A, also called copper-transporting ATPase 1, has a ubiquitous distribution and is found in multiple mammalian cell types, except for hepatocytes beyond the perinatal period. Its major roles are copper absorption in the small intestines and copper transport via the blood-cerebrospinal fluid and blood-brain barriers. In the liver, ATP7B is present uniformly and is responsible of metalation of ceruloplasmin and excretion of copper into the bile from the apical surface of hepatocytes.

Normal Human Copper Metabolism **Chapter | 2** 19

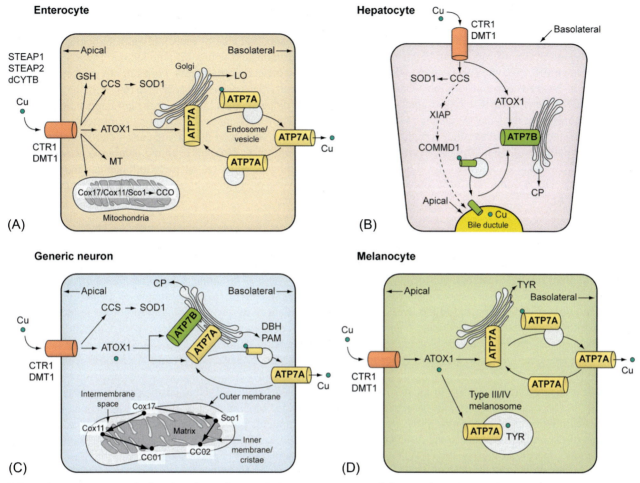

FIG. 1 Copper transport mechanisms in various cell types. (A) In enterocytes, copper (Cu$^+$) uptake is mediated by CTR1, possibly in concert with the metalloreductases STEAP1, STEAP2, and dCYTB. The roles of CTR2 (not shown) and DMT1 in this process are less certain. Within enterocytes, GSH and MT function in copper sequestration and storage. The chaperones CCS, ATOX1, Cox17, Cox11, and Sco1 ferry copper to specific proteins or organelles. With an increase in copper levels, ATP7A traffics to the basolateral surface and pumps copper into the blood. (B) In hepatocytes, ATOX1 provides copper to ATP7B for metalation of CP and traffics to the apical membrane to pump copper into the bile, the body's major mechanism for copper removal. CCS has been proposed to deliver copper to XIAP, which may interact *(dashed lines)* with COMMD1, a protein mutated in hepatic copper toxicosis of Bedlington terriers that may modulate ATP7B activity. This putative pathway is denoted by dashed lines. (C) In neuronal cells, ATP7A and ATP7B are expressed and required for maturation of CP, DBH, and PAM. CCO metalation is illustrated. (D) In melanocytes, TYR acquires copper within melanosomes via ATP7A, which also localizes to the *trans*-Golgi 4. *Abbreviations*: *ATOX1*, antioxidant 1 copper chaperone; *ATP7A*, copper-transporting ATPase 1; *ATP7B*, copper-transporting ATPase 2; *CCS*, copper chaperone for SOD1; *CCO*, cytochrome *c* oxidase; *COMMD1*, COMM domain-containing protein 1; *COX11*, cytochrome *c* oxidase assembly protein Cox11; *Cox17*, cytochrome *c* oxidase copper chaperone; *CP*, ceruloplasmin; *CTR*, copper transporter; *DBH*, dopamine-β-hydroxylase; *dCYTB*, cytochrome *b* reductase 1; *DMT1*, divalent metal transporter 1; *GSH*, glutathione; *LO*, lysyl oxidase; *MT*, metallothionein; *PAM*, peptidylglycine α-amidating monooxygenase; *Sco1*, protein SCO1 homologue; *SOD1*, superoxide dismutase; *STEAP*, six-transmembrane epithelial antigen of prostate; *TYR*, tyrosinase; and *XIAP*, X-linked inhibitor of apoptosis. *(Reprinted with permission from Kaler SG. ATP7A-related copper transport diseases-emerging concepts and future trends. Nat Rev Neurol 2011;7(1):15–29.)*

In the intestinal cells, ATP7A uses the energy derived from ATP hydrolysis, to transport copper from the enterocyte into the blood via the basolateral border. This final step of copper absorption is subject to regulation, as ATP7A levels are inversely correlated with systemic copper levels [25]. This is also explained by the response of ATP7A to intracellular copper concentrations. In normal copper states, ATP7A is usually localized in the *trans*-Golgi network, where it is responsible for the metalation of various cuproenzymes in the secretory pathway. When excess intracellular copper is present, ATP7A translocates to the plasma membrane and effluxes copper to the extracellular milieu, to prevent intracellular accumulation and toxicity. The latter is classically the situation in enterocytes, when copper is imported into the cellular cytoplasm by CTR1, leading to a higher intracellular copper concentration, which drives ATP7A to the basolateral membrane and permits transport of the micronutrient into the blood [26].

To sum up, luminal copper is reduced to Cu^+ and then imported via the apical side of the intestinal epithelial cell via CTR1. It is then loaded onto its chaperone ATOX1 that delivers it to ATP7A. The latter protein is the ultimate transporter responsible for its delivery into the portal circulation across the basolateral side of intestinal epithelial cells.

Intestinal absorption of copper can be affected by various modifiable and nonmodifiable factors. Demographic characteristics can influence copper absorption and reflect differences in copper requirements. For example, one report showed that gender influenced copper absorption, which was higher in women (71%) compared with men (64%), in the 20–59 years age group [27]. The efficiency of copper uptake can also depend on dietary copper content, with high-copper diets resulting in lower absorption of the metal, as mentioned previously. Bioavailability of copper can additionally vary depending on the food source, since copper absorption is higher in a protein-based diet compared with a vegetarian diet [23]. Finally, the intraluminal contents of the intestines can also interfere with copper uptake. Zinc, for example, inhibits copper absorption in the intestinal brush barrier [28], by induction of metal-chelating proteins called metallothioneins, and potentially by competing with a common Cu/Zn transporter in the enterocytes, or influencing activity of copper transporters or chaperones [18].

COPPER TRANSPORT AND EXCRETION

After absorption from the small intestine to the bloodstream, copper is predominantly bound to proteins like albumin [29] and transcuprein (alpha-2-macroglobulin) [30] and to amino acids such as histidine, glutamine, threonine, and cysteine [31]. These interactions with low-molecular-weight ligands are thought to facilitate copper uptake by the peripheral tissues. Complexed copper is then directed to the liver via the portal circulation. In this major organ of copper homeostasis, the metal may have three distinct destinations, depending on the body's needs and available copper pool. Copper can be stored in hepatocytes, secreted back into the bloodstream in the form of ceruloplasmin, or excreted into the bile. The latter is the major regulatory function of the liver in copper homeostasis, with the kidneys performing a secondary role in copper excretion [1].

On the molecular level, similar to its role in intestinal epithelial cells, CTR1 imports copper into the hepatocyte (Fig. 1B). In the cytosol, the ion can be retained intracellularly by binding to glutathione or small metallothioneins [32]. By neutralizing free copper ions in the cytoplasm, metallothioneins protect cells from toxicity [33]. Alternatively, CTR1 can direct imported copper to the ATOX1 chaperone, which then shuttles it to ATP7B. The latter transporter, also referred to as copper-transporting P-type ATPase 2, is the primary player in copper metabolism in the liver. Its roles in copper transport are similar to those of ATP7A. In normal copper states, ATP7B is localized to the *trans*-Golgi network and is responsible for copper incorporation into ceruloplasmin, which is then secreted into the circulation. As previously mentioned, this protein is implicated both in copper and iron metabolism. Ceruloplasmin is the major copper-containing protein in the blood, as 85%–95% of plasma copper is complexed within it. However, this protein does not seem to be directly implicated in copper uptake into cells [34]. This assumption is supported by the fact that copper metabolism is unaffected by mutations of ceruloplasmin gene that cause a rare autosomal recessive condition named aceruloplasminemia. Instead, iron accumulation in the liver and the brain is found, supporting the role of this ferroxidase in iron transport rather than copper transport [35]. Functionally, copper is therefore merely a cofactor for ceruloplasmin, just as for various other metalloenzymes. When hepatocyte copper levels rise, ATP7B translocates toward the apical hepatocyte membrane in a cytoplasmic vesicular component, to facilitate excretion of excess copper into the bile [36–38]. The precise mechanisms of ATP7B trafficking to the apical membrane and release of copper into bile are incompletely understood. One recently proposed mechanism is the movement of ATP7B and copper into lysosomes, which then undergo exocytosis via the canalicular membrane of hepatocytes [39]. ATP7B is therefore a crucial protein in copper excretion and homeostasis, and impairment of this vehicle causes Wilson disease, addressed elsewhere in this book.

CONCLUSION

Copper is a versatile divalent metal necessary as a cofactor for metalloenzymes but that can also induce toxic oxidative stress, for which reason its levels are tightly regulated. Multiple gene products intervene to maintain copper homeostasis at different levels and in various cell types. Copper importers, such as CTR1, are responsible for copper uptake into the cytoplasm of cells. In the intracellular milieu, chaperones like ATOX1 shuttle copper to designated cellular compartments and other copper transporters. Intracytoplasmic proteins such as metallothioneins and glutathione bind to Cu^+ and act as buffers to prevent the production of toxic reactive oxygen species. Finally, copper exporters such as ATP7A and ATP7B have dual functions depending on intracellular copper levels. In normal copper states, they transport copper into the secretory pathway for metalation of cuproenzymes. With excessive intracellular copper levels, they move to the plasma membrane and flush copper outside of the cell. Current efforts are directed at dissecting the specific mechanisms that govern the intracellular itineraries of copper transporters and analyzing their interactions with other protein partners. Such studies should allow further understanding of the function of these proteins and possibly contribute to the development of comprehensive treatments for copper-related diseases.

REFERENCES

[1] Institute of Medicine (US) Panel on Micronutrients. Dietary reference intakes for vitamin A, vitamin K, arsenic, boron, chromium, copper, iodine, iron, manganese, molybdenum, nickel, silicon, vanadium, and zinc—NCBI Bookshelf. National Academies Press (US); 2001. [Internet]. Available from, https://www.ncbi.nlm.nih.gov/pubmed.

[2] Nose Y, Wood LK, Kim B-E, Prohaska JR, Fry RS, Spears JW, et al. Ctr1 is an apical copper transporter in mammalian intestinal epithelial cells in vivo that is controlled at the level of protein stability. J Biol Chem 2010;285(42):32385–92.

[3] Gupta A, Lutsenko S. Human copper transporters: mechanism, role in human diseases and therapeutic potential. Future Med Chem 2009;1(6):1125–42.

[4] Desai V, Kaler SG. Role of copper in human neurological disorders. Am J Clin Nutr 2008;88(3):855S–8S.

[5] Polishchuk R, Lutsenko S. Golgi in copper homeostasis: a view from the membrane trafficking field. Histochem Cell Biol 2013;140(3):285–95.

[6] Hammond N, Wang Y, Dimachkie M, Barohn R. Nutritional neuropathies. Neurol Clin 2013;31(2):477–89.

[7] Kumar N, Crum B, Petersen RC, Vernino SA, Ahlskog JE. Copper deficiency myelopathy. Arch Neurol 2004;61(5):762–6.

[8] Zucconi GG, Cipriani S, Scattoni R, Balgkouranidou I, Hawkins DP, Ragnarsdottir KV. Copper deficiency elicits glial and neuronal response typical of neurodegenerative disorders. Neuropathol Appl Neurobiol 2007;33(2):212–25.

[9] Kwon M-J, Kim B, Lee YS, Kim T-Y. Role of superoxide dismutase 3 in skin inflammation. J Dermatol Sci 2012;67(2):81–7.

[10] Peña MMO, Lee J, Thiele DJ. A delicate balance: homeostatic control of copper uptake and distribution. J Nutr 1999;129(7):1251–60.

[11] Samanovic MI, Ding C, Thiele DJ, Darwin KH. Copper in microbial pathogenesis: meddling with the metal. Cell Host Microbe 2012;11(2):106–15.

[12] Weiss G, Carver PL. Role of divalent metals in infectious disease susceptibility and outcome. Clin Microbiol Infect 2018;24(1):16–23.

[13] Pennington JAT, Schoen SA, Salmon GD, Young B, Johnson RD, Marts RW. Composition of core foods of the U.S. food supply, 1982-1991: III. Copper, manganese, selenium, and iodine. J Food Compos Anal 1995;8(2):171–217.

[14] Vancampen DR, Mitchell EA. Absorption of Cu-64, Zn-65, Mo-99, and Fe-59 from ligated segments of the rat gastrointestinal tract. J Nutr 1965;86:120–4.

[15] Crampton RF, Matthews DM, Poisner R. Observations on the mechanism of absorption of copper by the small intestine. J Physiol 1965;178(1):111–26.

[16] Turnlund JR, Keyes WR, Anderson HL, Acord LL. Copper absorption and retention in young men at three levels of dietary copper by use of the stable isotope 65Cu. Am J Clin Nutr 1989;49(5):870–8.

[17] Turnlund JR. Human whole-body copper metabolism. Am J Clin Nutr 1998;67(5 Suppl.):960S–4S.

[18] Cousins RJ, Liuzzi JP. Trace metal absorption and transport. In: Said HM, editor. Physiology of the gastrointestinal tract. 6th ed. Academic Press; 2018. p. 1485–98. Available from, https://www.sciencedirect.com/science/article/pii/B978012809954400061X [cited 2018 Apr 26; Internet; Chapter 61].

[19] Gulec S, Collins JF. Molecular mediators governing iron-copper interactions. Annu Rev Nutr 2014;34:95–116.

[20] Gunshin H, Mackenzie B, Berger UV, Gunshin Y, Romero MF, Boron WF, et al. Cloning and characterization of a mammalian proton-coupled metal-ion transporter. Nature 1997;388(6641):482–8.

[21] Illing AC, Shawki A, Cunningham CL, Mackenzie B. Substrate profile and metal-ion selectivity of human divalent metal-ion transporter-1. J Biol Chem 2012;287(36):30485–96.

[22] Varada KR, Harper RG, Wapnir RA. Development of copper intestinal absorption in the rat. Biochem Med Metab Biol 1993;50(3):277–83.

[23] Berghe VD, Ve P, Klomp LW. New developments in the regulation of intestinal copper absorption. Nutr Rev 2009;67(11):658–72.

[24] Levy AR, Nissim M, Mendelman N, Chill J, Ruthstein S. Ctr1 intracellular loop is involved in the copper transfer mechanism to the Atox1 metallochaperone. J Phys Chem B 2016;120(48):12334–45.

[25] Chun H, Catterton T, Kim H, Lee J, Kim B-E. Organ-specific regulation of ATP7A abundance is coordinated with systemic copper homeostasis. Sci Rep 2017;7. [Internet; cited 2018 May 2]. Available from, http://www.ncbi.nlm.nih.gov/pmc/articles/PMC5607234/.

[26] Kaler SG. ATP7A-related copper transport diseases-emerging concepts and future trends. Nat Rev Neurol 2011;7(1):15–29.

[27] Johnson PE, Milne DB, Lykken GI. Effects of age and sex on copper absorption, biological half-life, and status in humans. Am J Clin Nutr 1992;56(5):917–25.

[28] Prasad AS, Brewer GJ, Schoomaker EB, Rabbani P. Hypocupremia induced by zinc therapy in adults. JAMA 1978;240(20):2166–8.

[29] Gordon DT, Leinart AS, Cousins RJ. Portal copper transport in rats by albumin. Am J Physiol 1987;252(3):E327–33. Pt 1.

[30] Liu N, Lo LS, Askary SH, Jones L, Kidane TZ, Trang T, et al. Transcuprein is a macroglobulin regulated by copper and iron availability. J Nutr Biochem 2007;18(9):597–608.

[31] Neumann PZ, Sass-Kortsak A. The state of copper in human serum: evidence for an amino acid-bound fraction. [Internet; cited 2018 May 3]. Available from, https://www.jci.org/articles/view/105566/pdf/render; 1967.

[32] Babula P, Masarik M, Adam V, Eckschlager T, Stiborova M, Trnkova L, et al. Mammalian metallothioneins: properties and functions. Metallomics 2012;4(8):739–50.

[33] Ruttkay-Nedecky B, Nejdl L, Gumulec J, Zitka O, Masarik M, Eckschlager T, et al. The role of metallothionein in oxidative stress. Int J Mol Sci 2013;14(3):6044–66.
[34] Hellman NE, Gitlin JD. Ceruloplasmin metabolism and function. Annu Rev Nutr 2002;22:439–58.
[35] Harris ZL, Takahashi Y, Miyajima H, Serizawa M, MacGillivray RT, Gitlin JD. Aceruloplasminemia: molecular characterization of this disorder of iron metabolism. Proc Natl Acad Sci U S A 1995;92(7):2539–43.
[36] Wang Y, Hodgkinson V, Zhu S, Weisman GA, Petris MJ. Advances in the understanding of mammalian copper transporters. Adv Nutr 2011;2(2):129–37.
[37] Roelofsen H, Wolters H, Van Luyn MJA, Miura N, Kuipers F, Vonk RJ. Copper-induced apical trafficking of ATP7B in polarized hepatoma cells provides a mechanism for biliary copper excretion. Gastroenterology 2000;119(3):782–93.
[38] Lim CM, Cater MA, Mercer JFB, Fontaine SL. Copper-dependent interaction of dynactin subunit p62 with the N terminus of ATP7B but not ATP7A. J Biol Chem 2006;281(20):14006–14.
[39] Polishchuk EV, Concilli M, Iacobacci S, Chesi G, Pastore N, Piccolo P, et al. Wilson disease protein ATP7B utilizes lysosomal exocytosis to maintain copper homeostasis. Dev Cell 2014;29(6):686–700.

Chapter 3

ATP7B Function

Hannah Pierson and Svetlana Lutsenko
Department of Physiology, Johns Hopkins University School of Medicine, Baltimore, MD, United States

THE BIOCHEMICAL FUNCTION OF ATP7B

Human ATP7B is a member of a large family of membrane-bound transporters called P-type ATPases or ATP-driven ion pumps. The members of this family use the energy of ATP hydrolysis to transport their substrates across various cellular membranes [1]. Human ATP7B belongs to a P_{1B}-ATPase subfamily, which includes evolutionarily conserved transporters involved in the homeostasis of transition metals (copper [Cu], zinc, cadmium, silver, and lead [2, 3]). Human cells express two structurally similar Cu-transporting ATPases: ATP7A and ATP7B. The biochemical function of these two proteins (i.e., an ATP-driven transmembrane transfer of Cu) is the same. However, ATP7A and ATP7B have distinct patterns of cell and tissue expression and differ in their regulatory responses. The physiological roles of ATP7A and ATP7B do not overlap, and neither ATP7A nor ATP7B compensates for the loss of each other's function, when mutated in disease.

ATP7B hydrolyzes ATP to translocate a reduced Cu(I) from the cytosol into the lumen of the *trans*-Golgi network (TGN) and into specialized vesicles of secretory pathway. In these compartments, Cu is used for functional activation of a Cu-dependent ferroxidase ceruloplasmin and for storage/export, respectively (Fig. 1). A small soluble protein Atox1, which binds and carries Cu in the cytosol, serves as a donor of Cu for ATP7B (Fig. 1).

Human ATP7B is a fairly large transmembrane protein (1465 amino acid residues, 165 kDa) with multiple domains (Fig. 2). The domains are involved in binding of ligands (Cu and ATP), regulation of conformational transitions, and determining the intracellular localization of ATP7B (for recent reviews, see Refs. [4–6]). Although the structure of the full-length ATP7B has not yet been solved, the high-resolution structures of several isolated domains have been determined [7–10]. These structures provide a useful template for studies of structural effects of the disease-causing mutations in corresponding domains [11, 12]. It has also become clear that the domains interact and affect each other's function [13–15]. Mutations or posttranslational modifications in one domain often have long-range effects causing significant changes in overall protein stability, activity, and localization (see Chapter 4), an important consideration for understanding of ATP7B function and regulation in normal cells and in disease.

The main molecular events underlying ATP7B-mediated Cu transport have been characterized (for recent review, see Ref. [4]). To perform its Cu transport function, ATP7B binds ATP within the cytosolic nucleotide-binding domain (N-domain). This step is obligatory, and the mutations that disrupt ATP binding cause the loss of ATP7B Cu transport activity (see Chapter 4). Cu, which is destined for transport, binds within the membrane portion of the protein. Simultaneous binding of both ligands (ATP and Cu) at the catalytic and transport sites, respectively, triggers ATP hydrolysis. During this process, the terminal γ-phosphate of ATP is transferred to an invariant aspartate residue D1027 forming transient phosphorylated intermediate (the so-called "catalytic" phosphorylation). D1027 is located within the highly conserved $D^{1027}KTG$ motif of the P-domain (Fig. 2); thus, when ATP binds, it bridges the N- and P-domains. D1027 is also only 42 residues away from the Cu transport site(s). Mutations that disrupt Cu binding to the transport sites also inhibit catalytic phosphorylation [16], despite separation between the Cu and ATP binding sites. This tight physical and functional coupling between different functional domains is a common property of all P-type ATPases. The interdomain coupling explains numerous effects that single-site disease-causing mutations often have on the activity, protein stability, and trafficking of ATP7B [17].

The ATP7B Cu transport sites are located within the transmembrane portion of the protein. The invariant $C^{983}PC^{985}$ motif contributes two cysteine residues to the binding/transport sites. Other amino acid residues are also likely to interact with Cu, although direct analysis of Cu coordination environment within the transport sites is still lacking. Transient phosphorylation of D1027 initiates conformational transitions within ATP7B, which allow Cu to dissociate from its intramembrane sites and exit into the lumen of secretory pathway [18]. There, Cu binds to Cu-dependent enzymes (such as

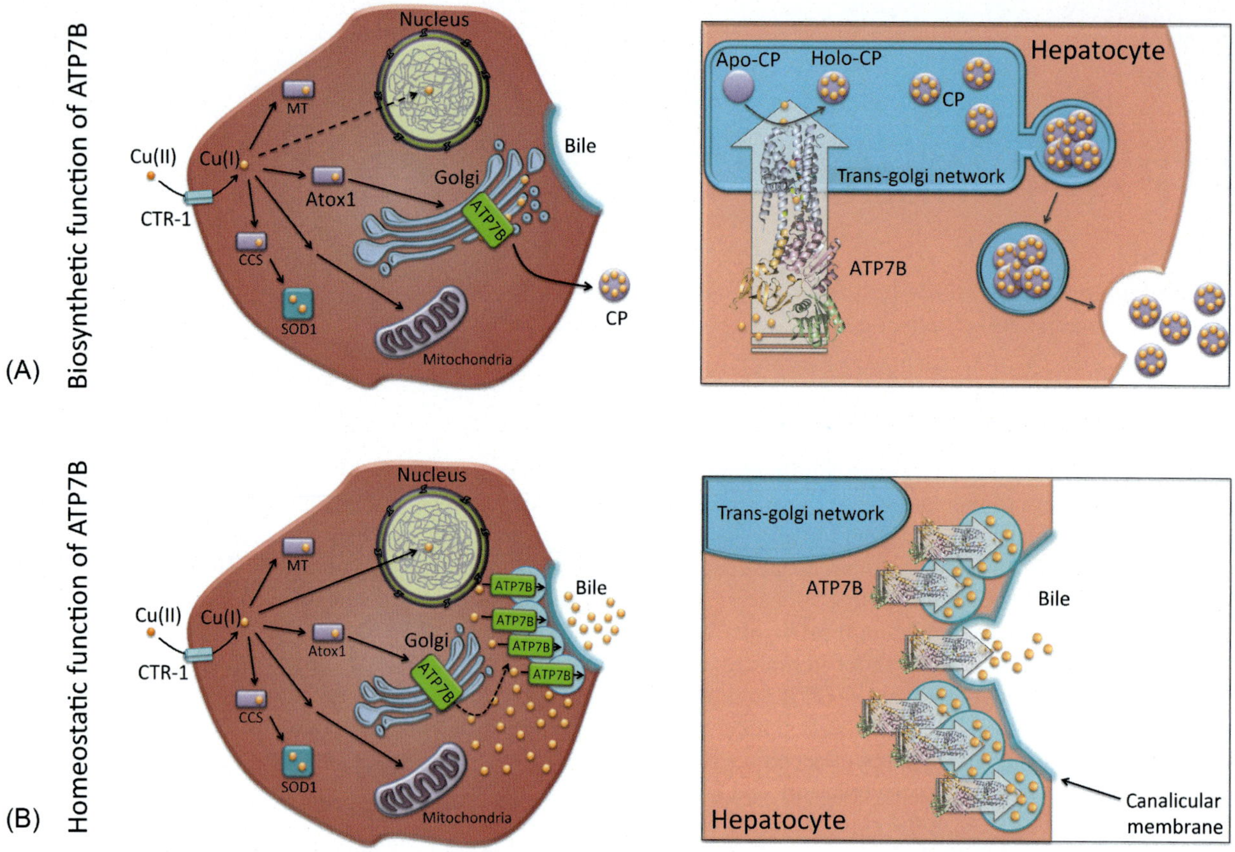

24 PART | II Molecular Mechanisms

FIG. 1 The dual function of ATP7B in hepatocytes. (A) Under basal conditions, Cu enters hepatocytes through the high-affinity Cu transporter CTR1 and is distributed to cytosolic chaperones. Atox1 delivers Cu to ATP7B. ATP7B pumps Cu into the lumen of the *trans*-Golgi network to facilitate metalation of ceruloplasmin. (B) When systemic Cu is elevated, hepatocytes become Cu overloaded. To export excess Cu, ATP7B traffics to vesicles in a subapical compartment and transports Cu into these vesicles, which then fuse with the canalicular membrane releasing Cu into the bile.

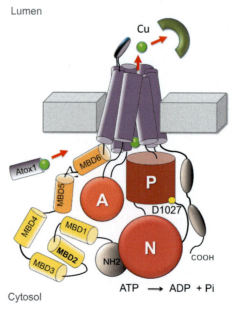

FIG. 2 Molecular architecture of ATP7B. ATP7B has a small (<30% by mass) transmembrane portion, which consists of eight transmembrane segments and forms a Cu-translocation pathway. The invariant CPC motif is located within the transmembrane domain. The rest of the protein is cytosolic and consists of a small N-terminal segment involved in apical trafficking, followed by six metal-binding domains (MBDs). MBDs are connected by loops with sites for phosphorylation by kinases. The A-, P-, and N-domains are central for catalytic activity and conformational transitions. The catalytic aspartate residue (D1027) is located within the P-domain and undergoes transient phosphorylation upon ATP hydrolysis.

ceruloplasmin) and, possibly, other molecules. Release of Cu from the protein is associated with the hydrolysis of the phosphorylated intermediate; the released energy is used to trigger further conformational changes and "reset" ATP7B for a new cycle of ATP hydrolysis and Cu transport. The ATPase activity for endogenous and recombinant ATP7B in microsomal membrane preparations was shown to be in a range of 1–1.8 µmol Pi/mg protein/h [16], although higher values (0.25–0.75 µmol Pi/mg of ATP7B protein per min) have also been suggested [4].

Despite significant progress in the characterization of ATP7B structure and function, it remains unclear how many Cu atoms bind within the transmembrane portion of ATP7B and how many Cu atoms are transported per one hydrolyzed ATP. Studies using bacterial orthologues suggest that there could be two intramembrane sites [19] but only one Cu(I) is transported per cycle of ATP hydrolysis [20]. In addition to the sites within the membrane domain, Cu binds to the cytosolic sites located within the large N-terminal "tail" of ATP7B, which comprises about 40% of the entire protein (Fig. 2). This region of ATP7B has important regulatory functions and modulates both the transport activity of ATP7B and its localization within the cell (in either TGN or vesicles).

CELLULAR FACTORS THAT AFFECT ATP7B FUNCTION

Intracellular Trafficking Allows ATP7B to Switch Between the Biosynthetic and Homeostatic Functions

In all cells and tissues characterized so far, ATP7B is targeted predominantly to intracellular compartments. The data on ATP7B imaging in cells are supported by quantitative mass-spectrometry analysis of proteomes of various cell compartments in hepatocytes [21]. The biochemical function of ATP7B in different cell compartments is the same, that is, ATP7B transfers Cu from the cytosol into the compartmental lumen. However, consequences of this transport activity differ depending on ATP7B localization. It is commonly believed that under basal or Cu-limiting conditions, ATP7B is targeted to the trans-Golgi network, TGN. This conclusion is supported by the colocalization of ATP7B with TGN38 or mannose-6-phosphate receptor [22–24] (Fig. 1). However, depending on the cell line, ATP7B was also found under basal conditions in the post-TGN compartment (identified by staining with Syntaxin 6 [25]) and late endosomal compartment, identified by Rab7 [26]. Only one high-resolution (electron microscopy) study of the endogenous ATP7B in liver tissues was published [27]. This study found endogenous ATP7B to be targeted to a subset of TGN vesicles at different stages of maturation [27], perhaps explaining the somewhat varied findings in cell lines. In the TGN/post-TGN compartment, ATP7B performs a biosynthetic function, that is, it delivers Cu to a Cu-dependent ferroxidase ceruloplasmin, completing the last step of biosynthesis and functional maturation of this protein [28].

ATP7B traffics between the TGN and vesicles in response to different signals (Cu elevation and hormonal signaling) to accommodate changing metabolic needs of a cell. In vesicles, ATP7B sequesters Cu for storage or further export out of the cell (for review, see Ref. [6]). This function is homeostatic, that is, the major outcome of ATP7B activity in vesicles is the maintenance of the appropriate Cu concentrations in the cytosol. In hepatocytes, ATP7B traffics from the TGN to vesicles that are distinct from the vesicles that carry ceruloplasmin to the plasma membrane (our data). Therefore, ATP7B in vesicles does not transfer Cu to ceruloplasmin. Instead, ATP7B sequesters excess Cu for eventual export via plasma membrane. Whether ATP7B transports Cu across the apical membrane or appears at the membrane transiently during vesicle fusion is unclear.

Under high Cu conditions, ATP7B can traffic to lysosomes [29]. This conclusion is based on the colocalization of ATP7B with the lysosomal marker LAMP1, and it led to a suggestion that lysosomal exocytosis of Cu could be a mechanism facilitating ATP7B-dependent Cu efflux [29]. However, this idea remains controversial [22]. It could be that trafficking of ATP7B to lysosomes occurs only when Cu is very high. Under more moderate Cu elevation, ATP7B moves (via basolateral membrane and transcytosis) to a distinct subapical compartment (Fig. 1). From this compartment, it can reach the apical membrane in a step that is sensitive to acidification and is inhibited by bafilomycin [30]. The short nine-amino acid sequence $F^{37}AFDNVGYE^{45}$ located in the N-terminal domain of ATP7B participates in the TGN retention and, in hepatocytes, is required for trafficking of ATP7B toward the apical membrane [31]. Myosin 5B, an important player in apical polarity and trafficking, regulates ATP7B delivery from the subapical compartment to the apical membrane [32].

Glutaredoxin 1 Interacts With ATP7B and Regulates ATP7B Trafficking

The N-terminus of ATP7B contains six metal-binding domains (MBDs), each with a characteristic Cys-x-x-Cys motif for the coordination of one Cu(I) per MBD (Fig. 2). In order for Cu to bind, the cysteine residues within the binding site must be reduced. Maintenance of the reduced state requires the activity of glutathione-dependent oxidoreductase glutaredoxin 1 (Grx1), which can directly bind to the N-terminal region of ATP7B (Singleton et al., 2010). Studies with the glutathione-detecting antibodies suggest that ATP7B could be glutathionylated, although whether MBDs or other Cys residues in ATP7B are glutathionylated remains unclear [33]. The suggestion that MBDs are glutathionylated is at odds with the in vitro studies of the MBDs redox potentials, which suggest that under physiologically relevant conditions, MBDs should be mostly reduced and unmodified [34]. It could be that glutathionylation is substoichiometric (i.e., only affects a fraction of protein). The role of glutathionylation could be to protect the metal-binding cysteines from oxidation under conditions of metabolic stress or other conditions associated with an oxidative shift in the cellular redox environment. The importance of maintaining cysteine in the reduced (or Cu-bound) state was illustrated in experiments where replacement of one metal-binding site Cys-x-x-Cys in MBD2 caused oxidation of several other MBDs [35].

Inactivation of Grx1 blocks the ability of ATP7B to move from the TGN to vesicles when Cu is elevated, presumably due to the disruption of Cu binding by MBDs [33]. Trafficking from the TGN to vesicles in response to Cu elevation is a step necessary for a subsequent Cu efflux out of the cell. Therefore, the loss of trafficking causes intracellular Cu accumulation and is equivalent to the loss of ATP7B Cu export function. It is also possible that Grx1 inhibition does not directly inactivate ATP7B but rather blocks Cu transfer from Atox1 (see below) or has both effects.

Atox1

The Cu chaperone Atox1 is a small (68 amino acid residues, 7.6 kDa) cytosolic protein, which is thought to carry Cu between the site of Cu entry into the cell and the secretory pathway, where Cu-transporting ATPases are localized. Similarly to MBDs of ATP7B, Atox1 has a Cys-x-x-Cys motif for the binding of one Cu atom. The metal-binding site of Atox1 is highly sensitive to changes in a cellular glutathione balance (which is determined by the ratio of reduced to oxidized glutathione, GSH-GSSG). When the GSH-GSSG ratio decreases, Atox1 undergoes reversible oxidation in vitro and in cells [34, 36, 37]. Grx1 is directly involved in the regulation of the oxidation state of Atox1 [34, 37]. The in vitro studies show that Atox1 is more sensitive to oxidation than ATP7B MBDs [34]. Therefore, Atox1 can respond to oxidative changes in the cytosol and adjust the amount of Cu transferred to the secretory pathway even if the oxidation state of Cu-transporting ATPases ATP7A and ATP7B is unchanged.

Physiological relevance of redox modulation of Atox1 was demonstrated in a cultured neuroblastoma cell line SH-SY5Y and in motor neurons of a developing spinal cord [36]. In both experimental systems, neuronal differentiation was associated with changes in the GSH-GSSG ratio. A higher ratio (a more reducing environment) produced a less oxidized Atox1 and therefore increased the ability of Atox1 to bind Cu, resulting in a higher Cu flow through the secretory pathway. The diminished Cu transfer to a secretory pathway in response to Atox1 oxidation was also directly demonstrated [34]. Whether how strongly Atox1 oxidation affects Cu transport to ATP7B or ATP7A or both transporters is still unclear. It is also uncertain whether Atox1 is an obligatory partner of ATP7B. The Atox1$^{-/-}$ mice show diminished Cu transfer to the secretory pathway [38, 39], but whether this effect is due to lower activity of ATP7A, ATP7B, or both ATP7A and ATP7B has not been explored. Survival of Atox1$^{-/-}$ mice into adulthood suggests that the loss of Atox1 function is compensated by other Cu-carrying proteins or low-molecular-weight carriers. It remains to be determined whether the Atox1$^{-/-}$ mice develop liver pathology similar to Wilson disease, which might be expected if Atox1 is essential for ATP7B activity.

Currently, the role of Atox1 in ATP7B function is based on the in vitro evidence. The Atox1-Cu complex exchanges Cu with the purified recombinant N-terminal domain of ATP7B and stimulates catalytic phosphorylation of the full-length ATP7B [40, 41]. In a metal-free form, Atox1 removes Cu from some, but not all, N-terminal MBDs of ATP7B and partially inhibits catalytic phosphorylation of ATP7B [41]. The mechanism of Cu transfer from Atox1 to individual MBDs of ATP7B has been extensively characterized (for review, see Ref. [42]). The studies revealed differences between MBDs in their interactions with Atox1 although in vitro all MBDs can be metallated by Atox1 [43, 44]. MBD1, MBD2, and MBD4 form more stable complexes with Atox1 compared with other MBDs [43]. MBD3, MBD5, and MBD6 do not form complexes, but can receive Cu from Atox1 if the Cu-binding sites in MBD1, MBD2 and MBD4 are mutated. When Cu transfer was analyzed using the full-length ATP7B or the entire N-terminal domain, MBD2 was the initial acceptor of Cu from Atox1 [40]. It was also suggested that the transfer of Cu from Atox1 to MBD2 disrupts transient interactions of MBD2 with MBD1 and MBD3, allowing further loading of Cu to MBDs [9, 14]. It is interesting that the CxxC-to-AxxA mutations within MBD2 or MBD3 increase susceptibility of others MBDs to oxidation [35]. This finding points to interactions between N-terminal MBDs that are important for proper ATP7B function. Disruption of the interdomain interactions also facilitates trafficking of ATP7B from TGN to vesicles [14, 23]. Altogether, the available data suggest that Atox1 regulates MBD Cu occupancy and may modulate the rate of ATP7B-dependent Cu transport and protein trafficking. Whether Atox1 is as an obligatory Cu donor for ATP7B and whether ATP7B can receive Cu from alternative sources need to be examined in vivo.

Kinase-Mediated Phosphorylation

In all cells studied so far (hepatocytes, HEK293, COS cells, and skin fibroblasts), ATP7B was shown to be phosphorylated by kinase(s) under basal growth conditions, and the level of phosphorylation increases in response to Cu elevation [15, 16, 45]. Studies with the ATP7B mutant trapped in the endoplasmic reticulum (ER) revealed that the basal phosphorylation is already present when ATP7B is transitioning from the ER to Golgi [23]. Significance of this phosphorylation is not clear; it has been suggested that it may protect ATP7B from degradation [46]. Upon Cu elevation, the extent of kinase-mediated phosphorylation of endogenous ATP7B increases approximately twofold [45]. This Cu-dependent kinase-mediated phosphorylation occurs in the TGN prior to ATP7B trafficking to vesicles [15]. Mutations of Ser340/Ser341 in the N-terminal domain of ATP7B decrease Cu-dependent phosphorylation [23]. The mutational studies also suggested that Cu-dependent phosphorylation could trigger changes in the interdomain interactions of ATP7B that are favorable for trafficking [23].

The N-terminal MBDs of ATP7B are connected by flexible linkers, which contain numerous sites for the kinase-mediated phosphorylation. Similarly, multiple phosphorylation sites were predicted in the long C-terminal "tail" of ATP7B. Mass-spectrometry analysis of the recombinant ATP7B revealed that the kinase-mediated phosphorylation occurs at as many as 24 sites and involves, predominantly, Ser/Thr residues [15]. Most sites are phosphorylated in either Cu-depleted or Cu-elevated conditions, and the extent of phosphorylation at each site (complete or partial) has been difficult to measure. Phosphorylation of Ser residues at the positions 340/341, 478, and 481 (N-terminus) and 1121 and 1453 (C-terminus) has been verified by independent studies [23, 46]. It should be noted that the mass-spectrometry study was done by expressing recombinant ATP7B in skin fibroblasts, where ATP7B is not normally expressed. Therefore, functional significance of numerous phosphorylation sites awaits further study. It could be that ATP7B is phosphorylated by different kinases in a cell-specific manner and/or in response to different signals, such as changes in Cu levels and hormonal signaling [45, 47, 48].

Interestingly, mutating Cu-binding cysteines in MBD6 inhibits both the catalytic phosphorylation and kinase-mediated phosphorylation, whereas inactivation of sites for Cu binding in the transmembrane domain only disrupts catalysis [4]. These results suggest that the catalytic activity of ATP7B is not required for kinase-mediated phosphorylation, whereas kinase-mediated modification may regulate ATP7B activity. Modulation of a Cu-dependent ATP hydrolysis in liver membranes by treatment with glucagon, insulin, and kinase inhibitors supports the role of kinases in the regulation of ATP7B activity [47, 48], although more direct evidence is needed. How many kinases regulate ATP7B is still uncertain. Inhibitor of protein kinase D, CID755673, inhibits the phosphorylation of ATP7B in vitro and decreases levels of the recombinant protein in COS-7 cells [46], suggesting that the protein kinase D regulates ATP7B function. Reagents activating PKC stimulate ATP7B activity in hepatocytes [47], providing evidence for the regulation of ATP7B by multiple kinases.

TISSUE SPECIFIC FUNCTIONS OF ATP7B
ATP7B Role in the Liver

ATP7B is expressed most highly in hepatocytes. The role of ATP7B in liver physiology is well established, and it is both biosynthetic and homeostatic [49]. Cu absorbed from the diet is directed first to the liver, where it is processed by hepatocytes. Under steady-state conditions, hepatic ATP7B exerts primarily its "biosynthetic function" (Fig. 1A), that is, ATP7B transports Cu into the TGN lumen and thus enables incorporation of Cu cofactor into ceruloplasmin [28, 50]. Holoceruloplasmin (Cu loaded) is then released into the bloodstream through the secretory pathway [51]. The liver is also primarily responsible for the maintenance of the systemic Cu balance, which is controlled through the "homeostatic function" of ATP7B (Fig. 1B). When systemic Cu levels increase, more Cu enters hepatocytes, and ATP7B undergoes vesicle-mediated trafficking to the canalicular membrane [52]. Vesicles containing ATP7B accumulate in a subapical compartment, where ATP7B facilitates the sequestration of excess Cu. ATP7B-containing vesicles fuse with the membrane [53] to release their contents into the bile. Whether this step is regulated or constitutive is unknown. ATP7B transiently appears at the plasma membrane and then endocytose into "vesicle recycling pool." When Cu levels decrease, ATP7B returns to the TGN through retrograde trafficking [25, 54]. The return is associated with the loss of a kinase-mediated phosphorylation.

Cu that is released into the bile is not bioavailable for the reuptake by the GI tract [55]. This observation provides some credence to the hypothesis of lysosome-mediated Cu exocytosis, as Cu within lysosome would be protected from reuptake (see above and [52]). It is also possible that Cu is secreted in a complex with other molecules, which cannot be reabsorbed, but identity of such molecules is currently unknown. Emerging data demonstrate that ATP7B-dependent Cu balance in the liver is tightly linked to the key metabolic pathways in the storage and utilization of carbohydrate and fat [56–58]. It is intriguing that Niemann-Pick C (NPC1) protein (associated with lipid storage disease) appears to modulate ATP7B function in the liver. The loss of NPC1 in humans and in mice was reported to result in Cu misbalance in the liver and serum [59–61].

As described above, the activity of hepatic ATP7B is stimulated by insulin and inhibited by glucagon through a pathway involved protein kinase A (PKA) [48, 62]. Protein kinase Cε (a di-acylglycerol/phospholipid responsive kinase) has also been implicated in modulating ATP7B function in hepatocytes [47]. Activation of PKCε by phorbol 12-myristate 13-acetate (PMA) significantly upregulates ATP7B activity. Targeted inactivation of ATP7B in hepatocytes (in Atp7b$^{\Delta Hep}$ mice) revealed additional aspects of ATP7B function [58]. Cu accumulation occurs within Atp7b$^{\Delta Hep}$ hepatocytes, similar to that observed in patients with Wilson disease or in mice with a globally inactivated Atp7b; however, liver disease manifests very differently [56]. The Atp7b$^{\Delta Hep}$ animals have an increased tolerance to hepatic Cu accumulation likely mediated by significant upregulation of metallothioneins and do not show significant inflammatory response. At advanced age (>40 weeks after birth), the ATP7B$^{\Delta Hep}$ mice develop hepatic steatosis, possibly as a result of the Cu-dependent changes in mitochondria, although the precise mechanism remains to be established. The ATP7B$^{\Delta Hep}$ mice also have higher amounts of adipose tissue and gain more weight on a standard diet compared with control animals [56]. Taken together, these data suggest that maintaining Cu balance in hepatocytes is essential for the organism lipid balance.

ATP7B in Intestine

Humans acquire Cu through dietary sources, and intestinal absorption is an important step in the maintenance of organismal Cu balance. Cu-transporting ATPase ATP7A (which is highly homologous to ATP7B) has a well-known role in dietary Cu acquisition: it functions as a basolateral Cu exporter in enterocytes delivering Cu to portal circulation (reviewed in Ref. [63]). ATP7B is also expressed in mouse intestine and in human colorectal cancer cells [64]. Functional characterization of ATP7B in the intestine has established a novel role for ATP7B: enteric ATP7B pumps Cu into vesicles for sequestration and storage in order to support normal absorptive processes (Fig. 3) [65].

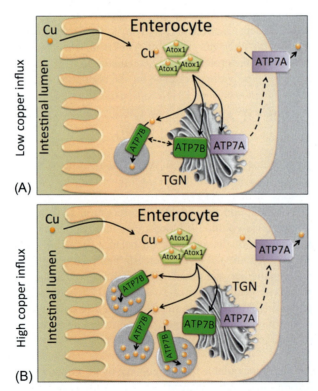

FIG. 3 ATP7B is an important player in intestinal Cu balance. (A) Cu is absorbed by the intestinal epithelium and delivered to the Golgi network. Under low dietary Cu loads, ATP7A delivers Cu to circulating pools, and ATP7B resides in both the Golgi and in cytoplasmic vesicles to buffer incoming Cu. (B) When dietary Cu load is high, ATP7A remains as an efflux pump delivering Cu into the blood stream. ATP7B functions to pump excess Cu into the lumen of small cytoplasmic vesicles for storage/sequestration.

Intestinal ATP7B may also deliver Cu to hephaestin. Hephaestin is a structural and functional homologue of ceruloplasmin [66] with an important role in dietary iron absorption. Inactivation of ATP7A in intestine has no effect on iron absorption [67], that is, the Cu-dependent ferroxidase hephaestin function appeared undisrupted. Since hephaestin receives its Cu in the secretory pathway, ATP7B is a likely candidate to fulfill this function. Recent studies by Petris and coworkers confirmed that the loss of both ATP7A/ATP7B in the intestine inactivates hephaestin and causes anemia. Lastly, disruption of the enteric ATP7B has profound consequences on a fat and cholesterol uptake [65]. How ATP7B couples the cytosolic Cu balance to lipid processing is currently unknown.

ATP7B in the Brain and Other Tissues

Expression of ATP7B in numerous regions of the brain has been well documented. ATP7B is abundantly expressed in choroid plexus, and Cu elevation triggers ATP7B trafficking to the basolateral membrane [68]. This result led to the suggestion that in the blood brain barrier, ATP7B exports excess of Cu from choroidal cells into the blood [68]. The functional importance of ATP7B in the CNS is also evident from pathological changes in the brains of Wilson disease patients. However, specific functions of ATP7B in either neurons or glial cells remain largely unclear. Studies in primary hippocampal neurons and cultured neuroblastoma cells using recombinant ATP7B have found that ATP7B is located in *trans*-Golgi network and upon Cu additions traffics to the plasma membrane of dendrites and cell body, but not the axons [69], presumably to facilitate Cu efflux. Cu delivery to such important Cu-dependent enzymes as peptidyl-alpha monooxygenase is mediated by ATP7A and not ATP7B [70], and the loss of ATP7A function is not compensated by ATP7B [71]. Studies in cultured differentiated noradrenergic cells SH-SY5Y suggest that ATP7B, instead, plays a regulatory role and modulates ATP7A activity (our studies, manuscript in preparation).

Similarly, scant information is available regarding the role of ATP7B in other organs or tissues. In mammary gland, ATP7B is expressed in luminal epithelial cells and contributes to Cu transport to the milk. Treating cultured mammary cells with a cocktail of lactation hormones or with Cu triggers trafficking of ATP7B toward the apical membrane and facilitates export of Cu from cell [72]. Inactivation of ATP7B is associated with lower Cu content in milk [73, 74]. In placenta, ATP7B is present in the syncytiotrophoblast, mostly in microvilli, and has a vesicular localization. Electron microscopy and studies of ATP7B trafficking behavior in placental cells indicate that ATP7B is involved in balancing Cu levels in the cytosol cells and may transport Cu back into maternal circulation [75, 76].

ATP7B is expressed in ciliated epithelial cells of airways [77], in hair cells [78], pineal glands [79], and the kidneys where it shows race- and age-dependent differences in expression [80]. The lack of ATP7B trafficking toward the plasma membrane in response to Cu elevation in cultured renal MDCK cells suggested that the renal ATP7B might play a role in Cu storage [81] rather than efflux. This suggestion is supported by the studies in $Atp7b^{-/-}$ mice, which accumulate Cu in the liver but not in the kidneys [82].

CONCLUSIONS

ATP7B is an essential player in systemic and cell-specific Cu homeostasis. The major biochemical characteristics of ATP7B, such as ATP dependence of Cu transport, the ability to form transiently phosphorylated intermediate, and the stimulation of ATPase activity by Cu have been experimentally demonstrated. However, many questions remain about intracellular regulation of Cu transport by ATP7B. Similarly, precise functions of ATP7B in specialized cells and tissues are largely unclear. Answering these questions is essential for understanding of a large spectrum of Wilson disease-causing mutations and their diverse phenotypic consequences.

ACKNOWLEDGMENT

The authors thank National Institute of Health for financial support (Grant R01DK071865).

REFERENCES

[1] Axelsen KB, Palmgren MG. Evolution of substrate specificities in the P-type ATPase superfamily. J Mol Evol 1998;46:84–101.
[2] Arguello JM, Eren E, Gonzalez-Guerrero M. The structure and function of heavy metal transport P1B-ATPases. Biometals 2007;20:233–48.
[3] Gupta A, Lutsenko S. Evolution of copper transporting ATPases in eukaryotic organisms. Curr Genomics 2012;13:124–33.
[4] Inesi G, Pilankatta R, Tadini-Buoninsegni F. Biochemical characterization of P-type copper ATPases. Biochem J 2014;463:167–76.
[5] Inesi G. Molecular features of copper binding proteins involved in copper homeostasis. IUBMB Life 2017;69:211–7.

[6] Polishchuk R, Lutsenko S. Golgi in copper homeostasis: a view from the membrane trafficking field. Histochem Cell Biol 2013;140:285–95.
[7] Banci L, Bertini I, Cantini F, Rosenzweig AC, Yatsunyk LA. Metal binding domains 3 and 4 of the Wilson disease protein: solution structure and interaction with the copper(I) chaperone HAH1. Biochemistry 2008;47:7423–9.
[8] Banci L, Bertini I, Cantini F, Migliardi M, Natile G, Nushi F, Rosato A. Solution structures of the actuator domain of ATP7A and ATP7B, the Menkes and Wilson disease proteins. Biochemistry 2009;48:7849–55.
[9] Yu CH, Dolgova NV, Dmitriev OY. Dynamics of the metal binding domains and regulation of the human copper transporters ATP7B and ATP7A. IUBMB Life 2017;69:226–35.
[10] Dmitriev O, Tsivkovskii R, Abildgaard F, Morgan CT, Markley JL, Lutsenko S. Solution structure of the N-domain of Wilson disease protein: distinct nucleotide-binding environment and effects of disease mutations. Proc Natl Acad Sci U S A 2006;103:5302–7.
[11] Dmitriev OY, Bhattacharjee A, Nokhrin S, Uhlemann EM, Lutsenko S. Difference in stability of the N-domain underlies distinct intracellular properties of the E1064A and H1069Q mutants of copper-transporting ATPase ATP7B. J Biol Chem 2011;286:16355–62.
[12] Kumar R, Arioz C, Li Y, Bosaeus N, Rocha S, Wittung-Stafshede P. Disease-causing point-mutations in metal-binding domains of Wilson disease protein decrease stability and increase structural dynamics. Biometals 2017;30:27–35.
[13] Rodriguez-Granillo A, Crespo A, Wittung-Stafshede P. Interdomain interactions modulate collective dynamics of the metal-binding domains in the Wilson disease protein. J Phys Chem B 2010;114:1836–48.
[14] Huang Y, Nokhrin S, Hassanzadeh-Ghassabeh G, Yu CH, Yang H, Barry AN, Tonelli M, Markley JL, Muyldermans S, Dmitriev OY, Lutsenko S. Interactions between metal-binding domains modulate intracellular targeting of Cu(I)-ATPase ATP7B, as revealed by nanobody binding. J Biol Chem 2014;289:32682–93.
[15] Braiterman LT, Gupta A, Chaerkady R, Cole RN, Hubbard AL. Communication between the N and C termini is required for copper-stimulated Ser/Thr phosphorylation of Cu(I)-ATPase (ATP7B). J Biol Chem 2015;290:8803–19.
[16] Pilankatta R, Lewis D, Adams CM, Inesi G. High yield heterologous expression of wild-type and mutant Cu$^+$-ATPase (ATP7B, Wilson disease protein) for functional characterization of catalytic activity and serine residues undergoing copper-dependent phosphorylation. J Biol Chem 2009;284:21307–16.
[17] Huster D, Kuhne A, Bhattacharjee A, Raines L, Jantsch V, Noe J, Schirrmeister W, Sommerer I, Sabri O, Berr F, Mossner J, Stieger B, Caca K, Lutsenko S. Diverse functional properties of Wilson disease ATP7B variants. Gastroenterology 2012;142:947–56. e945.
[18] Tadini-Buoninsegni F, Bartolommei G, Moncelli MR, Pilankatta R, Lewis D, Inesi G. ATP dependent charge movement in ATP7B Cu$^+$-ATPase is demonstrated by pre-steady state electrical measurements. FEBS Lett 2010;584:4619–22.
[19] Gonzalez-Guerrero M, Eren E, Rawat S, Stemmler TL, Arguello JM. Structure of the two transmembrane Cu$^+$ transport sites of the Cu$^+$-ATPases. J Biol Chem 2008;283:29753–9.
[20] Wijekoon CJ, Udagedara SR, Knorr RL, Dimova R, Wedd AG, Xiao Z. Copper ATPase CopA from *Escherichia coli*: quantitative correlation between ATPase activity and vectorial copper transport. J Am Chem Soc 2017;139:4266–9.
[21] Jadot M, Boonen M, Thirion J, Wang N, Xing J, Zhao C, Tannous A, Qian M, Zheng H, Everett JK, Moore DF, Sleat DE, Lobel P. Accounting for protein subcellular localization: a compartmental map of the rat liver proteome. Mol Cell Proteomics 2017;16:194–212.
[22] Lalioti V, Peiro R, Perez-Berlanga M, Tsuchiya Y, Munoz A, Villalba T, Sanchez C, Sandoval IV. Basolateral sorting and transcytosis define the Cu$^+$-regulated translocation of ATP7B to the bile canaliculus. J Cell Sci 2016;129:2190–201.
[23] Hasan NM, Gupta A, Polishchuk E, Yu CH, Polishchuk R, Dmitriev OY, Lutsenko S. Molecular events initiating exit of a copper-transporting ATPase ATP7B from the trans-Golgi network. J Biol Chem 2012;287:36041–50.
[24] Suzuki M, Gitlin JD. Intracellular localization of the Menkes and Wilson's disease proteins and their role in intracellular copper transport. Pediatr Int 1999;41:436–42.
[25] Guo Y, Nyasae L, Braiterman LT, Hubbard AL. NH2-terminal signals in ATP7B Cu-ATPase mediate its Cu-dependent anterograde traffic in polarized hepatic cells. Am J Physiol Gastrointest Liver Physiol 2005;289:G904–16.
[26] Harada M, Kumemura H, Sakisaka S, Shishido S, Taniguchi E, Kawaguchi T, Hanada S, Koga H, Kumashiro R, Ueno T, Suganuma T, Furuta K, Namba M, Sugiyama T, Sata M. Wilson disease protein ATP7B is localized in the late endosomes in a polarized human hepatocyte cell line. Int J Mol Med 2003;11:293–8.
[27] Huster D, Hoppert M, Lutsenko S, Zinke J, Lehmann C, Mossner J, Berr F, Caca K. Defective cellular localization of mutant ATP7B in Wilson's disease patients and hepatoma cell lines. Gastroenterology 2003;124:335–45.
[28] Hellman NE, Kono S, Mancini GM, Hoogeboom AJ, De Jong GJ, Gitlin JD. Mechanisms of copper incorporation into human ceruloplasmin. J Biol Chem 2002;277:46632–8.
[29] Polishchuk EV, Concilli M, Iacobacci S, Chesi G, Pastore N, Piccolo P, Paladino S, Baldantoni D, van ISC, Chan J, Chang CJ, Amoresano A, Pane F, Pucci P, Tarallo A, Parenti G, Brunetti-Pierri N, Settembre C, Ballabio A, Polishchuk RS. Wilson disease protein ATP7B utilizes lysosomal exocytosis to maintain copper homeostasis. Dev Cell 2014;29:686–700.
[30] Nyasae LK, Schell MJ, Hubbard AL. Copper directs ATP7B to the apical domain of hepatic cells via basolateral endosomes. Traffic 2014;15:1344–65.
[31] Braiterman L, Nyasae L, Guo Y, Bustos R, Lutsenko S, Hubbard A. Apical targeting and Golgi retention signals reside within a 9-amino acid sequence in the copper-ATPase, ATP7B. Am J Physiol Gastrointest Liver Physiol 2009;296:G433–44.
[32] Gupta A, Schell MJ, Bhattacharjee A, Lutsenko S, Hubbard AL. Myosin Vb mediates Cu+ export in polarized hepatocytes. J Cell Sci 2016;129:1179–89.
[33] Singleton WC, McInnes KT, Cater MA, Winnall WR, McKirdy R, Yu Y, Taylor PE, Ke BX, Richardson DR, Mercer JF, La Fontaine S. Role of glutaredoxin1 and glutathione in regulating the activity of the copper-transporting P-type ATPases, ATP7A and ATP7B. J Biol Chem 2010;285:27111–21.

[34] Hatori Y, Clasen S, Hasan NM, Barry AN, Lutsenko S. Functional partnership of the copper export machinery and glutathione balance in human cells. J Biol Chem 2012;287:26678–87.

[35] LeShane ES, Shinde U, Walker JM, Barry AN, Blackburn NJ, Ralle M, Lutsenko S. Interactions between copper-binding sites determine the redox status and conformation of the regulatory N-terminal domain of ATP7B. J Biol Chem 2010;285:6327–36.

[36] Hatori Y, Yan Y, Schmidt K, Furukawa E, Hasan NM, Yang N, Liu CN, Sockanathan S, Lutsenko S. Neuronal differentiation is associated with a redox-regulated increase of copper flow to the secretory pathway. Nat Commun 2016;7.

[37] Brose J, La Fontaine S, Wedd AG, Xiao Z. Redox sulfur chemistry of the copper chaperone Atox1 is regulated by the enzyme glutaredoxin 1, the reduction potential of the glutathione couple GSSG/2GSH and the availability of Cu(I). Metallomics 2014;6:793–808.

[38] Hamza I, Faisst A, Prohaska J, Chen J, Gruss P, Gitlin JD. The metallochaperone Atox1 plays a critical role in perinatal copper homeostasis. Proc Natl Acad Sci U S A 2001;98:6848–52.

[39] Ozumi K, Sudhahar V, Kim HW, Chen GF, Kohno T, Finney L, Vogt S, McKinney RD, Ushio-Fukai M, Fukai T. Role of copper transport protein antioxidant 1 in angiotensin II-induced hypertension: a key regulator of extracellular superoxide dismutase. Hypertension 2012;60:476–86.

[40] Walker JM, Huster D, Ralle M, Morgan CT, Blackburn NJ, Lutsenko S. The N-terminal metal-binding site 2 of the Wilson's disease protein plays a key role in the transfer of copper from Atox1. J Biol Chem 2004;279:15376–84.

[41] Walker JM, Tsivkovskii R, Lutsenko S. Metallochaperone Atox1 transfers copper to the NH2-terminal domain of the Wilson's disease protein and regulates its catalytic activity. J Biol Chem 2002;277:27953–9.

[42] Singleton C, Le Brun NE. Atx1-like chaperones and their cognate P-type ATPases: copper-binding and transfer. Biometals 2007;20:275–89.

[43] Banci L, Bertini I, Cantini F, Massagni C, Migliardi M, Rosato A. An NMR study of the interaction of the N-terminal cytoplasmic tail of the Wilson disease protein with copper(I)-HAH1. J Biol Chem 2009;284:9354–60.

[44] Yatsunyk LA, Rosenzweig AC. Cu(I) binding and transfer by the N terminus of the Wilson disease protein. J Biol Chem 2007;282:8622–31.

[45] Vanderwerf SM, Cooper MJ, Stetsenko IV, Lutsenko S. Copper specifically regulates intracellular phosphorylation of the Wilson's disease protein, a human copper-transporting ATPase. J Biol Chem 2001;276:36289–94.

[46] Pilankatta R, Lewis D, Inesi G. Involvement of protein kinase D in expression and trafficking of ATP7B (copper ATPase). J Biol Chem 2011;286:7389–96.

[47] Cardoso LH, Britto-Borges T, Vieyra A, Lowe J. ATP7B activity is stimulated by PKCvarepsilon in porcine liver. Int J Biochem Cell Biol 2014;54:60–7.

[48] Hilario-Souza E, Cuillel M, Mintz E, Charbonnier P, Vieyra A, Cassio D, Lowe J. Modulation of hepatic copper-ATPase activity by insulin and glucagon involves protein kinase A (PKA) signaling pathway. Biochim Biophys Acta 2016;1862:2086–97.

[49] Lutsenko S, Barnes NL, Bartee MY, Dmitriev OY. Function and regulation of human copper-transporting ATPases. Physiol Rev 2007;87:1011–46.

[50] Terada K, Nakako T, Yang XL, Iida M, Aiba N, Minamiya Y, Nakai M, Sakaki T, Miura N, Sugiyama T. Restoration of holoceruloplasmin synthesis in LEC rat after infusion of recombinant adenovirus bearing WND cDNA. J Biol Chem 1998;273:1815–20.

[51] Bissig KD, Honer M, Zimmermann K, Summer KH, Solioz M. Whole animal copper flux assessed by positron emission tomography in the Long-Evans Cinnamon rat—a feasibility study. Biometals 2005;18:83–8.

[52] Pena K, Coblenz J, Kiselyov K. Brief exposure to copper activates lysosomal exocytosis. Cell Calcium 2015;57:257–62.

[53] Cater MA, La Fontaine S, Shield K, Deal Y, Mercer JF. ATP7B mediates vesicular sequestration of copper: insight into biliary copper excretion. Gastroenterology 2006;130:493–506.

[54] Roelofsen H, Wolters H, Van Luyn MJ, Miura N, Kuipers F, Vonk RJ. Copper-induced apical trafficking of ATP7B in polarized hepatoma cells provides a mechanism for biliary copper excretion. Gastroenterology 2000;119:782–93.

[55] Linder MC, Wooten L, Cerveza P, Cotton S, Shulze R, Lomeli N. Copper transport. Am J Clin Nutr 1998;67:965S–971S.

[56] Huster D, Finegold MJ, Morgan CT, Burkhead JL, Nixon R, Vanderwerf SM, Gilliam CT, Lutsenko S. Consequences of copper accumulation in the livers of the Atp7b-/- (Wilson disease gene) knockout mice. Am J Pathol 2006;168:423–34.

[57] Huster D, Purnat TD, Burkhead JL, Ralle M, Fiehn O, Stuckert F, Olson NE, Teupser D, Lutsenko S. High copper selectively alters lipid metabolism and cell cycle machinery in the mouse model of Wilson disease. J Biol Chem 2007;282:8343–55.

[58] Muchenditsi A, Yang H, Hamilton JP, Koganti L, Housseau F, Aronov L, Fan H, Pierson H, Bhattacharjee A, Murphy RC, Sears CL, Potter JJ, Wooton-Kee CR, Lutsenko S. Targeted inactivation of copper-transporter Atp7b in hepatocytes causes liver steatosis and obesity in mice. Am J Physiol Gastrointest Liver Physiol 2017; https://doi.org/10.1152/ajpgi.00312.2016.

[59] Connemann BJ, Gahr M, Schmid M, Runz H, Freudenmann RW. Low ceruloplasmin in a patient with Niemann-Pick type C disease. J Clin Neurosci 2012;19:620–1.

[60] Vazquez MC, Martinez P, Alvarez AR, Gonzalez M, Zanlungo S. Increased copper levels in in vitro and in vivo models of Niemann-Pick C disease. Biometals 2012;25:777–86.

[61] Goez HR, Jacob FD, Fealey RD, Patterson MC, Ramaswamy V, Persad R, Johnson ES, Yager JY. An unusual presentation of copper metabolism disorder and a possible connection with Niemann-Pick type C. J Child Neurol 2011;26:518–21.

[62] Hilario-Souza E, Valverde RH, Britto-Borges T, Vieyra A, Lowe J. Golgi membranes from liver express an ATPase with femtomolar copper affinity, inhibited by cAMP-dependent protein kinase. Int J Biochem Cell Biol 2011;43:358–62.

[63] Pierson HL, Tumer Z. Copper metabolism, ATP7A, and Menkes disease. In: eLS. Chichester: John Wiley & Sons, Ltd.; 2015.

[64] Weiss KH, Wurz J, Gotthardt D, Merle U, Stremmel W, Fullekrug J. Localization of the Wilson disease protein in murine intestine. J Anat 2008;213:232–40.

[65] Pierson H, Muchenditsi A, Kim B, Ralle M, Zachos N, Huster D, Lutsenko S. The function of ATPase copper transporter ATP7B in intestine. Gastroenterology 2018;154:168–80.

[66] Vulpe CD, Kuo YM, Murphy TL, Cowley L, Askwith C, Libina N, Gitschier J, Anderson GJ. Hephaestin, a ceruloplasmin homologue implicated in intestinal iron transport, is defective in the sla mouse. Nat Genet 1999;21:195–9.

[67] Wang Y, Zhu S, Hodgkinson V, Prohaska JR, Weisman GA, Gitlin JD, Petris MJ. Maternofetal and neonatal copper requirements revealed by enterocyte-specific deletion of the Menkes disease protein. Am J Physiol Gastrointest Liver Physiol 2012;303:G1236–44.

[68] Fu X, Zhang Y, Jiang W, Monnot AD, Bates CA, Zheng W. Regulation of copper transport crossing brain barrier systems by Cu-ATPases: effect of manganese exposure. Toxicol Sci 2014;139:432–51.

[69] Jain S, Farias GG, Bonifacino JS. Polarized sorting of the copper transporter ATP7B in neurons mediated by recognition of a dileucine signal by AP-1. Mol Biol Cell 2015;26:218–28.

[70] El Meskini R, Culotta VC, Mains RE, Eipper BA. Supplying copper to the cuproenzyme peptidylglycine alpha-amidating monooxygenase. J Biol Chem 2003;278:12278–84.

[71] Niciu MJ, Ma XM, El Meskini R, Pachter JS, Mains RE, Eipper BA. Altered ATP7A expression and other compensatory responses in a murine model of Menkes disease. Neurobiol Dis 2007;27:278–91.

[72] Michalczyk A, Bastow E, Greenough M, Camakaris J, Freestone D, Taylor P, Linder M, Mercer J, Ackland ML. ATP7B expression in human breast epithelial cells is mediated by lactational hormones. J Histochem Cytochem 2008;56:389–99.

[73] La Fontaine S, Theophilos MB, Firth SD, Gould R, Parton RG, Mercer JF. Effect of the toxic milk mutation (tx) on the function and intracellular localization of Wnd, the murine homologue of the Wilson copper ATPase. Hum Mol Genet 2001;10:361–70.

[74] Buiakova OI, Xu J, Lutsenko S, Zeitlin S, Das K, Das S, Ross BM, Mekios C, Scheinberg IH, Gilliam TC. Null mutation of the murine ATP7B (Wilson disease) gene results in intracellular copper accumulation and late-onset hepatic nodular transformation. Hum Mol Genet 1999;8:1665–71.

[75] Hardman B, Michalczyk A, Greenough M, Camakaris J, Mercer J, Ackland L. Distinct functional roles for the Menkes and Wilson copper translocating P-type ATPases in human placental cells. Cell Physiol Biochem 2007;20:1073–84.

[76] Hardman B, Luff S, Ackland ML. Differential intracellular localisation of the Menkes and Wilson copper transporting ATPases in the third trimester human placenta. Placenta 2011;32:79–85.

[77] Ibricevic A, Brody SL, Youngs WJ, Cannon CL. ATP7B detoxifies silver in ciliated airway epithelial cells. Toxicol Appl Pharmacol 2010;243:315–22.

[78] Ding D, He J, Allman BL, Yu D, Jiang H, Seigel GM, Salvi RJ. Cisplatin ototoxicity in rat cochlear organotypic cultures. Hear Res 2011;282:196–203.

[79] Borjigin J, Payne AS, Deng J, Li X, Wang MM, Ovodenko B, Gitlin JD, Snyder SH. A novel pineal night-specific ATPase encoded by the Wilson disease gene. J Neurosci 1999;19:1018–26.

[80] Joseph S, Nicolson TJ, Hammons G, Word B, Green-Knox B, Lyn-Cook B. Expression of drug transporters in human kidney: impact of sex, age, and ethnicity. Biol Sex Differ 2015;6:4.

[81] Barnes N, Bartee MY, Braiterman L, Gupta A, Ustiyan V, Zuzel V, Kaplan JH, Hubbard AL, Lutsenko S. Cell-specific trafficking suggests a new role for renal ATP7B in the intracellular copper storage. Traffic 2009;10:767–79.

[82] Linz R, Barnes NL, Zimnicka AM, Kaplan JH, Eipper B, Lutsenko S. Intracellular targeting of copper-transporting ATPase ATP7A in a normal and Atp7b-/- kidney. Am J Physiol Renal Physiol 2008;294:F53–61.

Chapter 4

Biochemical and Cellular Properties of ATP7B Variants

Samuel Jayakanthan, Courtney McCann and Svetlana Lutsenko
Department of Physiology, Johns Hopkins University School of Medicine, Baltimore, MD, United States

INTRODUCTION

Cells employ copper (Cu) to activate oxygen and facilitate electron-transfer reactions. The ability of copper (Cu) to readily cycle between the Cu(I) and Cu(II) oxidation states makes this metal an indispensable cofactor for a variety of enzymes that are critical for organism function and survival. These enzymes participate in such processes as respiration, defense against oxygen radicals, biosynthesis of neuromodulators, iron transport, and other important processes [1, 2]. However, under conditions of Cu overload, observed in Wilson disease, redox properties of Cu became detrimental to cells causing metabolic abnormalities, inflammation, and cell death [3]. To maintain Cu homeostasis, human cells utilize the high-affinity Cu transporter CTR1 (for Cu uptake) and two Cu-transporting ATPases (Cu-ATPases) ATP7A and ATP7B (for Cu export) [4, 5]. These Cu transporters work together with Atox1, a small cytosolic protein, which has a metallochaperone (or Cu shuttle) function (Fig. 1). Atox1 binds Cu released from CTR1 at the basolateral plasma membrane and transfers Cu to ATP7B and ATP7A located in the *trans*-Golgi network (TGN) and vesicles of the endocytic pathway [9, 10]. ATP7A and ATP7B then transfer Cu into the lumen of the secretory pathway where Cu is utilized for activation of Cu-dependent enzyme or is exported out of the cells when present in excess. The Atox1-mediated transfer of Cu to the Cu-ATPases is a subject of redox regulation by the glutathione pair (GSH-GSSG) [7]. In neuronal cells, changes in the GSH-GSSG ratio modulate the ability of Atox1 to bind Cu and thus control the amount of Cu delivered to the secretory pathway [8]. ATP7A plays a "housekeeping" role in most tissues, except the liver, where its expression is low [11]. ATP7A is especially critical for the dietary Cu uptake in the intestine and for Cu entry into the brain [12]. Mutations in *ATP7A* result in systemic Cu deficiency and progressive and fatal neurodegenerative disorder known as Menkes' disease [13].

ATP7B is expressed in many tissues and is especially high in the liver, where ATP7B function is well understood. In hepatocytes, ATP7B transfers Cu to the Cu-dependent ferroxidase ceruloplasmin in the TGN and facilitates efflux of excess Cu across a canalicular membrane into the bile. In a normally functioning adult human liver, ATP7B maintains intracellular Cu levels at <50 µg Cu/g of the dry liver. Newborns have a significantly higher Cu concentration than adults [14]; in the postnatal livers, the excess Cu is sequestered within the yet-to-be-characterized vesicular compartments. Recent studies have demonstrated that ATP7B is also expressed in nonparenchymal cells of the liver, where ATP7B maintains Cu balance and may contribute to inflammatory response, when its function is lost [15]. Genetic mutations in ATP7B result in an impaired biliary efflux of Cu, an accumulation of Cu in hepatic tissues (>250–3000 µg Cu/g of the dry liver) [16], the loss of Cu incorporation into ceruloplasmin, and a spectrum of pathological changes known as Wilson disease, WD [5, 16–18]. Numerous WD-causing mutations have been identified (see the Human Gene Mutation Database (http://www.hgmd.cf.ac.uk/ac/all.php) and the Wilson Disease (WD) Mutation Database (http://www.wilsondisease.med.ualberta.ca/). The genetic studies have also uncovered higher frequency of some mutations in specific world populations and the absence of strong phenotype-genotype correlations (for recent examples, see Refs. [19–21]).

Phenotypic diversity (the variable time of onset, a spectrum of presenting manifestations, and the nonuniform course of the disease) is a characteristic feature of WD. The complexity of ATP7B structure and intracellular behavior may contribute to phenotypic diversity of WD. Mutations that completely prevent protein production or cause gross structural abnormalities result in a complete loss of copper transport activity. Other mutations, while deleterious, yield protein that has residual activity and retain some aspects of its function, thus leading to a slower development of the disease. ATP7B mediates its dual role in cofactor delivery to ceruloplasmin and Cu efflux by changing its intracellular localization (for details on ATP7B function, please see Chapter 3). Under basal conditions, ATP7B is located primarily in the late Golgi/*trans*-Golgi

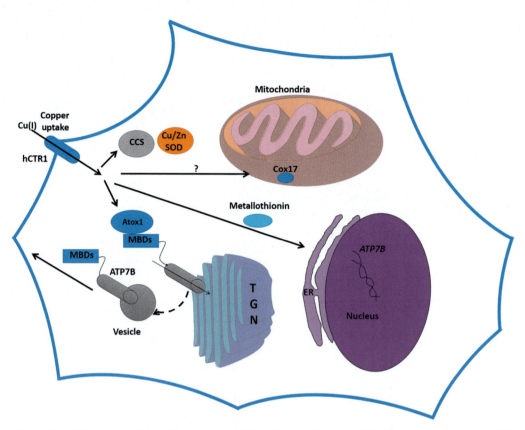

FIG. 1 Copper homeostasis in hepatocyte. Copper (Cu) is transported into hepatic cells by the high-affinity copper transporter CTR1 located at the basolateral membrane. The metal then binds to cytosolic shuttles or Cu chaperones. The Cu chaperone CCS delivers Cu to cytosolic Cu- and Zn-dependent superoxide dismutase, whereas ATOX1 transfers Cu to Cu-transporting ATPase ATP7B in the secretory pathway. The glutathione pair GSH-GSSG regulates the Cu uptake by CTR1 [6] and the Cu binding by Atox1 [7, 8]. When cytosolic Cu is elevated, ATP7B leaves the TGN and traffics in vesicles toward the canalicular membrane to facilitate export of excess Cu.

network (TGN) where it transfers Cu into the TGN lumen. When Cu is elevated, ATP7B undergoes a kinase-mediated phosphorylation [22, 23] and traffics from the TGN to the apically located vesicles [24]. This process depends on myosin-V, which facilitates docking of ATP7B vesicles near the apical membrane [25]. Excess Cu is sequestered in vesicles and then exported into the bile following vesicle fusion with the plasma membrane [5, 26, 27]. ATP7B is then endocytosed and, depending on the cellular Cu levels, either returns to the pool of recycling vesicles or traffics back to the TGN. The return to the TGN is associated with ATP7B dephosphorylation by phosphatase(s), which remain to be identified. Given this complex sequence of events, it is not surprising that the disease-causing mutations often modify more than one parameter of ATP7B behavior. The involvement of other proteins (kinases, phosphatases, and trafficking machinery) adds additional level of complexity and hence potential for variability. Altogether, mutations in *ATP7B* can be associated with a complete loss of ATP7B synthesis, misfolding, and instability/protein degradation. The mutation may markedly diminish ATP and/or Cu binding, interfere with the protein targeting/trafficking to proper intracellular compartments, or disrupt interaction with proteins involved in ATP7B regulation/modification. In this chapter, we provide examples of these various consequences.

MOLECULAR FEATURES OF ATP7B AND CATALYTIC CYCLE

ATP7B is a membrane protein, composed of multiple domains (Fig. 2A). The membrane portion of ATP7B has eight transmembrane α-helical segments. These segments form the Cu translocation pathway. The highly conserved $C^{983}PC^{985}$ motif within the sixth transmembrane segment contributes to Cu binding and translocation. Experimental mutations of these cysteines to serine residues produce the nonfunctional protein; the $Cys^{985}Tyr$ substitution is a WD-causing mutation. The cytosolic portion of ATP7B constitutes about 70% of protein mass and includes at least nine domains. The ATP binding and hydrolysis are mediated through the concerted action of the N (nucleotide binding) and P (phosphorylation) and A

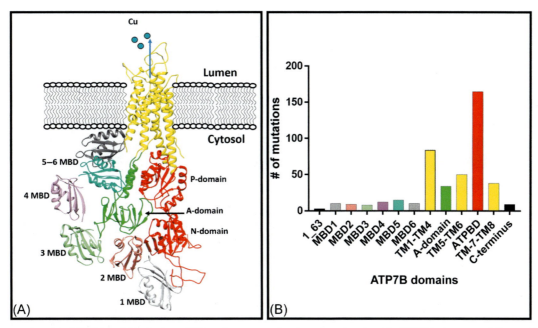

FIG. 2 Structural model of ATP7B and distribution of Wilson disease mutations in various domains. (A) ATP7B is shown as a ribbon model (generated by UCSF chimera). The cytosolic and transmembrane domains are as follows (MBD1, *light gray*; MBD2, *light orange*; MBD3, *light green*; MBD4, *purple*; MBD5, *cyan*; MBD6, *dark gray*; TM regions, *yellow*; A-domain, *green*; and ATP-binding domain, *red*). (B) Distribution of WD-causing mutations among the various domains of ATP7B is uneven.

(actuator) domains (Fig. 2A). ATP binds primarily within the N-domain; the conserved residues H1069, E1064, G1099, G1101, I1102, G1149, and N1150 contribute to ATP binding [40]. The substitution of invariant His1069 for glutamine (His^{1069}Gln) is one of the most common WD mutations in Caucasian and African American populations. Other WD mutations of ATP-coordinating residues include G1099S, G1101Arg, I1102T, and G1149A. The mutations in the vicinity to the nucleotide-coordinating residues may also significantly diminish ATP binding by altering the spatial organization of the binding pocket.

Following Cu binding to the transmembrane portion, the N-domain with bound ATP approaches the P-domain, and ATP phosphorylates the invariant catalytic aspartate D1027 by transferring its terminal phosphate to carboxyl side chain of the aspartate residue. The catalytic aspartate is located within the highly conserved D^{1027}KTGT motif in the P-domain. Formation of the phosphorylated intermediate and its subsequent hydrolysis by water supplies the energy necessary for conformational transitions within ATP7B and the expulsion of Cu from the transmembrane portion into the lumen of the secretory pathway. Experimental mutation D1027 > A completely eliminates ATP7B enzymatic activity and Cu transport, highlighting the essential role of this residue. The naturally occurring substitutions of conserved neighboring residues in the immediate vicinity of the catalytic aspartate (T1029A/I and T1031S/A) are also deleterious, as evidenced by the association of these mutations with WD.

The A-domain (which is also known as an "actuator" or a "transduction" domain) is thought to undergo rotation during the catalytic cycle and use its conserved T^{858}GE^{860} motif to facilitate dephosphorylation of the catalytic aspartate [28]. The movements of the N-, P-, and A-domains are coupled, and although direct evidence for ATP7B is still lacking, studies from other P-type ATPases and prokaryotic Cu-ATPases indicate that these domains and the membrane part of the protein are involved in a complex network of direct and long-range interactions. Significance of the individual N-, P-, and A-domains and transmembrane domain as well as their interactions is illustrated by the fact that the vast majority of WD-causing mutations are clustered within these domains and predicted contact regions (Fig. 2B). The ATP-binding domain (N- and P-domains) has by far the largest number of mutations followed by those in the transmembrane region and the A-domain.

PHYSICAL PROPERTIES OF ATP7B VARIANTS

While a great majority of the WD-causing mutations are yet to be fully characterized, available data indicate that the amino acid substitutions within ATP7B have direct effect on protein folding, stability, and function [29]. We have used calculations of solvent-accessible surface areas (SASA) and hydrophobicity indexes (Fig. 3A and B) of individual residues

36 PART | II Molecular Mechanisms

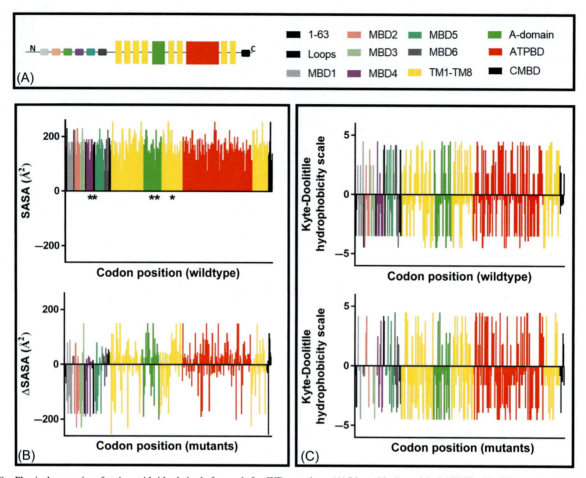

FIG. 3 Physical properties of amino acid side chains before and after WD mutations. (A) Linear block model of ATP7B with different domains shown in colors corresponding to the structural model in Fig. 2. (B) *Top*: solvent-accessible surface area (in Å²) for residues in the wild-type ATP7B. Each vertical line gives a value for the amino acid residue mutated in the corresponding position in WD (the domains are identified by color according to the block model). *Bottom*: change in the solvent-accessible area for the residues after the mutation for each position. Negative values indicate smaller area and the decrease in volume; positive values indicate increase in the surface area taken by the mutant. The discussed SNPs are marked by an asterisk (*). (C) *Top*: Kyte-Doolittle hydrophobicity values for the amino acids in wild-type ATP7B and (bottom) corresponding disease mutations (color-coded according to the block model). Altogether, the data illustrate that mutations within the regulatory N-terminal domain have a general trend of decreasing the volume of side chains and thus increasing the flexibility of this region, whereas mutations in the A- and ATP-binding domains are more often associated with an increased bulk of the side chains and therefore higher probability of significant structural change. Change in hydrophobic properties occurs in both directions; mutations in the transmembrane domain show more frequent decrease of hydrophobicity, which is likely to destabilize this domain.

to predict changes caused by each of the 457 WD-causing mutations at their respective positions. To do that, we placed in silico the correct amino acid residues or the corresponding WD mutant residues in the tripeptide G-X-G [30, 31], and the surface of the tripeptide was traced by a sphere [30]. Changes in solvent accessibility compared with the wild-type ATP7B (ΔSASA) illustrate whether mutation is associated with an increase or decrease in the local protein volume (Table 1); this in turn helps to predict the effect of each mutation on protein structure and stability within the immediate vicinity of the mutation. The changes in properties of the residues (hydrophobic vs hydrophilic) caused by mutation can be determined by comparing the Kyte-Doolittle Hydropathy (KDH) indexes of amino acids before and after substitution (Fig. 3B) [32]. The KDH index scale for amino acids in their native environment ranges from −4.5 (least hydrophobic) to +4.5 (very hydrophobic) (Table 1). Like changes in the solvent accessibility, changes in hydrophobicity could disrupt existing salt bridges and hydrogen-bond networks and destabilize interactions dependent on the hydrophobic contacts.

In the N-terminal metal-binding region, the mutations cause mostly decrease in protein volume in corresponding positions (Fig. 3A, bottom panel). These changes could increase protein flexibility, affect protein stability, and diminish the precision of interdomain interactions or interactions with proteins involved in ATP7B regulation [33, 34]. However, mutations of small and flexible Gly residues (such as G85 V in MBD1 and G591D in MBD6) have a marked destabilizing effect and decrease stability and folding or corresponding metal-binding domains [35, 36]. Lastly, the mutations may have only

TABLE 1 The Changes in Solvent-Accessible Surface and Hydrophobicity Index for Known WD Mutations

Mutation	SASA WT	Mutant (ΔSASA)	Hydrophobicity WT	Hydrophobicity Mutant	Mutation	SASA WT	Mutant (ΔSASA)	Hydrophobicity WT	Hydrophobicity Mutant	Mutation	SASA WT	Mutant (ΔSASA)	Hydrophobicity WT	Hydrophobicity Mutant
Ser15Gly	115	−40	−0.8	−0.4	Val456Leu	155	15	4.2	3.8	Cys703Tyr	135	95	2.5	−1.3
Asn41Ser	160	−45	−3.5	−0.8	Gln457Ter	180	−180	−3.5	0	Gln707Arg	180	45	−3.5	−4.5
Tyr44Asn	230	−70	−1.3	−3.5	Ala486Ser	115	0	1.8	−0.8	Leu708Pro	170	−25	3.8	−1.6
Cys72Ter	135	−135	2.5	0	Cys490Ter	135	−135	2.5	0	Gly710Ser	75	40	−0.4	−0.8
Gly85Val	75	80	−0.4	4.2	Leu492Ser	170	−55	3.8	−0.8	Gly710Arg	75	150	−0.4	−4.5
Gln95Ter	180	−180	−3.5	0	Gln511Ter	180	−180	−3.5	0	Gly710Ala	75	40	−0.4	1.8
Asp96Gly	150	−75	−3.5	−0.4	Gly515Val	75	80	−0.4	4.2	Gly710Val	75	80	−0.4	4.2
Ser105Ter	115	−115	−0.8	0	Leu523Ter	170	−170	3.8	0	Gly711Arg	75	150	−0.4	−4.5
Cys108Arg	135	90	2.5	−4.5	Glu529Gly	190	−115	−3.5	−0.4	Gly711Glu	75	115	−0.4	−3.5
Gln110Ter	180	−180	−3.5	0	Tyr532His	230	−35	−1.3	−3.2	Gly711Trp	75	80	−0.4	4.2
Gln111Ter	180	−180	−3.5	0	Tyr532Cys	230	−95	−1.3	2.5	Tyr713Cys	230	−95	−1.3	2.5
Gln115Ter	180	−180	−3.5	0	Val536Ala	155	−40	4.2	1.8	Tyr715His	230	−45	−1.3	−3.2
Ile116Thr	175	−35	4.5	−0.7	Pro539Leu	145	25	−1.6	3.8	Tyr715Ter	230	—	−1.3	—
Arg136Trp	225	30	−4.5	−0.9	Glu541Lys	190	10	−3.5	−3.9	Gln717Ter	180	−180	−3.5	0
Arg148Trp	225	30	−4.5	−0.9	Gln544Ter	180	−180	−3.5	0	Ala718Pro	115	30	1.8	−1.6
Gln155Ter	180	−180	−3.5	0	Leu549Pro	170	−25	3.8	−1.6	Ser721Pro	115	30	−0.8	−1.6
Cys157Phe	135	75	2.5	2.8	Ala553Glu	115	65	1.8	−3.5	Ala727Asp	115	35	1.8	−3.5
Gly170Val	75	80	−0.4	4.2	Gly591Ser	75	40	−0.4	−0.8	Ala727Val	115	40	1.8	4.2
Tyr187Ter	230	−230	−1.3	0	Gly591Asp	75	75	−0.4	−3.5	Met729Val	185	−30	1.9	4.2
Gln188Ter	180	−180	−3.5	0	Val597Ile	155	20	4.2	4.5	Val731Glu	155	35	4.2	−3.5
Gln193Ter	180	−180	−3.5	0	Ala604Pro	115	30	1.8	−1.6	Val731Ala	155	−40	4.2	1.8
Asp196Glu	150	30	−3.5	−3.5	Val606Gly	155	−80	4.2	−0.4	Leu732His	170	25	3.8	−3.2
Leu217Ter	170	−170	3.8	0	Glu611Lys	190	10	−3.5	−3.9	Leu732Pro	170	−25	3.8	−1.6
Cys271Ter	135	−135	2.5	0	Gly614Cys	75	60	−0.4	2.5	Val734Phe	155	55	4.2	2.8

Continued

38 PART | II Molecular Mechanisms

TABLE 1 The Changes in Solvent-Accessible Surface and Hydrophobicity Index for Known WD Mutations—cont'd

Mutation	SASA WT	Mutant (ΔSASA)	Hydrophobicity Index WT	Mutant	Mutation	SASA WT	Mutant (ΔSASA)	Hydrophobicity Index WT	Mutant	Mutation	SASA WT	Mutant (ΔSASA)	Hydrophobicity Index WT	Mutant
Gln286Ter	180	−180	−3.5	0	Arg616Gln	225	−45	−4.5	−3.5	Ala736Val	115	40	1.8	4.2
Gln289Ter	180	−180	−3.5	0	Arg616Trp	225	30	−4.5	−0.9	Thr737Arg	140	85	−0.7	−4.5
Tyr301Ter	230	−230	−1.3	0	Gly626Ala	75	40	−0.4	1.8	Thr737Ile	140	35	−0.7	4.5
Cys305Ter	135	−135	2.5	0	His639Tyr	195	35	−3.2	−1.3	Tyr741Ter	230	−230	−1.3	0
Asp329His	150	45	−3.5	−3.2	Leu641Ser	170	55	3.8	−0.8	Tyr741Cys	230	95	−1.3	2.5
Glu332Ter	190	−190	−3.5	0	Asp642His	150	45	−3.5	−3.2	Tyr743Cys	230	95	−1.3	2.5
Gly333Arg	75	150	−0.4	−4.5	Met645Arg	185	40	1.9	−4.5	Ser744Pro	115	30	−0.8	−1.6
Gln355Ter	180	−180	−3.5	0	Gln649Arg	180	45	−3.5	−4.5	Leu745Pro	170	−25	3.8	−1.6
Glu378Ter	190	−190	−3.5	0	Trp650Ter	255	−255	−0.9	0	Ile747Phe	175	35	4.5	2.8
Ile381Ser	175	−60	4.5	−0.8	Ser653Tyr	115	115	−0.8	−1.3	Leu748Met	170	15	3.8	1.9
Ser382Cys	115	20	−0.8	2.5	Cys656Ter	135	−135	2.5	0	Glu754Lys	190	10	−3.5	−3.9
Gln388Ter	180	−180	−3.5	0	Ser657Arg	115	110	−0.8	−4.5	Lys755Arg	200	25	−3.9	−4.5
Glu396Gln	190	−10	−3.5	−3.5	Met665Ile	185	−10	1.9	4.5	Ala756Gly	115	−40	1.8	−0.4
Glu396ter	190	−190	−3.5	—	Leu667ter	170	−170	3.8	0	Ala756Val	115	40	1.8	4.2
Ala399Pro	115	30	1.8	−1.6	Asn676Ile	160	15	−3.5	4.5	Arg758Met	225	−40	−4.5	1.9
Ser406Ala	115	0	−0.8	1.8	Gln680ter	180	−180	−3.5	0	Pro760Leu	145	25	−1.6	3.8
Pro410Leu	145	25	−1.6	3.8	Pro690Leu	145	25	−1.6	3.8	Phe763Tyr	210	20	2.8	−1.3
Glu412Ter	190	−190	−3.5	0	Gly691Arg	75	150	−0.4	−4.5	Asp765Asn	150	10	−3.5	−3.5
Glu418Lys	190	10	−3.5	−3.9	Ser693Pro	115	30	−0.8	−1.6	Asp765His	150	45	−3.5	−3.2
Asp419Asn	150	10	−3.5	−3.5	Ser693Cys	115	20	−0.8	2.5	Asp765Gly	150	−75	−3.5	−0.4
Leu436Val	170	−15	3.8	4.2	Ser693Tyr	115	115	−0.8	−1.3	Thr766Arg	140	85	−0.7	−4.5
Thr766Met	140	45	−0.7	1.9	Ser876Phe	115	95	−0.8	2.8	Thr991Met	140	45	−0.7	1.9
Pro767Ser	145	−30	−1.6	−0.8	Asn878Lys	160	40	−3.5	−3.9	Pro992His	145	50	−1.6	−3.2
Pro768His	145	50	−1.6	−3.2	Thr888Pro	140	5	−0.7	−1.6	Pro992Leu	145	25	−1.6	3.8
Pro768Leu	145	25	−1.6	3.8	Val890Met	155	30	4.2	1.9	Thr993Met	140	45	−0.7	1.9

Variant					Variant					Variant				
Met769Arg	185	40	1.9	-4.5	Gly891Asp	75	75	-0.4	-3.5	Val995Ala	155	-40	4.2	1.8
Met769Ile	185	-10	1.9	4.5	Gly891Val	75	80	-0.4	4.2	Met996Thr	185	-45	1.9	-0.7
Met769Val	185	-30	1.9	4.2	Gln898Arg	180	45	-3.5	-4.5	Gly998Asp	75	75	-0.4	-3.5
Leu770Pro	170	-25	3.8	-1.6	Ile899Phe	175	35	4.5	2.8	Gly1000Arg	75	150	-0.4	-4.5
Leu770Leu	170	0	3.8	3.8	Lys910Ter	200	-200	-3.9	0	Ala1003Thr	115	25	1.8	-0.7
Leu776Pro	170	-25	3.8	-1.6	Gln914Ter	180	-180	-3.5	0	Ala1003Val	115	40	1.8	4.2
Leu776Val	170	-15	3.8	4.2	Gln915Arg	180	45	-3.5	-4.5	Gln1004Pro	180	35	-3.5	-1.6
Arg778Gln	225	-45	-4.5	-3.5	Asp918Asn	150	10	-3.5	-3.5	Lys1010Thr	200	-60	-3.9	-0.7
Arg778Leu	225	55	-4.5	3.8	Arg919Gly	225	-150	-4.5	-0.4	Lys1010Arg	200	25	-3.9	-4.5
Arg778Gly	225	-150	-4.5	-0.4	Phe920Cys	210	-75	2.8	2.5	Gly1012Arg	75	150	-0.4	-4.5
Arg778Trp	225	30	-4.5	-0.9	Ser921Asn	115	45	-0.8	-3.5	Gly1012Val	75	80	-0.4	4.2
Trp779Gly	255	-180	-0.9	-0.4	Ser921Arg	115	110	-0.8	-4.5	Pro1014Leu	145	25	-1.6	3.8
Trp779Ter	255	–	-0.9	–	Ile929Val	175	-20	4.5	4.2	Met1017Ile	185	-10	1.9	4.5
Glu781gly	190	-115	-3.5	-0.4	Ser932Ter	115	-115	-0.8	0	Ala1018Val	115	40	1.8	4.2
Lys785Arg	200	25	-3.9	-4.5	Thr933pro	140	5	-0.7	-1.6	Lys1020Arg	200	25	-3.9	-4.5
Thr788Ile	140	35	-0.7	4.5	Thr935Met	140	45	-0.7	1.9	Ile1021Lys	175	25	4.5	-3.9
Leu795Arg	170	55	3.8	-4.5	Leu936Val	170	-15	3.8	4.2	Ile1021Val	175	-20	4.5	4.2
Ala803Thr	115	25	1.8	-0.7	Trp939Cys	255	-120	-0.9	2.5	Val1024Ala	155	-40	4.2	1.8
Glu810ter	190	-190	-3.5	0	Gly943Ser	75	40	-0.4	-0.8	Met1025Lys	185	15	1.9	-3.9
Arg816Ser	225	-110	-4.5	-0.8	Gly943Asp	75	75	-0.4	-3.5	Met1025Arg	185	40	1.9	-4.5
Arg827Trp	225	-80	-4.5	-1.6	Gly943Cys	75	60	-0.4	2.5	Thr1029Ile	140	35	-0.7	4.5
Arg827Pro	225	-80	-4.5	-1.6	Val949Gly	155	-80	4.2	-0.4	Thr1031Ala	140	-25	-0.7	1.8
Lys832Arg	200	25	-3.9	-4.5	Arg952Lys	225	-25	-4.5	-3.9	Thr1031Ser	140	-25	-0.7	-0.8
Gly836Glu	75	115	-0.4	-3.5	Gln963Ter	180	-180	-3.5	0	Thr1033Ser	140	-25	-0.7	-0.8
Pro840Leu	145	25	-1.6	3.8	Ile967Phe	175	35	4.5	2.8	Thr1033Ala	140	-25	-0.7	1.8
Asp842Asn	150	10	-3.5	-3.5	Arg969Gln	225	-45	-4.5	-3.5	His1034Pro	195	-50	-3.2	-1.6
Gly843Arg	75	150	-0.4	-4.5	Arg969Trp	225	30	-4.5	-0.9	Gly1035Val	75	80	-0.4	4.2
Thr850Ile	140	35	-0.7	4.5	Ala971val	115	40	1.8	4.2	Val1036Ile	155	20	4.2	4.5
Ser855Tyr	115	115	-0.8	-1.3	Thr974Met	140	45	-0.7	1.9	Arg1038Lys	225	-25	-4.5	-3.9
Leu856Arg	170	55	3.8	-4.5	Ser975Tyr	115	115	-0.8	-1.3	Arg1041Pro	225	-80	-4.5	-1.6

Continued

TABLE 1 The Changes in Solvent-Accessible Surface and Hydrophobicity Index for Known WD Mutations—cont'd

Mutation	SASA WT	SASA Mutant (ΔSASA)	Hydrophobicity WT	Hydrophobicity Mutant	Mutation	SASA WT	SASA Mutant (ΔSASA)	Hydrophobicity WT	Hydrophobicity Mutant	Mutation	SASA WT	SASA Mutant (ΔSASA)	Hydrophobicity WT	Hydrophobicity Mutant
Ile857Thr	175	−35	4.5	−0.7	Thr977Arg	140	85	−0.7	−4.5	Arg1041Trp	225	30	−4.5	−0.9
Thr858ala	140	−25	−0.7	1.8	Thr977Met	140	45	−0.7	1.9	Leu1043Pro	170	−25	3.8	−1.6
Glu860gly	190	−115	−3.5	−0.4	Leu979Gln	170	10	3.8	−3.5	Asp1047Val	150	5	−3.5	4.2
Ala861Thr	115	25	1.8	−0.7	Cys980Arg	135	90	2.5	−4.5	Pro1052Leu	145	20	−1.6	3.8
Val864Gly	155	−80	4.2	−0.4	Cys980Tyr	135	95	2.5	−1.3	Leu1057Pro	170	−25	3.8	−1.6
Gly869Arg	75	150	−0.4	−4.5	Ala982val	115	40	1.8	4.2	Ala1058Val	115	40	1.8	4.2
Gly869Val	75	80	−0.4	4.2	Cys985Tyr	135	95	2.5	−1.3	Gly1061Glu	75	115	−0.4	−3.5
Ala874Pro	115	30	1.8	−1.6	Ser986Phe	115	95	−0.8	2.8	Ala1063Val	115	40	1.8	4.2
Ala874Val	115	40	1.8	4.2	Leu987Pro	170	−25	3.8	−1.6	Glu1064Lys	190	10	−3.5	−3.9
Arg875Gly	225	−150	−4.5	−0.4	Gly988Arg	75	150	−0.4	−4.5	Glu1064Ala	190	−75	−3.5	1.8
Arg875Val	225	80	−4.5	4.2	Gly988Val	75	80	−0.4	4.2	Ala1065Pro	115	30	1.8	−1.6
Ser1067Asn	115	45	−0.8	−3.5	Met1169Thr	185	−45	1.9	−0.7	Gly1266Glu	75	115	−0.4	−3.5
Glu1068Gly	190	−115	−3.5	−0.4	Met1169Val	185	−30	1.9	4.2	Gly1266Val	75	150	−0.4	−4.5
His1069Gln	195	−15	−3.2	−3.5	Glu1173Lys	190	10	−3.5	−3.9	Gly1266Trp	75	150	−0.4	−4.5
His1069Asn	195	−35	−3.2	−3.5	Glu1173Gly	190	−115	−3.5	−0.4	Asp1267Asn	150	10	−3.5	−3.5
His1069Tyr	195	35	−3.2	−1.3	Gly1176Arg	75	150	−0.4	−4.5	Asp1267Ala	150	−35	−3.5	1.8
Pro1070Ser	145	−30	−1.6	−0.8	Gly1176Glu	75	115	−0.4	−3.5	Asp1267Val	150	5	−3.5	4.2
Ala1074Val	115	40	1.8	4.2	Thr1178Ala	140	−25	−0.7	1.8	Gly1268Arg	75	150	−0.4	−4.5
Cys1079Phe	135	75	2.5	2.8	Leu1181Pro	170	−25	3.8	−1.6	Asn1270Thr	160	−20	−3.5	−0.7
Glu1082Ter	190	−190	−3.5	0	Ala1183Thr	115	25	1.8	−0.7	Asp1271Asn	150	10	−3.5	−3.5
Leu1083Phe	170	40	3.8	2.8	Ala1183Gly	115	−40	1.8	−0.4	Pro1273Gln	145	35	−1.6	−3.5
Leu1088Ter	170	−170	3.8	0	Ile1184Thr	175	−35	4.5	−0.7	Pro1273Leu	145	25	−1.6	3.8
Gly1089Glu	75	115	−0.4	−3.5	Gly1186Ser	75	40	−0.4	−0.8	Pro1273Ser	145	−30	−1.6	−0.8
Gly1089Val	75	80	−0.4	4.2	Gly1186Cys	75	60	−0.4	2.5	Ala1274Thr	115	25	1.8	−0.7
Cys1091Tyr	135	95	2.5	−1.3	Ala1193Pro	115	30	1.8	−1.6	Ala1274Val	115	40	1.8	4.2

Biochemical and Cellular Properties of ATP7B Variants **Chapter | 4** **41**

Phe1094Leu	210	−40	2.8	3.8	Ala1197Thr	115	25	1.8	−0.7	Gln1277Ter	180	−180	−3.5	0
Gln1095Pro	180	35	−3.5	−1.6	Gln1200Pro	180	35	−3.5	−1.6	Ala1278Gly	115	−40	1.8	−0.4
Pro1098Arg	145	90	−1.6	−4.5	Gln1200Ter	180	−180	−3.5	0	Ala1278Val	115	40	1.8	4.2
Gly1099Ser	75	40	−0.4	−0.8	Ala1202Gly	115	−40	1.8	−0.4	Asp1279Gly	150	−75	−3.5	−0.4
Gly1101Arg	75	150	−0.4	−4.5	His1207Arg	195	30	−3.2	−4.5	Asp1279Tyr	150	80	−3.5	−1.3
Ile1102Thr	175	−35	4.5	−0.7	Gly1213Val	75	80	−0.4	4.2	Gly1281Asp	75	75	−0.4	−3.5
Cys1104Arg	135	90	2.5	−4.5	Asp1215Tyr	150	80	−3.5	−1.3	Gly1281Cys	75	60	−0.4	2.5
Cys1104Phe	135	75	2.5	2.8	Val1216Met	155	30	4.2	1.9	Gly1287Ser	75	40	−0.4	−0.8
Val1106Ile	155	20	4.2	4.5	Thr1220Met	140	45	−0.7	1.9	Gly1287Arg	75	150	−0.4	−4.5
Val1106Leu	155	15	4.2	3.8	Gly1221Glu	75	115	−0.4	−3.5	Thr1288Arg	140	85	−0.7	−4.5
Val1106Asp	155	−5	4.2	−3.5	Asp1222Asn	150	10	−3.5	−3.5	Thr1288Met	140	45	−0.7	1.9
Gly1111Asp	75	75	−0.4	−3.5	Asp1222Val	150	5	−3.5	4.2	Glu1293Lys	190	10	−3.5	−3.9
Gly1111Ala	75	40	−0.4	1.8	Asp1222Tyr	150	80	−3.5	−1.3	Ala1295Asp	115	35	1.8	−3.5
Leu1120Ter	170	−170	3.8	0	Arg1228Thr	225	85	−4.5	−0.7	Asp1296Asn	150	10	−3.5	−3.5
Val1140Ala	155	−40	4.2	1.8	Arg1228Ter	225	−225	−4.5	0	Val1297Ile	155	20	4.2	4.5
Gln1142His	180	15	−3.5	−3.2	Ile1230Val	175	−20	4.5	4.2	Val1297Asp	155	−5	4.2	−3.5
Gln1142Ter	180	−180	−3.5	0	Thr1232Pro	140	5	−0.7	−1.6	Val1298Ile	155	20	4.2	4.5
Val1146Met	155	30	4.2	1.9	Gln1233Pro	180	35	−3.5	−1.6	Val1298Leu	155	15	4.2	3.8
Ile1148Thr	175	−35	4.5	−0.7	Val1234Phe	155	55	4.2	2.8	Leu1299Phe	170	40	3.8	2.8
Gly1149Glu	75	115	−0.4	−3.5	Ile1236Thr	175	−35	4.5	−0.7	Leu1304Phe	170	40	3.8	2.8
Gly1149Ala	75	40	−0.4	1.8	Val1239Gly	155	−80	4.2	−0.4	Leu1305Pro	170	−25	3.8	−1.6
Arg1151His	225	−30	−4.5	−3.2	Ala1241Val	115	40	1.8	4.2	Ala1309Val	115	40	1.8	4.2
Arg1151Cys	225	−90	−4.5	2.5	Pro1245Thr	145	−5	−1.6	−0.7	Ser1310Arg	115	110	−0.8	−4.5
Trp1153Arg	255	−30	−0.9	−4.5	Lys1248Asn	200	−40	−3.9	−3.5	Ser1314Cys	115	20	−0.8	2.5
Trp1153Cys	255	−120	−0.9	2.5	Ala1250Gly	115	−40	1.8	−0.4	Arg1319Ter	225	−225	−4.5	0
Leu1154Pro	170	−25	3.8	−1.6	Val1252Ile	155	20	4.2	4.5	Arg1320Ser	225	−110	−4.5	−0.8
Arg1156His	225	−30	−4.5	−3.2	Leu1255Ile	170	5	3.8	4.5	Asn1324Ser	160	−80	−4.5	−1.6
Gly1158Val	75	80	−0.4	4.2	Gln1256Arg	180	45	−3.5	−4.5	Leu1327Val	170	−45	−3.5	−0.8
Asp1164Asn	150	10	−3.5	−3.5	Lys1258Ter	200	−200	−3.9	0	Ala1328Thr	115	−15	3.8	4.2
Ala1168Pro	115	30	1.8	−1.6	Val1262Phe	155	55	4.2	2.8	Leu1329Pro	170	25	1.8	−0.7
Ala1168Ser	115	0	1.8	−0.8	Gly1266Arg	75	150	−0.4	−4.5			−25	3.8	−1.6

Continued

TABLE 1 The Changes in Solvent-Accessible Surface and Hydrophobicity Index for Known WD Mutations—cont'd

Mutation	Solvent Accessible Surface Area (SASA) WT	Mutant (ΔSASA)	Hydrophobicity Index WT	Mutant	Mutation	Solvent Accessible Surface Area (SASA) WT	Mutant (ΔSASA)	Hydrophobicity Index WT	Mutant
Tyr1331Ser	230	−115	−1.3	−0.8	Ser1431Tyr	115	115	−0.8	−1.3
Tyr1331Cys	230	−95	−1.3	2.5	Ser1432Phe	115	95	−0.8	2.8
Asn1332Lys	160	40	−3.5	−3.9	Thr1434Met	140	45	−0.7	1.9
Asn1332Asp	160	−10	−3.5	−3.5	Ala1445Pro	115	30	1.8	−1.6
Val1334Asp	155	−5	4.2	−3.5	Ter1466Arg	—	225	—	−4.5
Gly1335Arg	75	150	−0.4	−4.5					
Ile1336Thr	175	−35	4.5	−0.7					
Gly1341Ser	75	40	−0.4	−0.8					
Gly1341Arg	75	150	−0.4	−4.5					
Gly1341Asp	75	75	−0.4	−3.5					
Gly1341Val	75	80	−0.4	4.2					
Ile1348Asn	175	−15	4.5	−3.5					
Gln1351Ter	180	−180	−3.5	0					
Pro1352Arg	145	90	−1.6	−4.5					
Pro1352Leu	145	25	−1.6	3.8					
Pro1352Ser	145	−30	−1.6	−0.8					
Trp1353Arg	255	−30	−0.9	−4.5					
Trp1353Ter	255	−255	−0.9	0					
Gly1355Ser	75	40	−0.4	−0.8					
Gly1355Asp	75	75	−0.4	−3.5					
Gly1355Cys	75	60	−0.4	2.5					
Ala1358Ser	115	0	1.8	−0.8					
Met1359Ile	185	−10	1.9	4.5					
Met1359Val	185	−30	1.9	4.2					
Ser1363Phe	115	95	−0.8	2.8					
Leu1368Pro	170	−25	3.8	−1.6					

Ser1369Leu	115	55	−0.8	3.8
Leu1373Pro	170	−25	3.8	−1.6
Leu1373Arg	170	55	3.8	−4.5
Gln1372Ter	180	−180	−3.5	0
Leu1373Pro	170	−25	3.8	−1.6
Cys1375Ser	135	−20	2.5	−0.8
Pro1379Ser	145	−30	−1.6	−0.8
Met1392Lys	185	15	1.9	−3.9
Trp1410Ter	255	−255	−0.9	0

Amino acid properties of 450 Wilson disease mutations are listed. The solvent-accessible surface area measurements were calculated for the residue X in the tripeptide G-X-G (as described in Refs. [30, 31]). The hydrophobicity index values for the wild-type and mutant residues are values calculated according to the hydrophobicity scale of Kyte and Doolittle [32].

minor, if any, effect on protein volume or hydrophobicity but alter the charge of the amino acid residue. Such mutations are likely to be less disruptive of protein function and may be associated with a milder disease phenotype. For example, N41S mutation, which does not significantly affect the volume occupied by this residue nor changes the hydrophilic nature of the residue in this position (Table 1), results in ATP7B variant that retains the Cu transport activity. The association of this mutation with Wilson disease could be due to the loss in the ability of ATP7B-N41S to faithfully traffic to the canalicular membrane and respond to changes in copper concentration (ATP7B-N41S traffics to basolaterally located vesicles [37]). The initial report on identification of this mutant does not describe whether the mutation was homozygous; therefore, it is also possible that the negative effect of ATP7B-N41S substitution is apparent only if another deleterious mutation is present on a second allele of *ATP7B* gene.

The transmembrane helical regions are largely hydrophobic owing to their embedding in the lipid bilayer. Wilson disease mutations in the TM region(s) cause reduction in solvent accessibility in a large number of codon positions (Fig. 3B, bottom). The mutants are also more hydrophilic in nature when compared with the wild type (Fig. 3C (bottom), which has a destabilizing effect on the hydrophobic helical segments in a lipid bilayer and in turn affects normal function of the pump or its intracellular sorting. The Cys985Y ($C^{983}PC^{985}$ motif in TM 6) substitution not only would prevent Cu(I) coordination but also could destabilize the helical bundle packing with its increased volume mutation. Mutations in the A-domain and the ATP-binding domains show visibly large-scale shifts in surface area but modest changes in the hydrophobicity scale (Fig. 3B and C). Compared with the wild type, the SASA values for the mutants indicate that the residues are more buried and less accessible to salt bridges and or other interactions. Since both A-domain and the ATP-binding domains undergo significant motions and changes in the interdomain contacts during ATP binding and hydrolysis, such mutations may have an adverse effect on catalytic phosphorylation.

WILSON DISEASE MUTATION IN VARIOUS DOMAINS

Yeast complementation system relies on the ability to restore the growth of Cu transport-deficient yeast strain by the expression of ATP7B or variants. This system has been used successfully to classify loss-of-function mutation from less severe mutations [38–41]. In combination with the studies of ATP7B variants in mammalian cells [41], this approach allows a fairly rapid initial characterization of a large number of mutants. A more laborious study directly measuring copper transport activity of 28 Wilson disease mutations along with the analysis of their enzymatic and cellular properties has also been published [29]. This study revealed that a significant fraction of ATP7B variants associated with Wilson disease has small but measurable transport activity (which also explains why some of the mutants may sustain cell growth in a yeast complementation assay). This finding is significant. Wilson disease is an autosomal recessive disorder (i.e., disorder, in which both copies of the *ATP7B* gene must be mutated for pathology to develop). Thus ATP7B proteins with various combinations of mutations will have various degrees of residual transport activity, which may in turn affect disease onset and progression.

Mutations in the N-Terminal Domain Are Rare But Have Significant Consequences

The N-terminal domain of human ATP7B is the most structurally diverse region of the protein, when ATP7B is compared with orthologues from other species. The N-terminal domain contains six metal-binding domains (MBDs), each with a conserved CxxC site for binding Cu(I) [28, 29, 42–47] (Fig. 2A). Each MBD has a ferredoxin fold and connected to other MBDs by loops of variable lengths. The loops allow the N-terminal domain flexibility necessary to rearrange upon copper transfer from the copper chaperone Atox1 [48]. The N-terminal domain also interacts with the neighboring cytosolic ATP-binding domain (and regulating the affinity for ATP) and A-domain, by perhaps forming a scaffold. Studies of the isolated domains have shown that interactions between MBDs and the ATP-binding domain are inhibitory [42, 49, 50]. While the atomic model of full length ATP7B remains to be unraveled, the solution structures of MBD1, MBD2, MBD3-MBD4, and MBD5-MBD6; the A-domain; and the nucleotide-binding domain of ATP7B have been determined [28, 50–53] contributing valuable insight. Indeed, structures of isolated domains from ATP7B and prokaryotic homologues have already been used as a model system to study the impact of Wilson disease mutations [28, 53–56].

The first 63 amino acids in the N-terminus of the ATP7B contain a short segment required for Cu-dependent trafficking of ATP7B from the TGN to the apical membrane [23, 45]. Mutations in this region lessen retention of ATP7B to the TGN under basal conditions and cause accumulation in vesicles in the vicinity of the basolateral membrane, instead of trafficking to the apical membrane [23, 45]. This abnormal ATP7B trafficking in the absence of Cu is likely to diminish incorporation of Cu into ceruloplasmin and prevent export of Cu into the bile. Three WD-causing mutations have been identified in this region—S15G, N41S, and Y44N. Biochemically, the mutations decrease the volume of the side chain and increase the hydrophilicity of the region (Fig. 3A and B). Since the region including these mutations is predicted to be unstructured, the likely explanation for the disruptive effect of mutations is the loss of interactions between this segment and other domains of ATP7B or regulatory proteins. Indeed, recent NMR studies have demonstrated that the peptide corresponding to N-terminal region interacts with the nucleotide-binding domain of ATP7B and remains associated with ATP7B even after the proteolytic cleavage within the N-terminal region [57].

The entire N-terminus, which encompasses all six MBDs, has 64 mutations, out of which 25 are codon termination mutants (not discussed) (Fig. 3A and B). Mutations of cysteine residues within the CxxC copper-binding sites would abolish copper binding and may negatively impact copper transfer to the TM regions. The C157F substitution in MBD2 is an example of such WD-causing mutation. The loss of copper binding by MBD2 was previously shown to prevent copper transfer from the cytosolic copper chaperone Atox1 to ATP7B and inhibit ATP7B activity [58]. Other disease mutations in MBDs are mostly located outside the CxxC metal-binding motifs and may alter folding of MBDs and destabilize the entire ATP7B. This idea is supported by the fact that the experimental deletion of the N-terminal segment, which includes up to five MBDs, does not inhibit ATP7B. Therefore, the negative effect of mutations within the N-terminal region is likely due to abnormal folding or disrupted regulation. Characterization of mutations G85V, L492S, R616W, G626A, and M645R revealed disruptions in normal ATP7B function [29]. Mutations G85V in MBD1 and L492S in MBD5 show neither catalytic nor copper transport activity [42]. The effect of G85V substitution can be easily explained. Gly85 is a conserved residue, and its location ensures flexibility of protein in this position (Fig. 3). Substitution of a small flexible glycine with a large bulky valine affects protein stability of MBD1 and the entire ATP7B [34, 35]; the unfolding/thermal instability of MBD1 would also alter interactions between MBD1, MBD2, and MBD3, which are important for Atox1-mediated copper transfer [36, 48, 57, 59]. The strong negative effect of L492S on ATP7B function is harder to understand. L492 is located seven residues away from the CxxC site of MBD5 and is unlikely to affect copper binding. The side chain of Leu492 is oriented inwardly, and the mutation to a smaller and more hydrophilic Ser may have some effect on protein structure/MBD folding. However, the ATP7B variant with this mutation shows high level of expression, arguing against significant protein unfolding/instability [29]. Therefore, why the marked negative effect on ATP7B activity is so dramatic is unclear. These uncertainties in predicting effect of mutations (even when structures of individual domains are available) highlight the need for the high-resolution structure of the entire ATP7B, which would reveal details of protein environment and more detailed functional analysis of the mutants. Mutation R616W (in MBD6) causes hyperphosphorylation, that is, the decreased ability to cleave phosphor-aspartate intermediate during ATP hydrolysis. The low transport activity of R616W variant is consistent with the idea that mutant ATP7B is trapped in one conformation and cannot proceed normally through the catalytic cycle. Similarly, the mutation G626A did not have any loss in catalytic phosphorylation, but had partial transport activity.

The M645R (in the segment connecting MBD6 and TM1) has been identified as a WD-causing mutant in several studies and is particularly common in Spanish population [60–62]. Yet, the in vitro analysis of this variant suggests that the copper transport activity of this variant is only slightly decreased. Lower expression of this variant in studies in mammalian cells suggests that the negative consequences of this mutation could be due to decreased protein abundance [29, 34]. As we discussed previously [63], it is also possible that M645 substitution has to be accompanied by a highly deleterious mutation on the second allele of ATP7B (such as deletions or stop codons) for disease phenotype to develop.

Actuator (A-Domain) and ATP Binding Domains (ATPBD) Are Major Targets of WD Mutations

ATP7B belongs to a large family of ATP-driven ion and lipid transporters called P-type ATPases. The calcium-transporting ATPase sarcoplasmic-endoplasmic reticulum calcium (SERCA) is the well-characterized member of this family; SERCA and ATP7B both hydrolyze ATP to transport their respective metals and show significant structural similarity for their A-, N-, and P-domains. In the SERCA, the actuator domain (A-domain) and ATP-binding domain (N- and P-domains together) interact with each other [64]. This interaction is necessary to facilitate conformational transitions of the protein during transport cycle. In ATP7B, A-domain and ATPBD are mutated in 34 and 165 positions, respectively. The P840L, I857T, and A874V substitutions in the A-domain have been characterized and found to inhibit transport activity and show high levels of phosphorylation of catalytic aspartate. These results are consistent with ATP7B being trapped in phosphorylated state and unable to hydrolyze the phosphate to yield energy for copper translocation across membrane. Changes in the interactions between the A-domain with the ATP-binding domain would explain the phenotype [29]. Analysis of the solvent-accessible surface area for two of these mutants and for several other mutations in the A-domain shows an increase in the surface area (a potentially destabilizing prospect, Table 1). Overall, the hydrophobicity of the 34 mutations is lower (Fig. 3A and B).

High abundance of the disease-causing mutants in the ATPBD highlights functional importance of this region (Fig. 2B). Several ATP7B mutants have been extensively characterized for structural defects, changes in transport activity, localization, and trafficking revealing broad spectrum of abnormalities [28, 29, 40, 65]. The P-domain mutants P992L, T1031S, and P1052L caused low transport activity and Cu uptake but did not fully eradicate the enzyme function [29]. The G1213V mutation is located in the "hinge" region, which connects the N- and P-domains, and inhibits copper transport activity. This effect is most likely caused by the disruption of a coordinated movement of the P- and N-domains, which is necessary for phosphorylation of the catalytic aspartate in the DKTGT motif [29]. In support of this hypothesis, molecular dynamics simulation on the G1213V mutant revealed marked changes of the normal mode motions around the hinge region [55]. Mutants D1222V and G1266R have low transport activity and render the enzyme completely inactive due to the inability of the enzyme to form a phosphorylated intermediate. These mutants are located close to the conserved D^{1027}KTGT site and could obstruct the phosphorylation site due to the presence of a large hydrophobic group (D1222>V) or an increase in volume and addition of a positively charged guanidine group (Fig. 3A and B).

Mutations in the N-domain affect protein folding and stability, ATP binding, and protein localization. The mutants H1069Q (discussed further below) and E1064K directly affect ATP binding. The negatively charged oxygen atoms in the glutamate E1064 coordinate the positive nitrogens in the adenosine moiety, whereas the histidine residue in position 1069 is involved in π-π stacking interactions with the adenosine ring stabilizing the ATP molecule [54]. NMR studies of the isolated N-domain with the E1064A mutation showed that the mutated protein is stable and retains the overall fold of the domain, but the positions of the α-helical segments were altered, creating a more open structure, misaligning the ATP-coordinating residues, and a complete loss of ATP binding [28]. Two other mutations R1151H that are located proximally to the H1069 site and C1104F that is located further away from the ATP-binding pocket were characterized using isothermal titration calorimetry studies. These studies showed that R1151H affected ATP-binding affinity and C1104F significantly affected protein folding [66]. The mutant L1083F is facing away from the ATP-binding pocket at the surface of the N-domain. In vitro, the mutant has only partial disruption to the transport activity and does not affect catalytic phosphorylation [29]. Yet, in cells, this mutant is trapped in the ER, and inability to reach the *trans*-Golgi network and vesicles would account for the WD phenotype associated with this mutation. These properties suggest that the residue L1083 may play an important role in the interdomain interactions necessary for structural integrity of ATP7B associated with its exit from the ER. Given recent finding that ATP7B forms a dimer, which persist when ATP7B traffics through the secretory pathway [67], it would be interesting to test whether the L1083F substitution may disrupt dimer formation.

The H1069Q Mutant

The H1069Q substitution is by far the best characterized WD mutation. This substitution causes decrease in ATP7B stability and retention in the ER [65, 68]. The mutation is most commonly found in Caucasian and African American patients [69–71]. Studies using yeast complementation showed that mutations to H1069 impaired the ATP7B function [65, 72, 73] and a significant inhibition of transport was confirmed by direct measurements of ATP7B copper transport activity [29]. The loss of this activity is likely due to the inability of the mutant to form a phosphorylated intermediate in the presence of ATP [74], which in turn could be a result of diminished affinity for ATP [66, 74]. Structural studies of the N-domain confirmed the role of H1069 in ATP binding [49, 50] but did not provide a clear answer why this mutation markedly decreases ATP7B stability. Molecular dynamics simulation studies using the NMR ensemble of the wild type and the mutant N-domain revealed differences in ATP orientation and coordination environment that may explain diminished binding [53]. These studies also showed that the histidine residue at position 1069 was vital for the proper orientation of ATP in the binding pocket. Further structural characterization of H1069Q showed large-scale destabilization of the mutant in vitro confirming that the structural instability of the N-domain is the primary reason for instability of entire ATP7B [28].

The Mutations Within the TM Domain Revealed Long-Range Interactions Between Different Parts of ATP7B

The transmembrane region of ATP7B contains eight transmembrane (TN) helical segments; they contain highly conserved residues that could coordinate Cu ions, especially TM4 and TM8; they also act as conduits for converting the ~32kJ/mol of energy acquired from ATP hydrolysis into movements associated with copper transport. Over 170 mutations have been identified within the TM domain, and they were found in all eight segments (Figs. 2B and 3). Only a handful has been well characterized. The mutation within predicted Cu-binding site cys-pro-cys (CPC) motif (C985Y) removes the side chain essential for copper binding replacing it with the bulky and hydrophobic residue further disrupting the copper translocation pathway; these changes account for the loss of ATP7B function. Mutations P760L, D765N, and M769V in TM4 decrease Cu transport activity but allow the formation of phosphorylated intermediate [29]. The hydrophobic nature of the replaced residues in TM domain is mostly preserved by mutations (Fig. 3A and B), but variation in the volumes of side chains can destabilize TM domain or preclude conformational transitions necessary for transport. No information is currently available of how the TM domain of ATP7B interacts with lipids and which mechanisms enable sorting of ATP7B within the TGN. Further studies are needed to examine these properties of ATP7B and the effect of mutations.

The S653Y mutant is an interesting example of a TM region mutation, which retains some Cu transport activity (and hence the delivery of Cu to ceruloplasmin) but lacks the ability to traffic from the TGN under elevated Cu conditions. The ATP7B-S653Y mutant is stable and is targeted normally to the TGN [37]. Experimental mutations in this region and molecular dynamics simulations revealed that replacing Ser653 with the bulky or charged residues modifies the local electrostatic environment and induces distortions within the TM1 and neighboring TM2 [37]. Although the distortion was local and did not significantly affect the TM segments directly involved in copper binding and transfer (TM4, TM7, and TM8), the structural change in TM1 altered the properties of the N-terminal domain, which is directly connected to TM1. Specifically, the mutation of N41, which in earlier studies was shown to cause constitutive trafficking (see above), in combination with the S653Y mutation did not have an effect. This is a very intriguing example of "canceling" effect of two WD-causing mutations and yet another demonstration of complexity and variety of effects. Although the presence of two or three mutations in the same allele is rear, such multiple mutations occur, and their effect may not necessarily be additive, as the above example demonstrates.

SINGLE NUCLEOTIDE POLYMORPHISMS MAY PLAY A LARGER ROLE IN DISEASE THAN PREVIOUSLY THOUGHT

Single nucleotide polymorphisms (SNPs) are the largest source of sequence variation in humans. However, contribution of SNPs to disease progression and manifestations is not well understood. SNPs are defined as genetic changes that are present in >1% of the general population [75]. Perhaps, due to this fact, SNPs have long been considered benign, and their contribution to disease phenotype has been understudied. Recently, SNPs have become a growing topic of interest. Dozens of SNPs have been identified in the *ATP7B* gene, and several of these are considered common, such as G2495A (Lys832Arg), C1366G (Val456Leu), and G1216T (Ala406Ser) (highlighted as (*) in Fig. 3B) [76, 77]. Interestingly, some of these SNPs frequently show up in WD patients. For example, the Lys832Arg SNP has been reported in WD patients in over 20 studies archived in the Wilson Disease Mutation Database. The first study addressing the biochemical and cellular properties of ATP7B SNPs was published in 2011. The studies illustrated a fine line between some polymorphisms and a disease mutation. The ATP7B Gly875Arg variant was considered an SNP but was also known as a WD-causing mutation in Indian population [78]. The protein folding of ATP7B875Arg (a less common variant) was very sensitive to cellular copper levels. In low copper, ATP7B875Arg was trapped in endoplasmic reticulum and unable to reach its normal Golgi localization and activate copper-dependent enzymes. By contrast, a more common ATP7B875Gly was less sensitive to copper depletion and targeted normally. Upon copper supplementation and excess, both ATP7B875Arg and ATP7B875Gly behave very similarly [78]. These results suggest that SNPs can have different "background" effects and exacerbate effects of WD mutation under certain conditions.

Direct functional studies of two other variants found that Val456Leu and Lys832Arg SNPs produce ATP7B with lower Cu transport activity [29]. More recently, Lys832Arg was studied in *Drosophila melanogaster* and was shown to be identified as a loss-of-function SNP. However, flies have only one copper transporter, ATP7, that shares homology with both ATP7B and ATP7A [79]. Expression of a Lys832Arg SNP construct in knockout *ATP7* flies did not rescue larval death. Although these data are intriguing, the relevance of findings to human cells is uncertain.

Four ATP7B SNPs were identified in the Indian population as useful predictive diagnosis markers for WD [80]. Additionally, a link has been suggested between two ATP7B SNPs (Lys832Arg and Arg952Lys) and increased the amount of bioavailable copper in Alzheimer's disease (AD) patients [81]. Interestingly, these SNPs are found at high frequencies in the normal population, but AD patients are more likely to be homozygous for these SNPs compared with healthy controls [82]. These studies have raised important questions about the contribution of SNPs in disease risk and progression. The effects of SNPs on the protein could be most significant when the SNP is located in a functionally significant domain, such as the ATP-binding domain in ATP7B (Fig. 2B, ATPBD shown in *red*). It is also possible that SNPs become significant under certain metabolic conditions or in response to aging when a decreasing protein quality control may exacerbate SNP-dependent changes of ATP7B properties. There is much to be learned about the impact of SNPs on ATP7B structure or function.

CONCLUSIONS

Wilson disease mutations affect all functional domains of ATP7B and have a broad spectrum of consequences. The mutational "landscape" highlights functional significance of different domains and identifies the ATP-binding domain as being most commonly targeted by substitutions. Specific location of WD mutations in ATP7B domains and changes in the volume, surface area, and hydrophobicity index of the side chains allow predictions of the potential impact of mutations on protein structure and activity. Nevertheless, direct and detailed characterization of mutations is needed, because the available data for currently characterized mutants show a broad spectrum of changes in stability, ligand-binding enzymatic and transport activity, and cellular behavior, all of which cannot be predicted by in silico analysis. Finally, emerging data on SNPs suggest that normal genetic variability could further modify properties of ATP7B and modulate effects of the disease-causing mutations.

REFERENCES

[1] Fraústro da Silva JJR, Williams RJP. The biological chemistry of the elements. 2nd ed. New York: Oxford University Press; 2001.
[2] Pena MM, Lee J, Thiele DJ. A delicate balance: homeostatic control of copper uptake and distribution. J Nutr 1999;129:1251–60.
[3] Gu M, Cooper JM, Butler P, Walker AP, Mistry PK, Dooley JS, Schapira AH. Oxidative-phosphorylation defects in liver of patients with Wilson's disease. Lancet 2000;356:469–74.
[4] Lutsenko S, Bhattacharjee A, Hubbard AL. Copper handling machinery of the brain. Metallomics 2010;2:596–608.
[5] Linz R, Lutsenko S. Copper-transporting ATPases ATP7A and ATP7B: cousins, not twins. J Bioenerg Biomembr 2007;39:403–7.
[6] Maryon EB, Molloy SA, Kaplan JH. Cellular glutathione plays a key role in copper uptake mediated by human copper transporter 1. Am J Physiol Cell Physiol 2013;304:C768–79.
[7] Hatori Y, Clasen S, Hasan NM, Barry AN, Lutsenko S. Functional partnership of the copper export machinery and glutathione balance in human cells. J Biol Chem 2012;287:26678–87.
[8] Hatori Y, Yan Y, Schmidt K, Furukawa E, Hasan NM, Yang N, Liu CN, Sockanathan S, Lutsenko S. Neuronal differentiation is associated with a redox-regulated increase of copper flow to the secretory pathway. Nat Commun 2016;7.
[9] Gray LW, Peng F, Molloy SA, Pendyala VS, Muchenditsi A, Muzik O, Lee J, Kaplan JH, Lutsenko S. Urinary copper elevation in a mouse model of Wilson's disease is a regulated process to specifically decrease the hepatic copper load. PLoS One 2012;7.
[10] Hatori Y, Lutsenko S. An expanding range of functions for the copper chaperone/antioxidant protein Atox1. Antioxid Redox Signal 2013;19:945–57.
[11] Lee J, Prohaska JR, Thiele DJ. Essential role for mammalian copper transporter Ctr1 in copper homeostasis and embryonic development. Proc Natl Acad Sci U S A 2001;98:6842–7.
[12] Lutsenko S, Tsivkovskii R, Walker JM. Functional properties of the human copper-transporting ATPase ATP7B (the Wilson's disease protein) and regulation by metallochaperone Atox1. Ann N Y Acad Sci 2003;986:204–11.
[13] Tumer Z, Horn N. Menkes disease: underlying genetic defect and new diagnostic possibilities. J Inherit Metab Dis 1998;21:604–12.
[14] Faa G, Liguori C, Columbano A, Diaz G. Uneven copper distribution in the human newborn liver. Hepatology 1987;7:838–42.
[15] Muchenditsi A, Yang H, Hamilton JP, Koganti L, Housseau F, Aronov L, Fan H, Pierson H, Bhattacharjee A, Murphy R, Sears C, Potter J, Wooton-Kee CR, Lutsenko S. Targeted inactivation of copper transporter Atp7b in hepatocytes causes liver steatosis and obesity in mice. Am J Physiol Gastrointest Liver Physiol 2017;313:G39–49.
[16] Danks DM. Copper and liver disease. Eur J Pediatr 1991;150:142–8.
[17] Roberts EA, Cox DW. Wilson disease. Bailliers Clin Gastroenterol 1998;12:237–56.
[18] Wilson SA. Kayser-Fleischer ring in cornea in two cases of Wilson's disease (progressive lenticular degeneration). Proc R Soc Med 1934;27:297–8.
[19] Gomes A, Dedoussis GV. Geographic distribution of ATP7B mutations in Wilson disease. Ann Hum Biol 2016;43:1–8.
[20] Cheng N, Wang H, Wu W, Yang R, Liu L, Han Y, Guo L, Hu J, Xu L, Zhao J, Han Y, Liu Q, Li K, Wang X, Chen W. Spectrum of ATP7B mutations and genotype-phenotype correlation in large-scale Chinese patients with Wilson disease. Clin Genet 2017;92:69–79.
[21] Ljubic H, Kalauz M, Telarovic S, Ferenci P, Ostojic R, Noli MC, Lepori MB, Hrstic I, Vukovic J, Premuzic M, Radic D, Ravic KG, Sertic J, Merkler A, Barisic AA, Loudianos G, Vucelic B. ATP7B gene mutations in Croatian patients with Wilson disease. Genet Test Mol Biomarkers 2016;20:112–7.
[22] Hasan NM, Gupta A, Polishchuk E, Yu CH, Polishchuk R, Dmitriev OY, Lutsenko S. Molecular events initiating exit of a copper-transporting ATPase ATP7B from the trans-Golgi network. J Biol Chem 2012;287:36041–50.
[23] Braiterman LT, Gupta A, Chaerkady R, Cole RN, Hubbard AL. Communication between the N and C termini is required for copper-stimulated Ser/Thr phosphorylation of Cu(I)-ATPase (ATP7B). J Biol Chem 2015;290:8803–19.
[24] Roelofsen H, Wolters H, Van Luyn MJ, Miura N, Kuipers F, Vonk RJ. Copper-induced apical trafficking of ATP7B in polarized hepatoma cells provides a mechanism for biliary copper excretion. Gastroenterology 2000;119:782–93.
[25] Gupta A, Schell MJ, Bhattacharjee A, Lutsenko S, Hubbard AL. Myosin Vb mediates Cu^+ export in polarized hepatocytes. J Cell Sci 2016;129:1179–89.
[26] de Bie P, Muller P, Wijmenga C, Klomp LW. Molecular pathogenesis of Wilson and Menkes disease: correlation of mutations with molecular defects and disease phenotypes. J Med Genet 2007;44:673–88.

[27] Gerbasi V, Lutsenko S, Lewis EJ. A mutation in the ATP7B copper transporter causes reduced dopamine beta-hydroxylase and norepinephrine in mouse adrenal. Neurochem Res 2003;28:867–73.

[28] Dmitriev OY, Bhattacharjee A, Nokhrin S, Uhlemann EM, Lutsenko S. Difference in stability of the N-domain underlies distinct intracellular properties of the E1064A and H1069Q mutants of copper-transporting ATPase ATP7B. J Biol Chem 2011;286:16355–62.

[29] Huster D, Kuhne A, Bhattacharjee A, Raines L, Jantsch V, Noe J, Schirrmeister W, Sommerer I, Sabri O, Berr F, Mossner J, Stieger B, Caca K, Lutsenko S. Diverse functional properties of Wilson disease ATP7B variants. Gastroenterology 2012;142:947–56. e945.

[30] Lesk AM, Chothia C. Solvent accessibility, protein surfaces, and protein folding. Biophys J 1980;32:35–47.

[31] Chothia C. The nature of the accessible and buried surfaces in proteins. J Mol Biol 1976;105:1–12.

[32] Kyte J, Doolittle RF. A simple method for displaying the hydropathic character of a protein. J Mol Biol 1982;157:105–32.

[33] Arioz C, Li Y, Wittung-Stafshede P. The six metal binding domains in human copper transporter, ATP7B: molecular biophysics and disease-causing mutations. Biometals 2017;30:823–40.

[34] de Bie P, van de Sluis B, Burstein E, van de Berghe PV, Muller P, Berger R, Gitlin JD, Wijmenga C, Klomp LW. Distinct Wilson's disease mutations in ATP7B are associated with enhanced binding to COMMD1 and reduced stability of ATP7B. Gastroenterology 2007;133:1316–26.

[35] Kumar R, Arioz C, Li Y, Bosaeus N, Rocha S, Wittung-Stafshede P. Disease-causing point-mutations in metal-binding domains of Wilson disease protein decrease stability and increase structural dynamics. Biometals 2017;30:27–35.

[36] Yu CH, Lee W, Nokhrin S, Dmitriev OY. The structure of metal binding domain 1 of the copper transporter ATP7B reveals mechanism of a singular Wilson disease mutation. Sci Rep 2018;8:581.

[37] Braiterman LT, Murthy A, Jayakanthan S, Nyasae L, Tzeng E, Gromadzka G, Woolf TB, Lutsenko S, Hubbard AL. Distinct phenotype of a Wilson disease mutation reveals a novel trafficking determinant in the copper transporter ATP7B. Proc Natl Acad Sci U S A 2014;111:E1364–73.

[38] Papur OS, Terzioglu O, Koc A. Functional characterization of new mutations in Wilson disease gene (ATP7B) using the yeast model. J Trace Elem Med Biol 2015;31:33–6.

[39] Lee BH, Kim JH, Lee SY, Jin HY, Kim KJ, Lee JJ, Park JY, Kim GH, Choi JH, Kim KM, Yoo HW. Distinct clinical courses according to presenting phenotypes and their correlations to ATP7B mutations in a large Wilson's disease cohort. Liver Int 2011;31:831–9.

[40] Hsi G, Cullen LM, Macintyre G, Chen MM, Glerum DM, Cox DW. Sequence variation in the ATP-binding domain of the Wilson disease transporter, ATP7B, affects copper transport in a yeast model system. Hum Mutat 2008;29:491–501.

[41] Park S, Park JY, Kim GH, Choi JH, Kim KM, Kim JB, Yoo HW. Identification of novel ATP7B gene mutations and their functional roles in Korean patients with Wilson disease. Hum Mutat 2007;28:1108–13.

[42] Tsivkovskii R, MacArthur BC, Lutsenko S. The Lys1010-Lys1325 fragment of the Wilson's disease protein binds nucleotides and interacts with the N-terminal domain of this protein in a copper-dependent manner. J Biol Chem 2001;276:2234–42.

[43] Lutsenko S, Barnes NL, Bartee MY, Dmitriev OY. Function and regulation of human copper-transporting ATPases. Physiol Rev 2007;87:1011–46.

[44] Ralle M, Lutsenko S, Blackburn NJ. Copper transfer to the N-terminal domain of the Wilson disease protein (ATP7B): X-ray absorption spectroscopy of reconstituted and chaperone-loaded metal binding domains and their interaction with exogenous ligands. J Inorg Biochem 2004;98:765–74.

[45] Guo Y, Nyasae L, Braiterman LT, Hubbard AL. NH2-terminal signals in ATP7B Cu-ATPase mediate its Cu-dependent anterograde traffic in polarized hepatic cells. Am J Physiol Gastrointest Liver Physiol 2005;289:G904–16.

[46] Tsivkovskii R, Eisses JF, Kaplan JH, Lutsenko S. Functional properties of the copper-transporting ATPase ATP7B (the Wilson's disease protein) expressed in insect cells. J Biol Chem 2002;277:976–83.

[47] DiDonato M, Narindrasorasak S, Forbes JR, Cox DW, Sarkar B. Expression, purification, and metal binding properties of the N-terminal domain from the Wilson disease putative copper-transporting ATPase (ATP7B). J Biol Chem 1997;272:33279–82.

[48] Yu CH, Yang N, Bothe J, Tonelli M, Nokhrin S, Dolgova NV, Braiterman L, Lutsenko S, Dmitriev OY. The metal chaperone Atox1 regulates the activity of the human copper transporter ATP7B by modulating domain dynamics. J Biol Chem 2017;292:18169–77.

[49] Dmitriev OY, Tsivkovskii R, Abildgaard F, Lutsenko S. NMR assignment of the Wilson disease associated protein N-domain. J Biomol NMR 2006;36(Suppl. 1):61.

[50] Dmitriev O, Tsivkovskii R, Abildgaard F, Morgan CT, Markley JL, Lutsenko S. Solution structure of the N-domain of Wilson disease protein: distinct nucleotide-binding environment and effects of disease mutations. Proc Natl Acad Sci U S A 2006;103:5302–7.

[51] Achila D, Banci L, Bertini I, Bunce J, Ciofi-Baffoni S, Huffman DL. Structure of human Wilson protein domains 5 and 6 and their interplay with domain 4 and the copper chaperone HAH1 in copper uptake. Proc Natl Acad Sci U S A 2006;103:5729–34.

[52] Banci L, Bertini I, Cantini F, Migliardi M, Natile G, Nushi F, Rosato A. Solution structures of the actuator domain of ATP7A and ATP7B, the Menkes and Wilson disease proteins. Biochemistry 2009;48:7849–55.

[53] Rodriguez-Granillo A, Sedlak E, Wittung-Stafshede P. Stability and ATP binding of the nucleotide-binding domain of the Wilson disease protein: effect of the common H1069Q mutation. J Mol Biol 2008;383:1097–111.

[54] Tsuda T, Toyoshima C. Nucleotide recognition by CopA, a Cu^+−transporting P-type ATPase. EMBO J 2009;28:1782–91.

[55] Jayakanthan S, Im H, McEvoy MM, Lutsenko S, Woolf TB. Molecular dynamic studies reveal conformation dynamics of Wilson disease mutations. J Mol Biol 2019 [in preparation].

[56] Jayakanthan S, Roberts SA, Weichsel A, Arguello JM, McEvoy MM. Conformations of the apo-, substrate-bound and phosphate-bound ATP-binding domain of the Cu(II) ATPase CopB illustrate coupling of domain movement to the catalytic cycle. Biosci Rep 2012;32:443–53.

[57] Corey H, Yang N, Bothe J, Tonelli M, Nokhrin S, Dolgova NV, Braiterman L, Lutsenko S, Dmitriev OY. The metal chaperone Atox1 regulates the activity of the human copper transporter ATP7B by modulating domain dynamics. J Biol Chem 2017;292(44):18169–77.

[58] Walker JM, Huster D, Ralle M, Morgan CT, Blackburn NJ, Lutsenko S. The N-terminal metal-binding site 2 of the Wilson's disease protein plays a key role in the transfer of copper from Atox1. J Biol Chem 2004;279:15376–84.

[59] Banci L, Bertini I, Cantini F, Massagni C, Migliardi M, Rosato A. An NMR study of the interaction of the N-terminal cytoplasmic tail of the Wilson disease protein with copper(I)-HAH1. J Biol Chem 2009;284:9354–60.

[60] Kalinsky H, Funes A, Zeldin A, Pel-Or Y, Korostishevsky M, Gershoni-Baruch R, Farrer LA, Bonne-Tamir B. Novel ATP7B mutations causing Wilson disease in several Israeli ethnic groups. Hum Mutat 1998;11:145–51.

[61] Brage A, Tome S, Garcia A, Carracedo A, Salas A. Clinical and molecular characterization of Wilson disease in Spanish patients. Hepatol Res 2007;37:18–26.

[62] Paradisi I, De Freitas L, Arias S. Most frequent mutation c.3402delC (p.Ala1135GlnfsX13) among Wilson disease patients in Venezuela has a wide distribution and two old origins. Eur J Med Genet 2015;58:59–65.

[63] Schushan M, Bhattacharjee A, Ben-Tal N, Lutsenko S. A structural model of the copper ATPase ATP7B to facilitate analysis of Wilson disease-causing mutations and studies of the transport mechanism. Metallomics 2012;4:669–78.

[64] Toyoshima C, Nomura H, Sugita Y. Crystal structures of Ca^{2+}-ATPase in various physiological states. Ann N Y Acad Sci 2003;986:1–8.

[65] Payne AS, Kelly EJ, Gitlin JD. Functional expression of the Wilson disease protein reveals mislocalization and impaired copper-dependent trafficking of the common H1069Q mutation. Proc Natl Acad Sci U S A 1998;95:10854–9.

[66] Morgan CT, Tsivkovskii R, Kosinsky YA, Efremov RG, Lutsenko S. The distinct functional properties of the nucleotide-binding domain of ATP7B, the human copper-transporting ATPase: analysis of the Wilson disease mutations E1064A, H1069Q, R1151H, and C1104F. J Biol Chem 2004;279:36363–71.

[67] Jayakanthan S, Braiterman LT, Hasan NM, Unger VM, Lutsenko S. Human copper transporter ATP7B (Wilson disease protein) forms stable dimers in vitro and in cells. J Biol Chem 2017;292:18760–74.

[68] Huster D, Hoppert M, Lutsenko S, Zinke J, Lehmann C, Mossner J, Berr F, Caca K. Defective cellular localization of mutant ATP7B in Wilson's disease patients and hepatoma cell lines. Gastroenterology 2003;124:335–45.

[69] Tanzi RE, Petrukhin K, Chernov I, Pellequer JL, Wasco W, Ross B, Romano DM, Parano E, Pavone L, Brzustowicz LM, et al. The Wilson disease gene is a copper transporting ATPase with homology to the Menkes disease gene. Nat Genet 1993;5:344–50.

[70] Petrukhin K, Fischer SG, Pirastu M, Tanzi RE, Chernov I, Devoto M, Brzustowicz LM, Cayanis E, Vitale E, Russo JJ, et al. Mapping, cloning and genetic characterization of the region containing the Wilson disease gene. Nat Genet 1993;5:338–43.

[71] Thomas GR, Roberts EA, Walshe JM, Cox DW. Haplotypes and mutations in Wilson disease. Am J Hum Genet 1995;56:1315–9.

[72] Iida M, Terada K, Sambongi Y, Wakabayashi T, Miura N, Koyama K, Futai M, Sugiyama T. Analysis of functional domains of Wilson disease protein (ATP7B) in Saccharomyces cerevisiae. FEBS Lett 1998;428:281–5.

[73] Hung IH, Suzuki M, Yamaguchi Y, Yuan DS, Klausner RD, Gitlin JD. Biochemical characterization of the Wilson disease protein and functional expression in the yeast *Saccharomyces cerevisiae*. J Biol Chem 1997;272:21461–6.

[74] Tsivkovskii R, Efremov RG, Lutsenko S. The role of the invariant His-1069 in folding and function of the Wilson's disease protein, the human copper-transporting ATPase ATP7B. J Biol Chem 2003;278:13302–8.

[75] Arias TD, Jorge LF, Barrantes R. Uses and misuses of definitions of genetic polymorphism: a perspective from population pharmacogenetics. Br J Clin Pharmacol 1990;31:117–9.

[76] Olsson C, Waldenstrom E, Westermark K, Landegre U, Syvanen AC. Determination of the frequencies of ten allelic variants of the Wilson disease gene (ATP7B), in pooled DNA samples. Eur J Hum Genet 2000;8:933–8.

[77] Kusuda Y, Hamaguchi K, Mori T, Shin R, Seike M, Sakata T. Novel mutations of the ATP7B gene in Japanese patients with Wilson disease. J Hum Genet 2000;45:86–91.

[78] Gupta A, Bhattacharjee A, Dmitriev OY, Nokhrin S, Braiterman L, Hubbard AL, Lutsenko S. Cellular copper levels determine the phenotype of the Arg875 variant of ATP7B/Wilson disease protein. Proc Natl Acad Sci U S A 2011;108:5390–5.

[79] Mercer SW, Wang J, Burke R. In vivo modeling of the pathogenic effect of copper transporter mutations that cause Menkes and Wilson diseases, motor neuropathy, and susceptibility to Alzheimer's disease. J Biol Chem 2017;292:4113–22.

[80] Gupta A, Maulik M, Nasipuri P, Chattopadhyay I, Das SK, Gangopadhyay PK, Indian Genome Variation C, Ray K. Molecular diagnosis of Wilson disease using prevalent mutations and informative single-nucleotide polymorphism markers. Clin Chem 2007;53:1601–8.

[81] Squitti R, Polimanti R, Siotto M, Bucossi S, Ventriglia M, Mariani S, Vernieri F, Scrascia F, Trotta L, Rossini PM. ATP7B variants as modulators of copper dyshomeostasis in Alzheimer's disease. Neuromolecular Med 2013;15:515–22.

[82] Bucossi S, Polimanti R, Mariani S, Ventriglia M, Bonvicini C, Migliore S, Manfellotto D, Salustri C, Vernieri F, Rossini PM, Squitti R. Association of K832R and R952K SNPs of Wilson's disease gene with Alzheimer's disease. J Alzheimers Dis 2012;29:913–9.

Chapter 5

Animal Models of Wilson Disease

Dominik Huster
Department of Gastroenterology and Oncology, Deaconess Hospital Leipzig, Academic Teaching Hospital, University of Leipzig, Leipzig, Germany

Mutations in ATP7B lead to Wilson disease (WD), a rare autosomal recessive disorder that is characterized by copper accumulation and toxicity predominantly in the liver and brain [1, 2]. The underlying genetic defect of WD was identified in the 1990s [3–6]. The defective gene, *ATP7B*, encodes a transmembrane copper-transporting P_1-type ATPase, essential for mammalian copper homeostasis [7]. The main function of ATP7B is the incorporation of copper into copper-dependent enzymes, such as ceruloplasmin, and export of excess copper into the bile [8]. ATP7B is also expressed in the brain; however, its specific function is still unknown. To a certain degree, ATP7B expression has been identified in several other tissues, such as the kidneys, lung, gut, placenta, and mammary gland [7, 9].

Meanwhile, our understanding of WD pathophysiology has increased substantially. Organ copper concentration, pathology, and clinical course of WD are highly variable [1], and unfortunately, a conclusive genotype–phenotype correlation has not been established [10]. The low incidence and the phenotypic variability of WD hamper identification of modifier genes and epigenetic factors involved in the pathogenesis. A deeper insight into the molecular physiology of the disease is hindered due to certain limitations in human research. Animal models for copper accumulation and toxicity are invaluable to increase our knowledge of the underlying pathophysiology of WD.

This review summarizes identification, properties, and novel insights derived from different animal models for WD and copper storage disorders. Animal models—mainly rodents—include natural/inbred or genetically generated models and provide profound insights into presymptomatic molecular events and mechanisms of disease development. The usefulness of animal models in identifying novel potential therapeutic strategies and agents for the treatment of WD is also addressed.

COPPER TOXICITY AND LIVER DISEASE IN RODENT MODELS OF WILSON DISEASE

Rodent models with *Atp7b* gene defects are meanwhile well described and reflect many parallels with copper metabolism and toxicity observed in humans. Rat or mouse models provide the most common experimental approach to study the process of hepatic copper accumulation to massive toxicity because of their low cost, ease of availability, and their usefulness for therapeutic or even genetic manipulation.

In contrast to what is observed in rodent models (see below), copper deficiency at birth is not clinically relevant in human WD. Another important difference to what is observed in human WD is the absence of an overt neurological phenotype—only slight abnormalities in behavioral tests were revealed in specific tests in some animals. Nevertheless, rodent models have been used to great advantage to study hepatic copper toxicity and evaluation of the therapeutic effectiveness of various compounds for the treatment of WD.

Long-Evans Cinnamon Rat

Long-Evans Cinnamon (LEC) rats are derived from an inbred strain of mutant Long-Evans agouti rats with a cinnamon-like fur color and were isolated in 1985 [11]. The LEC rat is widely accepted as a rodent model of Wilson disease and presents lots of common clinical features with human WD [12]. The rat Atp7b gene for *Atp7b* was cloned by Wu et al. in 1994 [13]. The gene defect was identified as a partial deletion of at least 900 bp of the coding region at the 3′ end and at least 400 bp of the downstream untranslated region. Further genetic experiments revealed the autosomal recessive pattern of this condition [14, 15]. LEC rats develop a severe acute hepatitis at ~4 months of age, with elevated bilirubin and liver enzymes and massive hepatocyte necrosis; mortality is between 30% and 40%. The surviving rats develop chronic hepatitis and fibrosis with malignant transformation (hepatocellular carcinoma or cholangiocarcinoma) by ~1 year of age [16–18]. The LEC rats spontaneously accumulate copper in the liver associated with a remarkable decrease in serum copper and ceruloplasmin levels comparable with human WD [19]. Hepatic copper accumulation reaches >1000 μg/g dry weight by 8 weeks [20].

LEC rats present a wide spectrum of histological changes in the liver: in LEC rats, slightly diseased at early ages, the liver histology reveals changes in nuclear size, hepatocyte ballooning, apoptosis, and increased number of Kupffer cells. With severe jaundice and elevated bilirubin, liver pathology is also characterized by cholestasis, hepatocyte necrosis, and erythrophagocytosis [21]. An upregulation of gene transcripts related to cytochrome P450, oxidative stress, DNA damage, and apoptosis was observed using microarray expression studies, and these accompany progression of liver disease in LEC rats [21]. Mori et al. summarized further histological characteristic features found in severely diseased animals such as a spotty coagulative necrosis of hepatocytes, appearance of giant hepatocytes, and enlarged polyploid nuclei. Preneoplastic lesions, including hyperplastic foci and nodules, developed in LEC rat livers with chronic hepatitis [18]. Furthermore, cholangiofibrosis is observed in stages of advanced liver disease and correlates with a potential for preneoplastic and neoplastic cellular transformation [22]. Hepatic mitochondrial changes are a characteristic phenomenon in LEC rat liver comparable with patients with WD. Sternlieb et al. described ultracellular changes in LEC rats characterized by mitochondrial pleomorphism, changes in matrix density, and abnormal morphology of cristae such as elongation, dilation, stacking, or matrix inclusions with electron-dense deposits [23]. Changes in LEC rat liver mitochondria resemble changes observed in human WD [24]. Mitochondria are a primary target of copper toxicity, leading to cross-linking and disintegration of mitochondrial membranes ultimately triggering hepatocyte death [25].

Changes in liver pathology and hepatic copper accumulation have been associated with metabolic changes, specifically in lipid metabolism. Levy et al. showed that 12-week-old LEC rats displayed increased hepatic triglycerides, free cholesterol, and cholesteryl ester [26].

Although liver tumors are not a frequent condition of WD [27], LEC rats have been studied as a model of hepatocellular carcinoma, although there are several shortcomings and differences compared with human HCC. The hepatocarcinogenesis in LEC rats is accompanied by reduced apoptosis, DNA strand breaks, and cellular proliferation [28]. The transformation from chronic hepatitis to hepatocellular carcinoma is characterized by overexpression and increased activity of proteins related to cell cycle and proliferation [29]. Src-related protein tyrosine kinase, associated with cell transformation, was overexpressed in LEC rat hepatocytes and correlates with the development of liver cancer [30]. Transcripts of DNA methyltransferases, enzymes also associated with liver cancer, were upregulated in liver tumors from LEC rats [31].

The brain pathology and neurological presentation in LEC rats has been less well studied than the liver pathology. Copper accumulates in all brain regions of LEC rats, and DNA single-strand breaks in the brain due to increased copper levels were observed [32]. Microarray analysis demonstrated changes in gene transcripts related to neuronal development, oxidative stress, apoptosis, and inflammation [33]. Basal ganglia showed a significant increase in superoxide dismutase (Mn-SOD) in LEC rats compared with controls, which mirrors increased oxidative stress [34]. In behavioral tests, LEC rats show higher locomotor activity, decreased habituation to startle response, and lower prepulse inhibition compared with controls [35]. Therefore, the usefulness of LEC rats for studying the neurological course of WD is rather limited.

LEC rats are extensively used to study therapeutic agents and strategies in WD comparable with treatment of human patients with WD. Zinc administration effectively prevented the development of hepatitis along with improved survival in LEC rats compared with untreated rats [20]. Jong-Hon et al. and Togashi et al. demonstrated the efficacy of oral penicillamine administration in preventing hepatitis in LEC rats shown by improvement of liver enzymes and liver histology [17, 36]. Copper chelation with trientine equally prevented the development of hepatitis, improved liver histology, and reduced the occurrence of hepatocellular carcinoma [37]. Tetrathiomolybdate was administered to LEC rats after onset of acute hepatitis, decreased hepatic copper concentration, and improved liver histology [38]. Further studies using LEC rats tested the implementation of new approaches to the treatment of WD based on hepatocyte transplantation and stem cell therapy as well as gene therapy [39, 40]. Chen et al. successfully treated LEC rats with *Atp7b*-transduced bone marrow mesenchymal stem cells administered through the portal vein. This treatment resulted in increased expression of *Atp7b*, increased ceruloplasmin concentration, lower hepatic copper, and decreased AST and ALT levels. Better liver engraftment and repopulation were achieved by preconditioning with radiation and ischemia–reperfusion performed before the infusion of *Atp7b*-transduced cells [41]. Gene therapy was tested in LEC rats by lentiviral *Atp7b* gene transfer [42]; this treatment resulted in decreased hepatic copper levels and improved liver fibrosis. Cell therapy with hepatocytes from the rat liver with normal copper metabolism was also successfully attempted in LEC rats; normal hepatocytes were transplanted intrasplenically. Treatment with bile salts before transplantation improved the efficacy of the procedure. This cell therapy was associated with decreased copper concentration and improved histology of LEC rat livers [43].

In summary, the LEC rat is a well-established animal model for WD used for many applications in basic and therapeutic research. Disadvantages include the lack of a well-defined control strain and the presence of several mutations in the LEC rat genome.

LPP Atp7b$^{-/-}$ Rat

A newly engineered rodent model for WD derived from the LEC rat but less well studied is the LPP Atp7b$^{-/-}$ rat. In order to segregate the three independent mutations identified in LEC rats, this new rat strain was generated by backcrossing of LEC rats to the Piebald Virol Glaxo (PVG) background [44]. These rats display a phenotype identical to that of the parental LEC rats but do not have the disadvantages associated with LEC rats: the LPP strain of rats does not exhibit the serotonin N-acyltransferase (NAT) defect and the cinnamon fur color found in LEC rats. In contrast to LEC rats with ~60% survival rate of the acute hepatitis [18], 100% of affected LPP rats die due to the development of fulminant hepatitis at about 100 days of age (Dr. H. Zischka, Helmholtz Center Munich, personal communication). LPP rats display severe mitochondrial changes due to copper overload comparable with LEC rats and human WD [25]. In a recent therapeutic animal trial using LPP rats, restoration of mitochondrial structure and function by the application of the bacterial peptide methanobactin has been compellingly demonstrated in vitro and in vivo along with reversion of acute liver failure in LPP rats [45]. LPP rats might be particularly beneficial for interventional studies for acute liver failure due to its rapid progression of hepatic disease, but further studies need to be performed to underline its usefulness as model for WD.

Toxic Milk Mouse

The first mouse model with a genetic defect of copper metabolism and hepatic copper accumulation comparable with human WD was first described by Rauch in 1983 [46]. While copper deficiency is not found in human WD, the offspring of mutant mice show signs of growth retardation, hypopigmentation, tremor, and death shortly after birth, attributable to copper deficit in the organism. Symptoms are exacerbated and accelerated after birth because of the greatly reduced copper content of their mother's breast milk. However, litters could be rescued by foster nursing on normal dams or by copper administration, hence the name of the model, toxic milk (tx) mouse. The genetic defect in *Atp7b* gene results in mislocalization of mutant Atp7b protein in the mammary gland of the tx mouse, leading to decreased secretion of copper with the milk [47]. In contrast to this and other mouse models with *Atp7b* gene defects, copper concentration in human breast milk is not affected in mothers with WD [48], and copper deficiency is not a clinical issue in newborns.

In 1995, shortly after the discovery of the human *ATP7B* gene, the linkage of the tx mutation to mouse chromosome 8 and the mouse homologue of human *ATP7B* was demonstrated [49, 50]. Theophilos et al. isolated cDNA clones encoding the murine homologue of WD gene *Atp7b* and identified the mutation (A4066G) causing the disease and established the tx mouse as the first mouse model for human WD [51]. The Met1356Val amino acid exchange in the eighth transmembrane domain of the Atp7b protein causes mislocalization and the loss of copper transport in in vitro model systems for copper transport and Atp7b localization [52, 53]. Comparable results have been shown for a number of human *ATB7B* variants both in vitro and in vivo [54, 55], which revealed similarities between human WD and the mouse model.

Young tx mice accumulate large amounts of copper in their liver followed by liver injury. Livers show nodular fibrosis at 6 months of age, bile duct hyperplasia, and infiltration with inflammatory cells. They exhibit many parallels in gross liver pathology and histological and ultrastructural abnormalities with liver pathology found in human WD [56]. A detailed investigation of age-dependent copper accumulation and influences on zinc and iron metabolism was provided by Allen et al. [57]. Copper-loading studies did not increase liver copper significantly (exceeding the copper levels in tx control mice without additional copper load), nor did they lead to significant acceleration of liver pathology or liver-related tumors, and the authors concluded that even tx mice have a remarkable ability to control excess copper intake without resulting in ill-health.

In agreement with an important finding frequently observed with human WD—decreased levels of holoceruloplasmin [58]—Rauch reported reduced ceruloplasmin activity in tx mice [46]. Mercer et al. confirmed reduced serum ceruloplasmin concentrations in tx mice and observed that mRNA levels of ceruloplasmin were unaffected or even higher in pregnant female tx mice [59]. They suggested that the low ceruloplasmin activity was a result of failed copper incorporation into apoceruloplasmin and concluded that the gene defect was not related to the ceruloplasmin gene.

Howel and Mercer provided a detailed analysis of trace element status and observed copper accumulation in the liver, kidney, spleen, muscle, brain, and red blood cells resulting in hemolysis, which is frequently found in human WD [60]. An important defense mechanism against elevated copper concentration in the liver, kidney, and spleen is the overproduction of metallothionein, which was also detected in tx mouse livers [61, 62].

Beside tremor in newborn mice, perhaps attributable to copper deficiency at this stage, no striking neurological symptoms could be detected in tx mice, as is observed in human WD. Several studies addressed copper concentration in the brain of tx mice [57, 60, 63]. Indeed, tx mice had significantly (~1.5–2-fold) higher copper concentrations in the cerebral cortex, corpus striatum, thalamus/hypothalamus, and brain stem compared with controls. Moreover, tx mice had significantly higher metallothionein content in all brain regions than controls, which might provide protection against copper toxicity and which could explain the lack of neurological symptoms related to copper toxicity [63].

As with other rodent models for WD, the tx mice have been utilized for various therapeutic interventions. Tx mice successfully underwent drug treatment with tetrathiomolybdate [64], a drug currently investigated in clinical trials in human WD. Tx mice were used to provide novel insight into effects of penicillamine treatment on brain copper handling. Penicillamine treatment increased free copper and enhanced oxidative stress at the beginning of administration, as shown by a reduced GSH/GSSG ratio and increased malondialdehyde in the cortex and basal ganglia. That observation could explain the worsening of neurological symptoms initially observed in penicillamine-treated patients [65].

Liver cell transplantation using hepatocytes isolated from 6- to 8-week male congenic DL mice and intrasplenically transplanted into 3- to 4-month-old *tx* mice partially corrected the *Atp7b* gene defect in tx mice and led to a significant decrease of copper concentration in the liver, kidney, and spleen in the transplanted mice; however, this approach failed to decrease brain copper concentration [66].

In a recent study by Shi et al. [67], hereditary abnormalities of copper metabolism in tx mice could be corrected partially by intrasplenic transplantation of homogeneous embryonic hepatocytes.

In another therapeutic approach utilizing the tx mice, Buck et al. performed bone marrow/stem cell transplantation and demonstrated initial correction of copper metabolism defects 5 months post transplantation [68]. Unfortunately, correction of copper metabolism was not sustained in the long term, that is, no significant improvement of copper parameters and liver histology was observed 9 months post transplantation.

Early-stage transplantation (at 2 months of age) of normal bone marrow cells has been found to be more effective in correcting liver function (in terms of lowering aminotransferase levels) and copper metabolism (decreased liver copper and rise of ceruloplasmin oxidase activity) in tx mice than late-stage transplantation (at 5 months of age) [69].

The tx mouse is a well-described model of WD with a natural mutation of *Atp7b* and many parallels to human WD providing lots of applications for basic and therapeutic research.

Toxic Milk Mouse From the Jackson Laboratory (tx-j Mouse)

Another naturally occurring genetic and phenotypic mouse model with parallels to human WD and the tx mouse is the toxic milk mouse from the Jackson laboratory (tx-j). Comparable with the tx mouse, the tx-j mouse's breast milk has low copper concentration leading to impaired growth and development. Therefore, tx-j pups have to be fostered to a control dam with normal copper metabolism. The tx-j mouse model has a spontaneous recessive point mutation that is different from the tx mouse, specifically at position 2135, in exon 8 of *Atp7b* gene, which leads to a G712D missense mutation, predicted to be in the second transmembrane region of the Atp7b [70].

The liver disease of the tx-j mouse is characterized by slow progression of liver histology to steatosis and cirrhosis. Liver histology is almost normal until 3–4 months of age [71, 72]. The presence of enlarged hepatocyte nuclei, about two times larger than controls, is among the first abnormalities to be observed. From 5 months onward, there is an increased inflammatory infiltrate and mild microvesicular steatosis. Mild fibrosis is present at 6–7 months of age, and ultimately, it will progress to cirrhosis [73]. Liver copper rapidly accumulates to approximately 40–50-fold above the normal level (controls) and remains elevated through the first year of life [72, 73]. Impairment of mitochondrial ultrastructure and function are typical changes associated with elevated copper. Electron microscopic images show cystic cristae dilation, increased citrate synthase activity, and decrease of complex IV activity [71]. Tx-j mice served as a model for studying interactions between copper accumulation and methionine metabolism. Tx-j mice showed a downregulation of *S*-adenosylhomocysteine hydrolase, responsible for metabolizing *S*-adenosylhomocysteine to homocysteine. *S*-adenosylhomocysteine elevation is associated with inhibition of DNA methylation, which was confirmed by global DNA hypomethylation in tx-j livers compared with controls [74]. As also observed in other rodent models, tx-j mice present changes in gene transcripts related to lipid metabolism. Fetal livers from tx-j mice showed downregulation of the expression of sterol regulatory element-binding transcription factor 1 (*Srebf1*), central to lipogenesis; carnitine palmitoyltransferase 1A (*Cpt1A*), important in fatty acid oxidation; and peroxisome proliferator-activated receptor alpha (*Ppara*), important in fatty acid oxidation [72].

Increased copper concentrations have been observed in the striatum, hippocampus, and cerebellum of tx-j mice brain at 12 months of age [73], whereas no significant changes in the cerebral cortex were detected in the same study. In contrast, another study showed approximately twofold higher copper concentrations in all the regions of the brain examined, including the cerebral cortex, hippocampus, striatum, and cerebellum [75]. Both studies revealed slight behavioral changes in tx-j mice compared with control mice. Tx-j mice showed abnormalities, specifically showing a preference in forelimb usage, and they were slower to reach maximal performance in the rotarod test than controls. Moreover, tx-j mice did not demonstrate any improvement in behavioral performance in the Morris water maze [73]. To dissect further the mechanism behind the behavioral differences, the authors investigated neuroinflammatory changes by immunohistochemistry. They found increased markers of astro- and microglial activation in the striatum and corpus callosum, but not in the cerebral cortex or the hippocampus. They also observed increased transcript levels of genes related to inflammation such as TNFα, iNOS, IL-1 β, and IL-4 [73]. Przybylkowski et al. utilized different behavioral tests on tx-j mice and demonstrated impaired locomotor performance in 12-month-old tx-j mice accompanied by increased copper and serotonin content [75]. The expression of dopaminergic-, noradrenergic-, and serotoninergic-specific enzymes to demonstrate neuronal loss was not different from controls.

As mentioned above, tx-j mice present with changes in methionine metabolism and DNA methylation potential [74]. In an interventional study, choline was administered to tx-j female mice prior to conception and during pregnancy [76]. Choline administration resulted in significantly lower hepatic copper levels and different expression of genes related to lipid and methionine metabolism. Moreover, an increase of DNA methylation was observed indicating that environmental and in utero factors might affect the WD phenotype. A follow-up study combined choline supplementation with penicillamine treatment after birth, which increased hepatic gene transcripts of methionine metabolism and genes related to oxidative phosphorylation [77]. The authors suggested these findings are relevant for future preventive treatments of WD.

Taken together, the tx-j mouse is a valid model of WD. It is characterized by relatively slow progression of liver damage over 1 year after birth with liver and brain metabolic changes accompanied by certain behavioral abnormalities.

Atp7b$^{-/-}$ Mouse

Advantages of a genetically engineered mouse model are the well-defined genetic background and the availability of a control strain making such models relevant for clinical and translational studies. The Atp7b null mouse was created by the introduction of an early termination codon in exon 2 of wild-type mouse Atp7b mRNA [78]. The Atp7b$^{-/-}$ (knockout) mouse features properties of both human WD and of the tx/tx-j mouse phenotype but displays liver pathology that exceeds observations made in other mouse models. Litters are born copper-deficient and accumulate copper in the liver, kidneys, and brain to a level up to 60-fold greater (the liver) than normal by 5 months of age [78], which is comparable with what is seen in tx and tx-j mice. Further investigation and description of liver pathology revealed more parallels with human liver disease [79]. Atp7b$^{-/-}$ mice have decreased serum holoceruloplasmin activity, increased urine copper excretion, intracellular and nuclear copper accumulation in the hepatocyte, and severe liver pathology, developing earlier and exceeding other mouse models. Hepatic pathologies include ultrastructural changes, steatosis, and mild inflammation at early stages (~6 weeks of age) and hepatitis, dysplasia, and necroinflammation in advanced stages (at ~12–20 weeks of age). Later (at ~36 weeks and older), bile duct proliferation, development of fibrosis, and even neoplastic proliferation of biliary cells are observed. The majority of mice survive severe hepatitis and show remarkable regeneration in large parts of the liver [79].

Free copper leads to generation of reactive oxygen species, which have deleterious effects on cellular membranes, proteins, DNA, and RNA molecules [80, 81]. Sternlieb described mitochondrial alterations in human WD livers [82]. Using mitochondria from Atp7b$^{-/-}$ mouse liver, mass spectrometry has revealed a potential underlying mechanism of mitochondrial injury and detected fragmentation of cardiolipin, an essential mitochondrial lipid, which is required for activity of certain mitochondrial proteins [83].

Metallothionein is significantly overexpressed in Atp7b$^{-/-}$ mouse livers and is potentially protective against copper toxicity, in agreement with observations in other species and human WD [84, 85]. While in healthy livers excess copper is excreted into the bile, WD patients have increased copper concentration in the urine suggesting an overflow or compensatory mechanism to release excess copper. In a study on Atp7b$^{-/-}$ mice, Gray et al. demonstrated a complex response to copper accumulation leading to elevated urinary copper excretion [86]. This response included downregulation of the Ctr1 copper transporter and the appearance of a *small copper carrier* in urine to excrete excess body copper.

Hepatic copper accumulation may affect lipid metabolism in mice and human WD as illustrated by hepatic steatosis. A combined microarray and real-time PCR study discovered downregulation of genes important for cholesterol and lipid metabolism in Atp7b$^{-/-}$ mice [85]. The rate-limiting enzyme HMG-CoA reductase and a variety of enzymes that involved cholesterol biosynthesis were downregulated in livers of knockout mice compared with wild-type mice at 6 weeks of age. Decreased cholesterol biosynthesis was observed and markedly lower levels of serum lipids (total cholesterol, HDL, VLDL cholesterol, and triglycerides; 1.2- to 3.6-fold reduction). These results were confirmed on studies with older mice [87] and were comparable with results found in LEC rats [26]. There is increasing evidence for a link between copper accumulation and hepatic cholesterol metabolism [85, 87–89].

Elevated copper levels in the liver due to the Atp7b gene defect also influence the cell cycle [85]. An upregulation of enzymes involved in cell cycle control, DNA replication, chromosome condensation and assembly, nuclear and cellular division, etc. occurs at early stages of the disease. At this stage, copper is highly elevated; however, cellular injury and inflammation in the liver are absent. These responses may trigger immune and inflammatory processes. Meanwhile, the upregulation of cell cycle machinery provides the basis for liver regeneration and other defense mechanisms observed in later phases of the hepatic disease. Data from expression studies were combined with analysis of the presymptomatic hepatic nuclear proteome and liver metabolites [89]. This approach revealed suppression of Farnesoid X nuclear receptor (FXR/NR1H4) and glucocorticoid receptor (GR/NR3C1) in copper-loaded livers, both involved in lipid metabolism, while molecules with a function in DNA repair and enzymes involved in oxidative defense were more abundant. Further studies revealed a decreased binding of the nuclear receptors FXR, RXR, HNF4α, and LRH-1 to promoter response elements and decreased mRNA expression of nuclear receptor target genes [90]. The disruption of nuclear receptor activity is a mechanism that may contribute to chronic hepatic copper accumulation and liver pathology seen in human WD.

Copper accumulates in $Atp7b^{-/-}$ mouse livers nonuniformly between intracellular compartments [79]. Early in the disease copper concentration reaches peak levels in the nucleus and cytosol, probably bound to metallothioneins. However, consequences of nuclear copper entry, alteration of the protein synthesis machinery, and correlation between copper elevation and specific nuclear responses remained puzzling. Burkhead et al. revealed that copper did not influence nuclear ion content, nor did it lead to significant protein oxidation, but it resulted in modification of nuclear proteins, associated with RNA processing [91].

$Atp7b^{-/-}$ mice have been used to test novel technologies such as laser ablation inductively coupled plasma mass spectrometry (LA-ICP-MS) to monitor hepatic copper distribution [92] and in vivo fluorescence detection of copper [93]. Positron emission tomography using new tracers to reveal trace element status has also provided insight into copper distribution [94].

In accordance with the other rodent models, no significant neurological symptoms or specific brain pathology has been observed in $Atp7b^{-/-}$ mice, and no detailed behavioral studies of $Atp7b^{-/-}$ mice have been published so far. However, in the initial report on $Atp7b^{-/-}$ mice, elevated brain copper levels (~2–2.5-fold increase at 5 months and older) along with slight neurological abnormalities (tremor, ataxia, and abnormal locomotor behavior in the second generation of pups) and growth retardation were observed [78]. LA-ICP-MS imaging of the brain revealed the copper accumulation in brain parenchyma increased by a factor of two in elder mice [95]. In contrast, another study [96] found only slight differences in total brain copper concentration and in different compartments (the basal ganglia, cerebral cortex, and cerebellum) compared with control mice. However, the same study revealed abnormal ultrastructural cell organelles (the mitochondria, deformed microtubules, and axons) in the neurons of $Atp7b^{-/-}$ mice that reflect a discrepancy between the absence of significant copper deposition and the presence of neuronal damage. The authors concluded that copper might not be the sole causative player and other unidentified pathogenetic factors might enhance toxic effects of copper on neurons.

High levels of copper in the liver followed by liver disease make the $Atp7b^{-/-}$ mouse a valuable model for novel therapeutic interventions. Pharmacological treatment of $Atp7b^{-/-}$ mice with the LXR agonist T0901317 ameliorated disease manifestations such as impaired liver function and histology, in line with improved markers of liver fibrosis, inflammation, and normalized lipid profiles [97]. A successful auxiliary liver transplantation using $Atp7b^{-/-}$ mice was undertaken by Cheng et al. [98]. In a gene therapy, trial prenatal gene transfer using lentiviral vectors containing the $Atp7b$ gene resulted in decreased liver copper content and improved hepatic copper metabolism and morphology in $Atp7b^{-/-}$ mice [99]. Gene therapy using an adeno-associated vector containing human ATP7B cDNA was successfully performed using $Atp7b^{-/-}$ mice in another recent study [100]. This treatment resulted in dose-dependent, long-term metabolic correction of copper imbalance as demonstrated by normalization of serum holoceruloplasmin, restoration of physiological (fecal) copper excretion, reduced urinary copper and serum transaminases, and prevention of liver injury without oncogenesis.

Compared with other mouse models of WD, liver pathology is most pronounced in the $Atp7b^{-/-}$ mouse model; however, the variability of the phenotype and the degree of regeneration in the liver are remarkably similar to human WD. The underlying cause of this variability has to be elucidated. An advantage of the $Atp7b^{-/-}$ mice is the availability of a control strain useful for functional and interventional studies and a better understanding of phenotypic diversity of WD.

OTHER ANIMAL MODELS FOR COPPER STORAGE DISORDERS

Other mammalian species with inherited defects in copper metabolism and consecutive hepatic copper accumulation have been described; however, the underlying genetic defect is not the same as in WD or is in many cases unknown.

Dogs With Copper-Associated Hepatitis

Bedlington terriers accumulate hepatic copper, which results in liver fibrosis and cirrhosis [101, 102]; however, the underlying genetic defect was identified in the *COMMD1* gene (formerly known as *MURR1*, exon 2 deletion) and is thus different from *Atp7b* gene [101, 103, 104]. This limits its usefulness as a WD model. However, hepatic copper accumulation and liver pathology resemble human WD. Overt neurological defects have not been observed, and the diseased dogs have normal serum concentrations of ceruloplasmin. Comparable with humans with WD, Bedlington terriers can be treated successfully with copper-chelating agents such as D-penicillamine [101].

Labrador retrievers have recently been found to develop canine copper toxicosis resulting in liver cirrhosis, but the underlying genetic defect is different to that found in WD and has an unknown and complex inheritance pattern [101]. However, a recent genome-wide association study revealed the involvement of ATP7B mutations and could therefore be considered as a first natural, nonrodent model for ATP7B-associated copper toxicosis [105].

North Ronaldsay Sheep

The copper-sensitive North Ronaldsay sheep is another animal model for hepatic and neurological copper accumulation, however clearly clinically and genetically different from WD [106, 107]. The genetic defect is still unknown, but this model may nevertheless have importance for the investigation of copper-induced liver fibrogenesis and for studying copper accumulation in the brain, the role of the blood–brain barrier, and the influence of environmental copper [107–109].

CONCLUSION AND RESEARCH PERSPECTIVES

Over the past decades, animal models of WD have increased our understanding of this disease and allowed initial exploration of established and novel therapies. Each model has specific properties, similarities, and dissimilarities to human WD. A common but unexplained feature of all models with human WD is the exceptional diversity in phenotype, clinical course, and outcome of the disease. Important properties in comparison with human WD are summarized in Table 1.

Animal data must be always scrutinized rigorously before drawing final conclusions about the relatedness to human disease and consequent implementation of novel therapies. Although WD has become a treatable disorder with an overall fair prognosis, several problems such as drug side effects, neurological deterioration after therapy initiation, residual symptoms, the lack of adherence, or failure of drug treatment underline the importance of developing novel therapies. The availability of valid model organisms enables innovative research strategies such as cell-based therapies and gene therapy [39, 40] and drug development [118, 119].

TABLE 1 Overview of Animal Models for WD

Model	Gene Defect	Similarities	Differences	Therapeutic Interventions	Reference
Rat models					
LEC rat	*Atp7b* deletion of 900 bp of the coding region at the 3′ end and ~400 bp of the downstream untranslated region	Hepatic copper accumulation Liver disease Increased brain copper Mitochondrial abnormalities Lipid metabolism affected Low ceruloplasmin levels Increased metallothionein	Only slight neurological symptoms Liver tumors (hepatocellular carcinoma and cholangiocarcinoma) common in LEC rats but uncommon in human WD	Penicillamine Zinc Trientine Tetrathiomolybdate Hepatocyte transplantation Stem cell therapy Gene therapy	[11, 13–17, 19–22, 26, 28–32, 35–42, 110–114]
LPP rat	Same as LEC rat (position n2110) of *Atp7b*, other gene mutations corrected by backcrossing	Hepatic copper accumulation Severe liver disease Mitochondrial abnormalities	No neurological symptoms	Methanobactin	[25, 44, 45]

Continued

TABLE 1 Overview of Animal Models for WD—cont'd

Model	Gene Defect	Similarities	Differences	Therapeutic Interventions	Reference
Mouse models					
Tx mouse	Met1386Val mutation (A4066G, exon 20)	Hepatic copper accumulation Liver disease Increased brain copper Mitochondrial abnormalities Low ceruloplasmin levels Increased metallothionein	Copper-deficient breast milk, relevant copper deficiency at birth Tremor in new born mice due to copper deficiency No neurological symptoms	Penicillamine Tetrathiomolybdate Hepatocyte transplantation Stem cell transplantation	[46, 47, 49–53, 56, 60, 61, 64–68]
Tx-j mouse	Gly712Asp mutation (G2135A, exon 8)	Hepatic copper accumulation Liver disease Increased brain copper Mitochondrial abnormalities Slight neurological symptoms, abnormal behavioral tests	Copper-deficient breast milk, relevant copper deficiency at birth	Choline and penicillamine (preventive)	[70–77]
Atp7b$^{-/-}$ mouse	Multiple stop codons in exon 2	Hepatic copper accumulation Liver disease Increased brain copper Mitochondrial abnormalities Lipid metabolism affected Low ceruloplasmin levels Increased metallothionein	Copper-deficient breast milk, relevant copper deficiency at birth No neurological disease	LXR agonist T0901317 Liver transplantation Gene therapy	[78, 79, 85, 95, 97–99, 115]
Dog models [116]					
Bedlington terrier	*COMMD 1*$^{del/del}$ (exon 2)	Hepatic copper accumulation Liver disease	Gene defect in *COMMD 1* Ceruloplasmin normal No neurological disease	Penicillamine Tetramine cupruretic agents	[101, 103, 117]
Labrador retriever	Unknown, complex ATPB: Arg1453Gln	Hepatic copper accumulation Liver disease	Gene defect unknown No neurological disease	Penicillamine	[101, 105]
Sheep model					
North Ronaldsay sheep	Unknown	Hepatic copper accumulation Liver disease	Gene defect unknown No neurological disease Influence of environmental copper	–	[106–109]

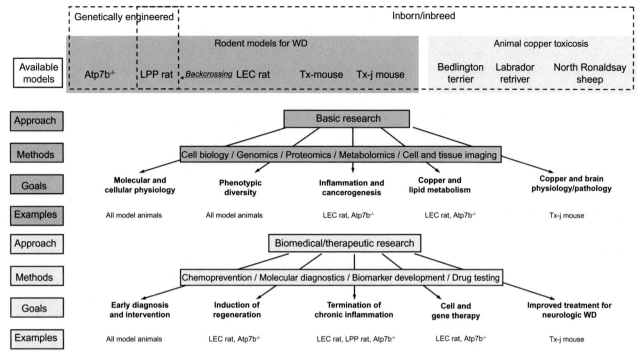

FIG. 1 Sources and research targets of animal models for WD.

Another unsolved problem in WD is the unknown pathophysiology of the neurological involvement. Currently, little is known about the specific mechanism and the diversity of symptoms. Any novel and targeted therapy is based on a better understanding of metal homeostasis in the brain.

An overview of the important research goals, utilizing mouse models for basic and therapeutic research for a better understanding and treatment of WD, is shown in Fig. 1.

In conclusion, mammalian models will continue to provide important basic knowledge about copper metabolism in the liver, the brain, and other organs. This will pave the way for the discovery of novel therapies to support recovery from copper toxicity due to the *ATP7B* gene defect in WD and other modifying factors.

Further refinement and development of specifically designed animal models for unmet needs such as better understanding of neurological WD, influence of environmental factors on phenotype, or improved targeted therapies are a complex task that requires close collaboration between clinical practice and experimental model research. This scientific interaction is essential to allow the continuous refinement of the existing animal models of WD with the aim to improve the clinical care for patients with WD.

ACKNOWLEDGMENT

The author thanks Dr. G. Sawers for the critical reading of the manuscript and helpful advice.

REFERENCES

[1] Huster D. Wilson disease. Best Pract Res Clin Gastroenterol 2010;24(5):531–9.

[2] Lorincz MT. Neurologic Wilson's disease. Ann N Y Acad Sci 2010;1184:173–87.

[3] Petrukhin K, Fischer SG, Pirastu M, Tanzi RE, Chernov I, Devoto M, et al. Mapping, cloning and genetic characterization of the region containing the Wilson disease gene. Nat Genet 1993;5(4):338–43.

[4] Tanzi RE, Petrukhin K, Chernov I, Pellequer JL, Wasco W, Ross B, et al. The Wilson disease gene is a copper transporting ATPase with homology to the Menkes disease gene. Nat Genet 1993;5(4):344–50.

[5] Bull PC, Thomas GR, Rommens JM, Forbes JR, Cox DW. The Wilson disease gene is a putative copper transporting P-type ATPase similar to the Menkes gene. Nat Genet 1993;5(4):327–37.

[6] Yamaguchi Y, Heiny ME, Gitlin JD. Isolation and characterization of a human liver cDNA as a candidate gene for Wilson disease. Biochem Biophys Res Commun 1993;197(1):271–7.

[7] Lutsenko S, Barnes NL, Bartee MY, Dmitriev OY. Function and regulation of human copper-transporting ATPases. Physiol Rev 2007; 87(3):1011–46.

[8] Lutsenko S, Efremov RG, Tsivkovskii R, Walker JM. Human copper-transporting ATPase ATP7B (the Wilson's disease protein): biochemical properties and regulation. J Bioenerg Biomembr 2002;34(5):351–62.

[9] Weiss KH, Wurz J, Gotthardt D, Merle U, Stremmel W, Fullekrug J. Localization of the Wilson disease protein in murine intestine. J Anat 2008; 213(3):232–40.

[10] Ferenci P. Phenotype-genotype correlations in patients with Wilson's disease. Ann N Y Acad Sci 2014;1315:1–5.

[11] Sasaki M, Yoshida MC, Kagami K, Takeichi N, Kobayashi H, Dempo K, et al. Spontaneous hepatitis in a inbred strain of Long-Evans rats. Rat News Lett 1985;14:4–6.

[12] Terada K, Sugiyama T. The Long-Evans Cinnamon rat: an animal model for Wilson's disease. Pediatr Int 1999;41(4):414–8.

[13] Wu J, Forbes JR, Chen HS, Cox DW. The LEC rat has a deletion in the copper transporting ATPase gene homologous to the Wilson disease gene. Nat Genet 1994;7(4):541–5.

[14] Yoshida MC, Masuda R, Sasaki M, Takeichi N, Kobayashi H, Dempo K, et al. New mutation causing hereditary hepatitis in the laboratory rat. J Hered 1987;78(6):361–5.

[15] Masuda R, Yoshida MC, Sasaki M, Dempo K, Mori M. Hereditary hepatitis of LEC rats is controlled by a single autosomal recessive gene. Lab Anim 1988;22(2):166–9.

[16] Okuda K. Hepatocellular carcinoma: recent progress. Hepatology 1992;15(5):948–63.

[17] Jong-Hon K, Togashi Y, Kasai H, Hosokawa M, Takeichi N. Prevention of spontaneous hepatocellular carcinoma in Long-Evans Cinnamon rats with hereditary hepatitis by the administration of D-penicillamine. Hepatology 1993;18(3):614–20.

[18] Mori M, Hattori A, Sawaki M, Tsuzuki N, Sawada N, Oyamada M, et al. The LEC rat: a model for human hepatitis, liver cancer, and much more. Am J Pathol 1994;144(1):200–4.

[19] Li Y, Togashi Y, Sato S, Emoto T, Kang JH, Takeichi N, et al. Spontaneous hepatic copper accumulation in Long-Evans Cinnamon rats with hereditary hepatitis. A model of Wilson's disease. J Clin Invest 1991;87(5):1858–61.

[20] Medici V, Sturniolo GC, Santon A, D'Inca R, Bortolami M, Cardin R, et al. Efficacy of zinc supplementation in preventing acute hepatitis in Long-Evans Cinnamon rats. Liver Int 2005;25(4):888–95.

[21] Klein D, Lichtmannegger J, Finckh M, Summer KH. Gene expression in the liver of Long-Evans Cinnamon rats during the development of hepatitis. Arch Toxicol 2003;77(10):568–75.

[22] Schilsky ML, Quintana N, Volenberg I, Kabishcher V, Sternlieb I. Spontaneous cholangiofibrosis in Long-Evans Cinnamon rats: a rodent model for Wilson's disease. Lab Anim Sci 1998;48(2):156–61.

[23] Sternlieb I, Quintana N, Volenberg I, Schilsky ML. An array of mitochondrial alterations in the hepatocytes of Long-Evans Cinnamon rats. Hepatology 1995;22(6):1782–7.

[24] Sternlieb I. Mitochondrial and fatty changes in hepatocytes of patients with Wilson's disease. Gastroenterology 1968;55(3):354–67.

[25] Zischka H, Lichtmannegger J, Schmitt S, Jagemann N, Schulz S, Wartini D, et al. Liver mitochondrial membrane crosslinking and destruction in a rat model of Wilson disease. J Clin Invest 2011;121(4):1508–18.

[26] Levy E, Brunet S, Alvarez F, Seidman E, Bouchard G, Escobar E, et al. Abnormal hepatobiliary and circulating lipid metabolism in the Long-Evans Cinnamon rat model of Wilson's disease. Life Sci 2007;80(16):1472–83.

[27] Pfeiffenberger J, Mogler C, Gotthardt DN, Schulze-Bergkamen H, Litwin T, Reuner U, et al. Hepatobiliary malignancies in Wilson disease. Liver Int 2015;35(5):1615–22.

[28] Jia G, Tohyama C, Sone H. DNA damage triggers imbalance of proliferation and apoptosis during development of preneoplastic foci in the liver of Long-Evans Cinnamon rats. Int J Oncol 2002;21(4):755–61.

[29] Masaki T, Shiratori Y, Rengifo W, Igarashi K, Matsumoto K, Nishioka M, et al. Hepatocellular carcinoma cell cycle: study of Long-Evans Cinnamon rats. Hepatology 2000;32(4 Pt 1):711–20.

[30] Masaki T, Tokuda M, Shiratori Y, Shirai M, Matsumoto K, Nishioka M, et al. A possible novel src-related tyrosine kinase in cancer cells of LEC rats that develop hepatocellular carcinoma. J Hepatol 2000;32(1):92–9.

[31] Miyoshi E, Jain SK, Sugiyama T, Fujii J, Hayashi N, Fusamoto H, et al. Expression of DNA methyltransferase in LEC rats during hepatocarcinogenesis. Carcinogenesis 1993;14(4):603–5.

[32] Hayashi M, Fuse S, Endoh D, Horiguchi N, Nakayama K, Kon Y, et al. Accumulation of copper induces DNA strand breaks in brain cells of Long-Evans Cinnamon (LEC) rats, an animal model for human Wilson disease. Exp Anim 2006;55(5):419–26.

[33] Lee BH, Kim JH, Kim JM, Heo SH, Kang M, Kim GH, et al. The early molecular processes underlying the neurological manifestations of an animal model of Wilson's disease. Metallomics 2013;5(5):532–40.

[34] Kim DW, Ahn TB, Kim JM, Jeon GS, Seo JH, Jeon BS, et al. Enhanced Mn-SOD immunoreactivity in the dopaminergic neurons of Long-Evans Cinnamon rats. Neurochem Res 2005;30(4):475–8.

[35] Fujiwara N, Iso H, Kitanaka N, Kitanaka J, Eguchi H, Ookawara T, et al. Effects of copper metabolism on neurological functions in Wistar and Wilson's disease model rats. Biochem Biophys Res Commun 2006;349(3):1079–86.

[36] Togashi Y, Li Y, Kang JH, Takeichi N, Fujioka Y, Nagashima K, et al. D-penicillamine prevents the development of hepatitis in Long-Evans Cinnamon rats with abnormal copper metabolism. Hepatology 1992;15(1):82–7.

[37] Sone K, Maeda M, Wakabayashi K, Takeichi N, Mori M, Sugimura T, et al. Inhibition of hereditary hepatitis and liver tumor development in Long-Evans Cinnamon rats by the copper-chelating agent trientine dihydrochloride. Hepatology 1996;23(4):764–70.

[38] Klein D, Arora U, Lichtmannegger J, Finckh M, Heinzmann U, Summer KH. Tetrathiomolybdate in the treatment of acute hepatitis in an animal model for Wilson disease. J Hepatol 2004;40(3):409–16.
[39] Gupta S. Cell therapy to remove excess copper in Wilson's disease. Ann N Y Acad Sci 2014;1315:70–80.
[40] Filippi C, Dhawan A. Current status of human hepatocyte transplantation and its potential for Wilson's disease. Ann N Y Acad Sci 2014;1315:50–5.
[41] Chen S, Shao C, Dong T, Chai H, Xiong X, Sun D, et al. Transplantation of ATP7B-transduced bone marrow mesenchymal stem cells decreases copper overload in rats. PLoS One 2014;9(11).
[42] Merle U, Encke J, Tuma S, Volkmann M, Naldini L, Stremmel W. Lentiviral gene transfer ameliorates disease progression in Long-Evans Cinnamon rats: an animal model for Wilson disease. Scand J Gastroenterol 2006;41(8):974–82.
[43] Joseph B, Kapoor S, Schilsky ML, Gupta S. Bile salt-induced pro-oxidant liver damage promotes transplanted cell proliferation for correcting Wilson disease in the Long-Evans Cinnamon rat model. Hepatology 2009;49(5):1616–24.
[44] Ahmed S, Deng J, Borjigin J. A new strain of rat for functional analysis of PINA. Brain Res Mol Brain Res 2005;137(1–2):63–9.
[45] Lichtmannegger J, Leitzinger C, Wimmer R, Schmitt S, Schulz S, Kabiri Y, et al. Methanobactin reverses acute liver failure in a rat model of Wilson disease. J Clin Invest 2016;126(7):2721–35.
[46] Rauch H. Toxic milk, a new mutation affecting cooper metabolism in the mouse. J Hered 1983;74(3):141–4.
[47] Michalczyk AA, Rieger J, Allen KJ, Mercer JF, Ackland ML. Defective localization of the Wilson disease protein (ATP7B) in the mammary gland of the toxic milk mouse and the effects of copper supplementation. Biochem J 2000;352(Pt 2):565–71.
[48] Dorea JG. Iron and copper in human milk. Nutrition 2000;16(3):209–20.
[49] Rauch H, Wells AJ. The toxic milk mutation, tx, which results in a condition resembling Wilson disease in humans, is linked to mouse chromosome 8. Genomics 1995;29(2):551–2.
[50] Reed V, Williamson P, Bull PC, Cox DW, Boyd Y. Mapping of the mouse homologue of the Wilson disease gene to mouse chromosome 8. Genomics 1995;28(3):573–5.
[51] Theophilos MB, Cox DW, Mercer JF. The toxic milk mouse is a murine model of Wilson disease. Hum Mol Genet 1996;5(10):1619–24.
[52] La Fontaine SS, Theophilos MB, Firth SD, Gould R, Parton RG, Mercer JF. Effect of the toxic milk mutation (tx) on the function and intracellular localization of Wnd, the murine homologue of the Wilson copper ATPase. Hum Mol Genet 2001;10(4):361–70.
[53] Voskoboinik I, La Fontaine S, Mercer JF, Camakaris J. Functional studies on the Wilson copper P-type ATPase and toxic Milk mouse mutant. Biochem Biophys Res Commun 2001;281(4):966–70.
[54] Huster D, Hoppert M, Lutsenko S, Zinke J, Lehmann C, Mossner J, et al. Defective cellular localization of mutant ATP7B in Wilson's disease patients and hepatoma cell lines. Gastroenterology 2003;124(2):335–45.
[55] Huster D, Kuhne A, Bhattacharjee A, Raines L, Jantsch V, Noe J, et al. Diverse functional properties of Wilson disease ATP7B variants. Gastroenterology 2012;142(4):947–56. e5.
[56] Biempica L, Rauch H, Quintana N, Sternlieb I. Morphologic and chemical studies on a murine mutation (toxic milk mice) resulting in hepatic copper toxicosis. Lab Invest 1988;59(4):500–8.
[57] Allen KJ, Buck NE, Cheah DM, Gazeas S, Bhathal P, Mercer JF. Chronological changes in tissue copper, zinc and iron in the toxic milk mouse and effects of copper loading. Biometals 2006;19(5):555–64.
[58] Gitlin JD. Wilson disease. Gastroenterology 2003;125(6):1868–77.
[59] Mercer JF, Grimes A, Danks DM, Rauch H. Hepatic ceruloplasmin gene expression is unaltered in the toxic milk mouse. J Nutr 1991;121(6):894–9.
[60] Howell JM, Mercer JF. The pathology and trace element status of the toxic milk mutant mouse. J Comp Pathol 1994;110(1):37–47.
[61] Koropatnick J, Cherian MG. A mutant mouse (tx) with increased hepatic metallothionein stability and accumulation [see comments]. Biochem J 1993;296(Pt 2):443–9.
[62] Deng DX, Ono S, Koropatnick J, Cherian MG. Metallothionein and apoptosis in the toxic milk mutant mouse. Lab Invest 1998;78(2):175–83.
[63] Ono S, Koropatnick DJ, Cherian MG. Regional brain distribution of metallothionein, zinc and copper in toxic milk mutant and transgenic mice. Toxicology 1997;124(1):1–10.
[64] Czachor JD, Cherian MG, Koropatnick J. Reduction of copper and metallothionein in toxic milk mice by tetrathiomolybdate, but not deferiprone. J Inorg Biochem 2002;88(2):213–22.
[65] Chen DB, Feng L, Lin XP, Zhang W, Li FR, Liang XL, et al. Penicillamine increases free copper and enhances oxidative stress in the brain of toxic milk mice. PLoS One 2012;7(5).
[66] Allen KJ, Cheah DM, Wright PF, Gazeas S, Pettigrew-Buck NE, Deal YH, et al. Liver cell transplantation leads to repopulation and functional correction in a mouse model of Wilson's disease. J Gastroenterol Hepatol 2004;19(11):1283–90.
[67] Shi Z, Liang XL, Lu BX, Pan SY, Chen X, Tang QQ, et al. Diminution of toxic copper accumulation in toxic milk mice modeling Wilson disease by embryonic hepatocyte intrasplenic transplantation. World J Gastroenterol 2005;11(24):3691–5.
[68] Buck NE, Cheah DM, Elwood NJ, Wright PF, Allen KJ. Correction of copper metabolism is not sustained long term in Wilson's disease mice post bone marrow transplantation. Hepatol Int 2008;2(1):72–9.
[69] Chen X, Xing S, Feng Y, Chen S, Pei Z, Wang C, et al. Early stage transplantation of bone marrow cells markedly ameliorates copper metabolism and restores liver function in a mouse model of Wilson disease. BMC Gastroenterol 2011;11:75.
[70] Coronado V, Nanji M, Cox DW. The Jackson toxic milk mouse as a model for copper loading. Mamm Genome 2001;12(10):793–5.
[71] Roberts EA, Robinson BH, Yang S. Mitochondrial structure and function in the untreated Jackson toxic milk (tx-j) mouse, a model for Wilson disease. Mol Genet Metab 2008;93(1):54–65.

[72] Le A, Shibata NM, French SW, Kim K, Kharbanda KK, Islam MS, et al. Characterization of timed changes in hepatic copper concentrations, methionine metabolism, gene expression, and global DNA methylation in the Jackson toxic milk mouse model of Wilson disease. Int J Mol Sci 2014;15(5):8004–23.

[73] Terwel D, Loschmann YN, Schmidt HH, Scholer HR, Cantz T, Heneka MT. Neuroinflammatory and behavioural changes in the Atp7B mutant mouse model of Wilson's disease. J Neurochem 2011;118(1):105–12.

[74] Medici V, Shibata NM, Kharbanda KK, LaSalle JM, Woods R, Liu S, et al. Wilson's disease: changes in methionine metabolism and inflammation affect global DNA methylation in early liver disease. Hepatology 2013;57(2):555–65.

[75] Przybylkowski A, Gromadzka G, Wawer A, Bulska E, Jablonka-Salach K, Grygorowicz T, et al. Neurochemical and behavioral characteristics of toxic milk mice: an animal model of Wilson's disease. Neurochem Res 2013;38(10):2037–45.

[76] Medici V, Shibata NM, Kharbanda KK, Islam MS, Keen CL, Kim K, et al. Maternal choline modifies fetal liver copper, gene expression, DNA methylation, and neonatal growth in the tx-j mouse model of Wilson disease. Epigenetics 2014;9(2):286–96.

[77] Medici V, Kieffer DA, Shibata NM, Chima H, Kim K, Canovas A, et al. Wilson disease: epigenetic effects of choline supplementation on phenotype and clinical course in a mouse model. Epigenetics 2016;11(11):804–18.

[78] Buiakova OI, Xu J, Lutsenko S, Zeitlin S, Das K, Das S, et al. Null mutation of the murine ATP7B (Wilson disease) gene results in intracellular copper accumulation and late-onset hepatic nodular transformation. Hum Mol Genet 1999;8(9):1665–71.

[79] Huster D, Finegold MJ, Morgan CT, Burkhead JL, Nixon R, Vanderwerf SM, et al. Consequences of copper accumulation in the livers of the atp7b-/- (Wilson disease gene) knockout mice. Am J Pathol 2006;168(2):423–34.

[80] Valko M, Morris H, Cronin MT. Metals, toxicity and oxidative stress. Curr Med Chem 2005;12(10):1161–208.

[81] Tapiero H, Townsend DM, Tew KD. Trace elements in human physiology and pathology. Copper. Biomed Pharmacother 2003;57(9):386–98.

[82] Sternlieb I. Copper and the liver. Gastroenterology 1980;78(6):1615–28.

[83] Yurkova IL, Arnhold J, Fitzl G, Huster D. Fragmentation of mitochondrial cardiolipin by copper ions in the Atp7b-/- mouse model of Wilson's disease. Chem Phys Lipids 2011;164(5):393–400.

[84] Evans GW, Bubois RS, Hambidge KM. Wilson's disease: identification of an abnormal copper-binding protein. Science 1973;181(4105):1175–6.

[85] Huster D, Purnat TD, Burkhead JL, Ralle M, Fiehn O, Stuckert F, et al. High copper selectively alters lipid metabolism and cell cycle machinery in the mouse model of Wilson disease. J Biol Chem 2007;282(11):8343–55.

[86] Gray LW, Peng F, Molloy SA, Pendyala VS, Muchenditsi A, Muzik O, et al. Urinary copper elevation in a mouse model of Wilson's disease is a regulated process to specifically decrease the hepatic copper load. PLoS One 2012;7(6).

[87] Ralle M, Huster D, Vogt S, Schirrmeister W, Burkhead JL, Capps TR, et al. Wilson disease at a single cell level: intracellular copper trafficking activates compartment-specific responses in hepatocytes. J Biol Chem 2010;285(40):30875–83.

[88] Huster D, Lutsenko S. Wilson disease: not just a copper disorder. Analysis of a Wilson disease model demonstrates the link between copper and lipid metabolism. Mol Biosyst 2007;3(12):816–24.

[89] Wilmarth PA, Short KK, Fiehn O, Lutsenko S, David LL, Burkhead JL. A systems approach implicates nuclear receptor targeting in the Atp7b(-/-) mouse model of Wilson's disease. Metallomics 2012;4(7):660–8.

[90] Wooton-Kee CR, Jain AK, Wagner M, Grusak MA, Finegold MJ, Lutsenko S, et al. Elevated copper impairs hepatic nuclear receptor function in Wilson's disease. J Clin Invest 2015;125(9):3449–60.

[91] Burkhead JL, Ralle M, Wilmarth P, David L, Lutsenko S. Elevated copper remodels hepatic RNA processing machinery in the mouse model of Wilson's disease. J Mol Biol 2011;406(1):44–58.

[92] Boaru SG, Merle U, Uerlings R, Zimmermann A, Flechtenmacher C, Willheim C, et al. Laser ablation inductively coupled plasma mass spectrometry imaging of metals in experimental and clinical Wilson's disease. J Cell Mol Med 2015;19(4):806–14.

[93] Hirayama T, Van de Bittner GC, Gray LW, Lutsenko S, Chang CJ. Near-infrared fluorescent sensor for in vivo copper imaging in a murine Wilson disease model. Proc Natl Acad Sci U S A 2012;109(7):2228–33.

[94] Peng F, Lutsenko S, Sun X, Muzik O. Positron emission tomography of copper metabolism in the Atp7b-/- knock-out mouse model of Wilson's disease. Mol Imaging Biol 2012;14(1):70–8.

[95] Boaru SG, Merle U, Uerlings R, Zimmermann A, Weiskirchen S, Matusch A, et al. Simultaneous monitoring of cerebral metal accumulation in an experimental model of Wilson's disease by laser ablation inductively coupled plasma mass spectrometry. BMC Neurosci 2014;15:98.

[96] Dong Y, Shi SS, Chen S, Ni W, Zhu M, Wu ZY. The discrepancy between the absence of copper deposition and the presence of neuronal damage in the brain of Atp7b(-/-) mice. Metallomics 2015;7(2):283–8.

[97] Hamilton JP, Koganti L, Muchenditsi A, Pendyala VS, Huso D, Hankin J, et al. Activation of liver X receptor/retinoid X receptor pathway ameliorates liver disease in Atp7B(-/-) (Wilson disease) mice. Hepatology 2016;63(6):1828–41.

[98] Cheng Q, He SQ, Gao D, Lei B, Long X, Liang HF, et al. Early application of auxiliary partial orthotopic liver transplantation in murine model of Wilson disease. Transplantation 2015;99(11):2317–24. https://doi.org/10.1097/TP.0000000000000787.

[99] Roybal JL, Endo M, Radu A, Gray L, Todorow CA, Zoltick PW, et al. Early gestational gene transfer with targeted ATP7B expression in the liver improves phenotype in a murine model of Wilson's disease. Gene Ther 2012;19(11):1085–94.

[100] Murillo O, Luqui DM, Gazquez C, Martinez-Espartosa D, Navarro-Blasco I, Monreal JI, et al. Long-term metabolic correction of Wilson's disease in a murine model by gene therapy. J Hepatol 2016;64(2):419–26.

[101] Fieten H, Penning LC, Leegwater PA, Rothuizen J. New canine models of copper toxicosis: diagnosis, treatment, and genetics. Ann N Y Acad Sci 2014;1314:42–8.

[102] Favier RP, Spee B, Schotanus BA, van den Ingh TS, Fieten H, Brinkhof B, et al. COMMD1-deficient dogs accumulate copper in hepatocytes and provide a good model for chronic hepatitis and fibrosis. PLoS One 2012;7(8).
[103] van De Sluis B, Rothuizen J, Pearson PL, van Oost BA, Wijmenga C. Identification of a new copper metabolism gene by positional cloning in a purebred dog population. Hum Mol Genet 2002;11(2):165–73.
[104] Forman OP, Boursnell ME, Dunmore BJ, Stendall N, van den Sluis B, Fretwell N, et al. Characterization of the COMMD1 (MURR1) mutation causing copper toxicosis in Bedlington terriers. Anim Genet 2005;36(6):497–501.
[105] Fieten H, Gill Y, Martin AJ, Concilli M, Dirksen K, van Steenbeek FG, et al. The Menkes and Wilson disease genes counteract in copper toxicosis in Labrador retrievers: a new canine model for copper-metabolism disorders. Dis Model Mech 2016;9(1):25–38.
[106] Haywood S, Muller T, Mackenzie AM, Muller W, Tanner MS, Heinz-Erian P, et al. Copper-induced hepatotoxicosis with hepatic stellate cell activation and severe fibrosis in North Ronaldsay lambs: a model for non-Wilsonian hepatic copper toxicosis of infants. J Comp Pathol 2004;130(4):266–77.
[107] Haywood S, Paris J, Ryvar R, Botteron C. Brain copper elevation and neurological changes in north ronaldsay sheep: a model for neurodegenerative disease? J Comp Pathol 2008;139(4):252–5.
[108] Haywood S, Vaillant C. Overexpression of copper transporter CTR1 in the brain barrier of North Ronaldsay sheep: implications for the study of neurodegenerative disease. J Comp Pathol 2014;150(2–3):216–24.
[109] Haywood S, Muller T, Muller W, Heinz-Erian P, Tanner MS, Ross G. Copper-associated liver disease in North Ronaldsay sheep: a possible animal model for non-Wilsonian hepatic copper toxicosis of infancy and childhood. J Pathol 2001;195(2):264–9.
[110] Suzuki KT, Ogra Y, Ohmichi M. Molybdenum and copper kinetics after tetrathiomolybdate injection in LEC rats: specific role of serum albumin. J Trace Elem Med Biol 1995;9(3):170–5.
[111] Terada K, Kawarada Y, Miura N, Yasui O, Koyama K, Sugiyama T. Copper incorporation into ceruloplasmin in rat livers. Biochim Biophys Acta 1995;1270(1):58–62.
[112] Terada K, Nakako T, Yang XL, Iida M, Aiba N, Minamiya Y, et al. Restoration of holoceruloplasmin synthesis in LEC rat after infusion of recombinant adenovirus bearing WND cDNA. J Biol Chem 1998;273(3):1815–20.
[113] Irani AN, Malhi H, Slehria S, Gorla GR, Volenberg I, Schilsky ML, et al. Correction of liver disease following transplantation of normal rat hepatocytes into Long-Evans Cinnamon rats modeling Wilson's disease. Mol Ther 2001;3(3):302–9.
[114] Malhi H, Irani AN, Volenberg I, Schilsky ML, Gupta S. Early cell transplantation in LEC rats modeling Wilson's disease eliminates hepatic copper with reversal of liver disease. Gastroenterology 2002;122(2):438–47.
[115] Lutsenko S. Atp7b-/- mice as a model for studies of Wilson's disease. Biochem Soc Trans 2008;36(Pt 6):1233–8.
[116] Fieten H, Leegwater PA, Watson AL, Rothuizen J. Canine models of copper toxicosis for understanding mammalian copper metabolism. Mamm Genome 2012;23(1–2):62–75.
[117] van de Sluis B, Groot AJ, Wijmenga C, Vooijs M, Klomp LW. COMMD1: a novel protein involved in the proteolysis of proteins. Cell Cycle 2007;6(17):2091–8.
[118] Gateau C, Delangle P. Design of intrahepatocyte copper(I) chelators as drug candidates for Wilson's disease. Ann N Y Acad Sci 2014;1315:30–6.
[119] Weiss KH, Stremmel W. Clinical considerations for an effective medical therapy in Wilson's disease. Ann N Y Acad Sci 2014;1315:81–5.

Chapter 6

Cellular Copper Toxicity: A Critical Appraisal of Fenton-Chemistry-Based Oxidative Stress in Wilson Disease

Hans Zischka[*,†] and Josef Lichtmannegger[*]

[*]Institute of Molecular Toxicology and Pharmacology, Helmholtz Center Munich, German Research Center for Environmental Health, Neuherberg, Germany, [†]Institute of Toxicology and Environmental Hygiene, Technical University of Munich, Munich, Germany

ABBREVIATIONS

εdA	1,N^6-ethenodeoxyadenosine
εdC	3,N^4-ethenodeoxycytidine
2,3-DHAB	2,3-dihydroxybenzoic acid (hydroxyl radical reacted SA)
8-OHdG	8-oxo-2′-deoxyguanosine
Acr-dG	acrolein deoxyguanosine
Akr1A1	aldo-keto reductase 1 family member A1
Akr1B7	aldo-keto reductase 1 family B7
ALP	alkaline phosphatase
ALT	(GPT) alanine aminotransferase
AST	(GOT) aspartate aminotransferase
ATP7B	ATPase copper transporting beta
BER	base excision repair of oxidized bases in DNA
BGD	N-benzyl-D-glucamine dithiocarbamate
BILI	bilirubin
BUN	blood urea nitrogen
BW	body weight
CAT	catalase
Cp	ceruloplasmin
CPK	creatine phosphokinase
Cro-dG	crotonaldehyde deoxyguanosine
CTR1	copper transport protein 1
Cu(I)-MT	copper (+1)-containing metallothionein
Cu/Zn-SOD	superoxide dismutase 1
D-PA	D-penicillamine
DTT	dithiothreitol
EM	electron microscopy
ER	endoplasmic reticulum
ESR	electron spin resonance
ETC	electron transport chain
F334	Fischer rats
FFA	free fatty acids
G6PD	glucose-6-phosphate dehydrogenase
GAPDH	glyceraldehyde-3-phosphate dehydrogenase
GGT	gamma-glutamyltransferase
GI	gastrointestinal (tract)
GPX	glutathione peroxidase

GSH	glutathione
GSSG-R	glutathione reductase
GST	glutathione *S*-transferase
H&E	hematoxylin-eosin
Hb	hemoglobin
HCC	hepatocellular carcinoma
HNE	4-hydroxy-2-nonenal
HO-1	heme oxygenase-1
HPLC	high-performance liquid chromatography
i.p.	intraperitoneal injection
i.v.	intravenous injection
IHC	immunohistochemistry
Ile	isoleucine
LDH	lactate dehydrogenase
LE	Long-Evans rat
LEA	Long-Evans Agouti rat
LEC	Long-Evans Cinnamon rat
Leu	leucine
LPP	LEC descendant, homozygotes deficient in Atp7b
MB	methanobactin
MDA	malondialdehyde
Mn-SOD	superoxide dismutase 2
MT	metallothionein
mtDNA	mitochondrial DNA
nanoLC-NSI/MS2	nanoflow liquid chromatography/nanospray ionization tandem mass spectrometry
NO	nitric oxide
OGG1	8-oxoguanine glycosylase
OH-Pro	hydoxyproline
p.o.	orally
PDH	pyruvate dehydrogenase
PDI	protein disulfide isomerase
Pro	proline
PUFA	polyunsaturated fatty acids
RIA	radioimmunoassay
ROS	reactive oxygen species
S9	postmitochondrial supernatant
SA	salicylic acid
s.c.	subcutaneous injection.
SN	supernatant
TBARS	thiobarbituric acid reactive substances
Trx	thioredoxine
TTM	tetrathiomolybdate
TUNEL	terminal deoxynucleotidyl transferase-mediated fluorescein-dUTP nick-end labeling
tx mouse	toxic milk mutant mouse
Val	valine
VitC	ascorbic acid
VitE	α-tocopherol
WD	Wilson disease

INTRODUCTION

Copper is a redox-active trace element, characterized by its ability to shift between the oxidized (cupric (Cu^{2+})) and the reduced state (cuprous (Cu^{1+})) [1, 2]. Several redox-active enzymes use this feature by tight incorporation of copper into their active sites [3, 4]. Recently, 54 copper-binding or copper-transporting proteins have been identified in the human proteome [5], which account for a very small amount of the total proteome, but nevertheless are of vital importance. Consequently, copper deficiency results in severe pathologies, anemia, neurological defects, connective tissue defects, oxidative damages, hypopigmentation, and hypothermia [6]. Conversely, copper overload can be highly detrimental as well. In humans, Wilson disease (WD) is the prime example for such toxicity due to excessive copper accumulation. This autosomal recessive genetic disorder is caused by mutations in the gene coding for the copper-transporting ATPase, ATP7B [7, 8]. ATP7B is primarily expressed in the liver, where it maintains body copper homeostasis. With ~2 mg daily dietary intake, copper is delivered via the portal vein from the GI tract to the liver and taken up into hepatocytes by the copper transporter CTR1 [2]. Within hepatocytes, copper is distributed by copper chaperones to copper redox enzymes, delivered systemically via ceruloplasmin, or excreted into the bile by ATP7B [9]. Consequently, dysfunctional ATP7B in WD leads to massive copper overload not only in patients' livers (10–27-fold) [10, 11] but also in patients' brains (up to eightfold) [12, 13], the latter most likely due to systemic copper spillover. Untreated, WD is fatal, due to liver or kidney failure or hemolysis [14, 15].

While the primary toxin in WD is (excess) copper, a matter of constant debate is, which of its adverse effects are causative for hepatocyte death in WD. One major hypothesis has been that aqueous free copper (and iron) catalyzes the formation of hydroxyl radicals (OH•) via Fenton- and Haber-Weiss-based chemistry [16] according to the following:

$$Fe^{3+}/Cu^{2+} + O_2^{\bullet-} \leftrightarrow Fe^{2+}/Cu^{1+} + O_2$$

$$Fe^{2+}/Cu^{1+} + H_2O_2 \rightarrow Fe^{3+}/Cu^{2+} + OH^{\bullet} + OH^{-} \quad \text{Fenton reaction}$$

$$O_2^{\bullet-} + H_2O_2 \leftrightarrow OH^{\bullet} + OH^{-} + O_2 \quad \text{Haber-Weiss reaction}$$

Thus, minute amounts of free copper ions lead to hydroxyl radicals, which are (with a one-electron reduction potential of +2.33 V) most powerful oxidants reacting at diffusion control rates with organic matter [17]. Oxidative stress by OH• directly attacks amino acids and DNA bases or abstracts hydrogen from double bonds in fatty acids, thereby generating reactive aldehydes or lipid hydroperoxides [18, 19]. The resulting aldehydes (e.g., acrolein, malondialdehyde, and 4-hydroxynonenal) in turn can form adducts with the DNA bases or amino acids in proteins [20]. Such damage may not only cause inflammation, triggering progressing fibrosis and cirrhosis, but also lead to hepatocyte death or dysplasia (e.g., Refs. [21, 22]).

However, a Fenton-chemistry-based copper toxicity as prime deleterious mechanism in WD has been challenged by thorough calculations demonstrating the lack of free, that is, unbound, copper in cells [23]. It was calculated that, assuming a cell volume of 10^{-14} L, the mean free copper concentration would correspond to 10^{-33} mol or 10^{-9} atoms [24]. Thus, cellular pools of free copper are virtually nonexistent, and pathologically, the cellular copper load would need to rise several orders of magnitude before free copper would appear, even if cellular settings would not change. Moreover, it has been demonstrated that copper is throughout tightly bound to proteins or low-molecular-weight ligands within cells [25, 26]. In situations of copper overload, elevated levels of metallothionein are known to capture and safely store large copper amounts [27]. In addition, accumulating cellular copper can be balanced to quite high amounts without clinical manifestations [28, 29]. These findings clearly argue against copper catalyzed hydroxyl radical emergence as early and immediately occurring toxic mechanism in WD.

Driven by these conflicting observations, we review here relevant studies concerning oxidative stress in WD. We start with a collective noncomprehensive survey of studies on oxidative stress markers in WD patients and related animal models. Thereafter, we focus on intervening therapies that aimed at targeting different aspects of copper toxicity. For the sake of conciseness, we restrict ourselves here to human studies and animal studies mainly in the LEC rat and its descendant the LPP rat. In a final paragraph, we provide our major conclusions from these studies.

OXIDATIVE STRESS IN WD PATIENTS AND RELATED ANIMAL MODELS

A relatively limited number of studies have assessed oxidative stress markers directly in WD patient samples (Table 1). About half of these studies have concentrated on plasma and the other half on liver, the primarily affected tissue in WD. Nevertheless, in plasma, a reduction in the antioxidants GSH and vitamin E and in total antioxidant capacity has been reported together with an increase in markers indicative of lipid peroxidation [31, 36–38]. These features prominently appeared in the context of WD patients with neurological symptoms (Table 1). In WD patient liver samples, mostly from explanted, that is, severely damaged, livers, decreased GSH and antioxidative enzyme activities (SOD, GPX, and GST) and increased DNA adducts and signs of lipid peroxidation have been reported [30, 33, 35]. In addition, severe mitochondrial damages (reviewed in Ref. [67]) and signs of ER stress have been observed [39]. Thus, in severe WD cases associated with liver failure or neurological disease progression, altered lipid peroxidation and diminished antioxidative enzymatic activity have been demonstrated.

TABLE 1 Oxidative Stress in WD Patients and Animal Models

Citation	Patients	Disease Stage	Marker	Sample(s)	Results
Humans					
Summer and Eisenburg (1985) [30]	WD versus normal	Chronic liver disease	Oxidative defense	Liver	GSH and GST activity severely ↓ in WD
Von Herbay et al. (1994) [31]	WD versus normal		Vitamin E	Plasma	Vitamin E ↓, correlating with free copper in serum
Ogihara et al. (1995) [32]	WD versus normal		Antioxidants, TBARS	Plasma	Antioxidants ↑ upon treatment with D-PA
Nair et al. (1998) [33]	WD versus normal	WD liver failure	DNA damage (εdA and εdC) caused by lipid peroxidation	Liver explants	Promutagenic DNA adducts threefold elevated in livers from patients undergoing orthotopic liver transplantation
Gu et al. (2000) [34]	WD versus normal	WD liver failure	Enzyme damage	Liver explants	Massive decrease in mitochondrial respiratory complex activities and in redox-sensitive aconitase activity
Nagasaka et al. (2006) [35]	WD versus normal	Liver damage: mild, strong, fulminant	Antioxidants, lipid peroxidation, oxidative defense	Liver	GSH ↓, TBARS ↑, Cu/Zn-SOD and Mn-SOD ↓, GPX ↓ especially in strong and fulminant WD but not in asymptomatic WD
Bruha et al. (2012) [36]	WD hepatic versus neurologic		Total antioxidant capacity, cytokines	Plasma	Lower total antioxidant capacity in the neurological versus hepatic form, proinflammatory cytokines ↑
Kalita et al. (2014) [37]	WD versus normal	Neurological versus asymptomatic WD	Antioxidants, lipid peroxidation	Plasma	GSH and total antioxidative capacity ↓, lipid peroxidation ↑ (untreated neurological > treated neurological > asymptomatic = controls)
Kalita et al. (2015) [38]	Neurological WD	6 months D-PA-treated WD	Antioxidants, lipid peroxidation	Plasma	Neurological worsening upon D-PA treatment is associated with GSH and antioxidative capacity ↓ but lipid peroxidation ↑
Oe et al. (2016) [39]	WD		Endoplasmic reticulum stress	Liver	Dilated and disorganized ER indicates abnormal protein folding

Citation	Model	Age (weeks)	Marker	Sample(s)	Results
Rodents					
Yamada et al. (1992) [40]	F344, LEC	4, 8	Lipid peroxidation (TBARS)	Liver	No elevated lipid peroxidation in asymptomatic LEC (8 weeks). Doubled lipid peroxidation in diseased LEC (8 weeks) >200 μg Cu/g liver
Yamamoto et al. (1993) [41]	LEA, LEC	5, 10, 15, 26, 40, 52, 104	DNA damage (8-OHdG) (HPLC)	Liver, kidney, brain	Significant 8-OHdG peak in the liver and kidney at acute hepatitis (15 weeks); no elevation at 5, 10 weeks, decline after acute phase

TABLE 1 Oxidative Stress in WD Patients and Animal Models—cont'd

Citation	Model	Age (weeks)	Marker	Sample(s)	Results
Stephenson et al. (1994) [42]	tx mouse	28–40	Lipid peroxidation (TBARS)	Liver microsomes	No lipid peroxidation of microsomes by Cu(I)-MT from mouse liver. Further addition of peroxide or Cu^{2+} causes lipid peroxidation
Ohhira et al. (1995) [43]	LEA, LEC	8, 14–15, 16–19	Radical-metabolizing enzymes, lipid peroxidation	Liver	Increased GSSG-R, decreased GPX, and unaltered Cu/Zn-SOD in LEC. Mn-SOD and lipid peroxidation significantly increased in hepatitis
Nair et al. (1996) [44]	LEA, LEC	7, 18, 30, 87	DNA damage (ϵdA and ϵdC) caused by lipid peroxidation	Liver	DNA adducts correlate with hepatic copper levels and massively peak at age fulminant hepatitis (18 weeks)
Yamamoto et al. (1997) [45]	LEA, LEC	6, 14, 18, 24	Ascorbate, ubiquinol/ubiquinone %, PUFA/FFA %, 16:1/FFA %	Plasma	Ascorbate significantly ↓ at all ages in LEC. Ubiquinol significantly ↑ at 6, 14, 18 weeks in LEC. PUFA/FFA % massively ↑ and 16:1/FFA % ↓ at 24 weeks in LEC
Suzuki et al. (1997) [46]	LEC	10–15 weeks (before jaundice) 15–17 weeks (jaundice)	Ascorbate versus hydroxyl radicals (ESR)	Liver (heated SN)	Hydroxyl radicals are produced in LEC with the onset of jaundice (hepatitis)
Nakamura et al. (1997) [47]	LEC	15	Hydroxyl radicals (ESR)	Liver (Cu(I)-MT)	No hydroxyl radicals emerge from Cu(I)-MT, only after liberation of copper by mercury ions
Rui and Suzuki (1997) [48]	LEC	9, 10, 11, 12, 13, 14, 15	Lipid peroxidation (MDA)	Liver, serum	Doubled MDA levels in the liver and serum in jaundiced rats
Chung et al. (1999) [49]	LEA, LEC	18	DNA damage (Acr-dG, Cro-dG) caused by PUFA-oxidation	Liver	26-fold and 7–20-fold elevated DNA levels in Acr-dG and Cro-dG, respectively, in LEC versus LEA livers
Yamamoto et al. (1999) [50]	Wistar, LEC	Healthy 9 weeks Jaundiced 14–24 weeks	Lipid peroxidation (TBARS) catalase and GPX activities	Liver (S-9)	3–4-fold increased H_2O_2 generation and significant losses in antioxidative enzymes in jaundiced LEC versus healthy LEC and controls
Yamamoto et al. (2001) [51]	Wistar, LEC	Healthy 8 weeks Jaundiced 12 weeks	Hydroxyl radicals (2,3-DHBA/SA)	Plasma, liver	Doubling of 2,3-DHAB/SA in the plasma and livers of jaundiced LEC versus healthy LEC and controls
Shishido et al. (2001) [52]	LEC	13	Hydroxyl radicals (ESR)	Liver Cu(I)-MT	Cu(I)-MT from LEC livers just before hepatitis still acts as antioxidant
Klein et al. (2003) [53]	LEA, LEC	10	Gene expression (array)	Liver	DNA damage repair enzymes ↑ and pro-inflammatory cytokines ↑
Choudhury et al. (2003) [54]	LEA, LEC	8, 14, 16, 17, 18, 24, 40	DNA damage repair enzymes	Liver	Activity and expression of BER enzymes significantly ↓ in acute (16–18 weeks) and early chronic (24 weeks) hepatitis

Continued

TABLE 1 Oxidative Stress in WD Patients and Animal Models—cont'd

Citation	Model	Age (weeks)	Marker	Sample(s)	Results
Samuele et al. (2005) [55]	LE, LEC	11, 14	Thiol status, NO metabolites, lipid peroxidation, SOD	Liver, brain	Significantly lower protein thiols and GSH/GSSG ratios in LEC versus LE. Protective nitrite massively ↓ in LEC versus LE. Lipid peroxidation up only in diseased LEC
Nair et al. (2005) [56]	LEC ± curcumin	6, 8, 12, 16, 32	nDNA and mtDNA (εdA and εdC) damage by lipid peroxidation	Liver	nDNA damage peaks at 8 weeks, mtDNA damage peaks at 12 weeks. Curcumin enhances oxidative DNA damage
Yasuda et al. (2006) [57]	LEC	Healthy (10–14) Subtle hepatitis (19–21) Acute hepatitis (26–31) Chronic hepatitis (44–50)	H_2O_2, lipid peroxides, oligosaccharide cleavage from glycoproteins	Serum	H_2O_2, lipid peroxides, and copper massively ↑ in acute hepatitis in LEC serum associated with glycoprotein damage, but not in earlier and later disease stages
Jia et al. (2006) [58]	LEC, LEC/(F344 × LEC)	24, 48	Gene expression (array and RT-PCR)	Liver	Significant mRNA increase of Akr1B7 but not of HO-1, Trx, Akr1A1, and G6PDH in LEC versus controls
Huster et al. (2006) [59]	Atp7b$^{-/-}$ mouse		Histopathology, IHC, EM	Serum, liver	Serum copper and serum ALT ↑. Structural mitochondrial alterations
Marquez-Quinones et al. (2007) [60]	LEC	6, 9, 11–13, D-PA-treated controls	Gene expression (array, RT-PCR), lipid peroxidation	Liver, urine	8-Isoprostane (8-IsoPGF$_{2\alpha}$) elevated upon jaundice in urine. Protein-, proteasome-, and nucleotide metabolism involved genes significantly ↑ in diseased animals
Roberts et al. (2008) [61]	tx mouse, C3H control	4–24	Antioxidant proteins, mtDNA	Liver	Structural mitochondrial alterations occur early. Thioredoxin-2 and peroxiredoxin-3 rise progressively. No mtDNA depletion
Marquez-Quinones et al. (2010) [62]	LEC	6, 9, 11–13, D-PA-treated controls	HNE-protein adducts, 8-isoprostaglandine in urine	Liver, urine	8-Isoprostane (8-IsoPGF$_{2\alpha}$) ↑ upon jaundice in urine. HNE-protein adducts appear in the first stages of hepatitis and are abundant during hepatitis
Wang et al. (2011) [63]	LEA, LEC$^{+/-}$, LEC$^{-/-}$	4, 12, 24	DNA damage (LC-MS/MS/MS)	Liver, brain	Massive liver damage in LEC$^{-/-}$ (12 weeks) coincides with DNA lesions. LEC$^{+/-}$ presents with higher levels of DNA lesions than LEA
Zischka et al. (2011) [29]	LEC, LPP	7, 9, 13, 15	Lipid peroxidation, oxidative enzyme damage	Liver	Diseased animals present with significant mitochondrial oxidative damages. Structural mitochondrial alterations occur early and are caused by copper interaction with protein thiols

TABLE 1 Oxidative Stress in WD Patients and Animal Models—cont'd

Citation	Model	Age (weeks)	Marker	Sample(s)	Results
Karmahapatra et al. (2013) [64]	LEA, LEC	8–40	DNA damage, repair enzymes, GSH/GSSG ratio, GAPDH activity	Liver	In acute hepatitis, DNA damage significantly ↑ and GSH/GSSG ↓. Decreased enzyme activities restorable with DTT
Yu et al. (2016) [65]	LEA, LEC	14–16	DNA damage (nanoLC-NSI/MS^2)	Liver, brain	In LEC liver, twofold higher levels of direct ROS-induced DNA damage and 50% higher lipid peroxidation-related DNA damage
Kumar et al. (2016) [66]	Wistar	4, 8, 12	CuSO$_4$ intoxication	Brain, liver, kidney	Oral copper overload causes total antioxidant capacity ↓, GSH ↓, and lipid peroxidation ↑ in all tissues

2,3-DHAB, 2,3-dihydroxybenzoic acid (hydroxyl radical reacted SA); *8-OHdG*, 8-oxo-2′-deoxyguanosine; *Acr-dG*, acrolein deoxyguanosine; *Akr1A1*, aldo-keto reductase 1 family member A1; *Akr1B7*, aldo-keto reductase 1 family B7; *ALT*, alanine aminotransferase; *BER*, base excision repair of oxidized bases in DNA; *Cro-dG*, crotonaldehyde deoxyguanosine; *Cu(I)-MT*, copper (+1)-containing metallothionein; *Cu/Zn-SOD*, superoxide dismutase 1; *D-PA*, D-penicillamine; *DTT*, dithiothreitol; *εdA*, 1,N^6-ethenodeoxyadenosine; *εdC*, 3,N^4-ethenodeoxycytidine; *EM*, electron microscopy; *ER*, endoplasmic reticulum; *ESR*, electron spin resonance; *F334*, Fischer rats; *FFA*, free fatty acids; *G6PDH*, glucose-6-phosphate dehydrogenase; *GAPDH*, glyceraldehyde-3-phosphate dehydrogenase; *GPX*, glutathione peroxidase; *GSH*, glutathione; *GSSG-R*, glutathione reductase; *GST*, glutathione S-transferase; *HNE*, 4-hydroxy-2-nonenal; *HO-1*, heme oxygenase 1; *HPLC*, high-performance liquid chromatography; *IHC*, immunohistochemistry; *LE*, Long-Evans rat; *LEA*, Long-Evans Agouti rat; *LEC*, Long-Evans Cinnamon rat; *LPP*, LEC descendant, homozygotes deficient in Atp7b; *MDA*, malondialdehyde; *Mn-SOD*, superoxide dismutase 2; *mtDNA*, mitochondrial DNA; *nanoLC-NSI/MS2*, nanoflow liquid chromatography/nanospray ionization tandem mass spectrometry; *NO*, nitric oxide; *PUFA*, polyunsaturated fatty acids; *S9*, postmitochondrial supernatant; *SA*, salicylic acid; *SN*, supernatant; *TBARS*, thiobarbituric acid reactive substances; *Trx*, thioredoxin; *tx mouse*, toxic milk mutant mouse; *WD*, Wilson disease.

In contrast to the few WD patient studies, a larger number of studies have tackled the issue of copper-related oxidative stress in animal models of WD (Table 1). Foremost, these studies were done not only in the LEC rat and its descendant the LPP$^{-/-}$ rat but also in Atp7b$^{-/-}$ and tx mice. For a detailed description of these WD animal models, see Chapter 5 or excellent articles of Burkhead et al. [68] and Medici and Huster [69]. Importantly, these animals lack a functional ATP7B and are thus true-to-life WD animal models that can be compared with nonaffected Wistar, LEA, and LPP$^{+/-}$ rats or wild-type mice, respectively.

As a result, these studies have amply demonstrated the occurrence of oxidative stress in WD animal livers (Table 1). Lipid peroxidation, DNA damage, oxidative protein adducts, reduced antioxidants, and DNA repair enzyme activities as well as hydrogen peroxide (H$_2$O$_2$) increases were repeatedly observed. Thus, oxidative stress is an unequivocal feature in WD animals. Importantly, however, most if not all of these studies have demonstrated the appearance of these markers in diseased animals with clear signs of liver damage and jaundice. Moreover, many of these studies have emphasized the absence of such markers in still healthy WD animals (Table 1). This also includes the formation of hydroxyl radicals from copper-containing metallothionein (MT) isolated from LEC livers [47, 52]. OH• was only found to emerge after liberation of copper from MT by added peroxides, Cu^{2+}, or mercury ions [42, 47]. Further, it has been reported that, just before the onset of severe hepatitis, copper-loaded MT still largely acts as antioxidant [52]. Therefore, a fully blown oxidative stress appears late in the disease progression and may be regarded as concurrent feature with clinically apparent hepatitis and jaundice in WD animals. In contrast, in nondiseased WD animals, enzymatic impairments, mitochondrial alterations, lower protein thiol amounts, and reductions in the repair systems against thiol modifications have been reported [29, 43, 53, 55, 61]. Thus, a progressive copper burden is not completely neutralized by copper scavengers like MT, and there are early occurring impairments in still healthy WD animals that are distinct from Fenton-chemistry-based copper insults.

TREATING COPPER TOXICITY IN WD RATS

Many studies have tested various therapeutic approaches against copper-induced impairments in WD rats (Tables 2–4). Their main rationale was to elucidate the contributing role of an increasing copper burden and its diverse consequences to the disease progression in WD animal livers.

TABLE 2 Therapeutic Strategies in LEC (LPP) Rats Aiming at a Reduction of the Hepatic Copper Burden

Citation	Treatment/Drug	Regimen	Measured Variables	Outcome
Togashi et al. (1992) [70]	D-PA	100 mg/kg (for 12 weeks)	GOT, GPT, histology, Cu (liver, urine, serum), Cp	No hepatitis, liver Cu ↓ (40%–50%), urine Cu ↑ (fourfold)
Jong-Hon et al. (1993) [71]	D-PA	100 mg/kg (from 11 to 70 weeks)	GOT, GPT, histology, serum Cu, GST (IHC), H&E, 8-OHdG	No hepatitis, no carcinoma, GST positive cells ↓, 8-OHdG ↓ (50%)
Suzuki et al. (1993) [72]	TTM	5–10 mg/kg (i.p., 8 days)	Cu/Zn/Fe (liver, kidney, spleen, serum) Cytosolic MT, Cu/Zn-SOD	No hepatitis Liver Cu ↓ but spleen, kidney, serum Cu ↑ No effect on liver Fe, MT decoppering, residual Cu in nonsoluble fraction
Sugawara and Sugawara (1994) [73]	Cu-deficient diet	30–65 days of age	Serum AST, ALT, GGT, Cp, Cu Liver Cu-MT, cytosol-MT	No hepatitis, Cu-MT 80% ↓
Sone et al. (1996) [74]	Trientine	5 mg/kg (s.c., 2× per week)	GOT, GPT, H&E, Cu/Fe/Zn (liver and urine)	No hepatitis, liver Cu ↓ (33%), liver Fe ↑ (12%–16%), urine Cu ↑ (77%), urine Fe ↓ (45%) HCC ↓ (33%)
Sugawara et al. (1999) [75]	TTM	2 × 10 mg/kg (s.c. on day 101 and 105, terminated on day 125)	Serum AST, ALT, LDH, BILI, BUN, Cu, Cp, Mo Cu (urine, bile, liver, kidney, testis, and brain)	Cures hepatitis, bile Cu ↑, urine Cu ↓, liver Cu ↑, testis Cu ↑, kidney Cu ↓, brain Cu unaltered
Sugawara et al. (1999) [76]	TTM	5 mg/kg (s.c., 2× per week, 40–105 days of age)	Serum AST, ALT, TBARS, BUN, Hb, Cu, Cp Cu/Fe/Mo (liver, kidney, testis, spleen, and brain), MT	TBARS ↑ (90%), liver Cu ↓ (65%), liver Fe ↑ (104%), Cu ↑ in kidney (threefold), spleen (10-fold) and testis (threefold), liver MT ↓ (95%)
Sugawara et al. (1999) [77]	Zinc acetate Zn-L-carnosine	1 g/kg (30–100 days of age)	Serum AST, ALT, Cp, BILI, TBARS Cu/Fe/Zn (serum, liver, kidney, spleen, intestine, and brain), MT	No hepatitis, TBARS ↓ (40%), serum Cu ↓ (95%), liver Cu ↓ (32%), intestine Cu ↓ (54%), kidney Cu ↓ (97%), liver Fe ↓ (38%), intestine MT ↑
Klein et al. (2000) [78]	D-PA	100 mg/kg (9–15 weeks) Or 9 × 500 mg/kg by gastric intubation	Serum AST, BILI, histology, liver Cu (total, cytosolic, noncytosolic, MT-bound, and lysosomes),	D-PA does not decopper MT but avoids Cu accumulation in noncytosolic fractions like lysosomes
Santon et al. (2004) [79]	Zinc acetate	50 mg/mL (daily gavage for 60 days)	Liver MT, oxidized MT, TUNEL	No difference in total liver MT, oxidized MT in zinc treated liver ↓ (80%), TUNEL staining in zinc treated ↓ (93%)
Klein et al. (2004) [80]	TTM	1 × 10 mg/kg (i.p.)	Serum AST, BILI, Cu, Mo Liver Cu/Fe/Zn, MT	Single treatment restores hepatic integrity in 11/12 rats, serum Cu ↑, liver Cu ↓ (30%), Cu-MT ↓ (25%)
Medici et al. (2005) [81]	Zinc acetate	80 mg/mL (daily gavage for 1 or 2 weeks) 50 mg/mL (daily gavage for 8 weeks)	Serum/urine Hb, AST, ALT, BILI, Cu Liver Cu/Fe/Zn, 8-OHdG	Serum: liver damage markers significantly down, Cu ↓ Liver: Cu ↓ (34% at 2 weeks, 11% at 8 weeks), Fe ↓ (75% at 8 weeks), 8-OHdG 25% ↓ at 8 weeks
Gonzales et al. (2005) [82]	Zinc acetate	1 g/kg diet (from 4 to 20 weeks)	BW, liver Cu, histology, H&E, MT, intestinal Cu/Zn/MT	Liver Cu ↓ (≤ 30%) at 20 weeks: lower hepatic damage,

TABLE 2 Therapeutic Strategies in LEC (LPP) Rats Aiming at a Reduction of the Hepatic Copper Burden—cont'd

Citation	Treatment/Drug	Regimen	Measured Variables	Outcome
				liver MT ↑ (30%), apoptosis ↓ (43%), intestinal Cu equal and MT ↑ (376%)
Shimada et al. (2005) [83]	BGD D-PA Trientine	2 mmol/kg BGD (i.p., 2×/week, 11–19 weeks) 1 mmol/kg D-PA or trientine (i.p.)	Cu (urine and bile; single dose), BW, ALT, BILI, H&E Cu/Fe/Zn (liver, kidney, spleen, brain, and serum)	Cu excretion into bile (BGD) or urine (D-PA and trientine) BGD: no hepatitis, liver Cu ↓ (33%), no kidney Cu, no serum Cu, liver Fe ↓ (64%)
Summer et al. (2011) [84]	MB	6×/13× 200 mg/kg (i.p., daily or 3× per week)	Serum AST, histology, bile Cu/Fe/Zn, liver Cu, MT	MB elevates bile Cu and restores hepatic integrity, liver Cu ↓ (56%/75%)
Zischka et al. (2011) [29]	MB D-PA (LPP rat)	62.5 mg/kg MB (i.p., 3× per week, 5 weeks) 100 mg/kg D-PA (in tap water, 5 weeks)	Liver Cu (total, cytosol, mitochondria, and lysosomes), mitochondrial activity, serum AST, BILI, BW, H&E	No hepatitis if treatment starts before liver damage onset, Cu depletion in mitochondria (75%) and lysosomes (25%) but not in cytosol; restored mitochondria
Lichtmannegger et al. (2016) [28]	MB D-PA Trientine (LPP rat)	150 mg/kg MB (IP/IV) 100 mg/kg D-PA (p.o.) 480 mg/kg trientine (p.o.)	Liver Cu (total and mitochondria), mitochondrial structure (EM), activity, AST, BILI, BW, H&E, Cp	One-week MB (but not D-PA/trientine) treatment restores hepatic and mitochondrial integrity, Cu depletion in mitochondria (50%) by MB; two treatment phases (5 days each) avoid hepatitis for ≥170 days

8-OHdG, 8-oxo-2′-deoxyguanosine; *ALT*, (GPT) alanine aminotransferase; *AST*, (GOT) aspartate aminotransferase; *BGD*, N-benzyl-D-glucamine dithiocarbamate; *BILI*, bilirubin; *BUN*, blood urea nitrogen; *BW*, body weight; *Cp*, ceruloplasmin; *D-PA*, D-penicillamine; *EM*, electron microscopy; *GGT*, gamma-glutamyltransferase; *GST*, glutathione S-transferase; *H&E*, hematoxylin-eosin; *Hb*, hemoglobin; *HCC*, hepatocellular carcinoma; *IHC*, immunohistochemistry; *i.p.*, intraperitoneal injection; *i.v.*, intravenous injection; *LDH*, lactate dehydrogenase; *MB*, methanobactin; *MT*, metallothionein; *p.o.*, orally; *s.c.*, subcutaneous injection; *Cu/Zn-SOD*, superoxide dismutase; *TBARS*, thiobarbituric acid reactive substances; *TTM*, tetrathiomolybdate; *TUNEL*, terminal deoxynucleotidyl transferase-mediated fluorescein-dUTP nick-end labeling.

TABLE 3 Therapeutic Antioxidant Strategies in LEC Rats

Citation	Treatment/Drug	Regimen		Measured Variables	Outcome
Yamazaki et al. (1993) [85]	Dietary VitE deficiency versus control versus surplus	3–28 weeks Defi: ≤0.1 ppm total tocopheryls Cont: 20 ppm α-tocopheryl acetate Surp: 59 ppm α-tocopheryl nicotinate		BW, serum GOT/GPT, α-tocopherol, lipid peroxidation (MDA, serum, and liver)	Females: no significant shift in hepatitis onset or death Males: earlier (VitE-deficient) or later onset (VitE surplus) hepatitis and death Lipid peroxides in serum and liver: def > cont > surp (comparable in males and females)
Hawkins et al. (1995) [86]	VitC, VitE, Pro, β-carotene, Trx, Leu/Ile/Val	Diet:	0.8% VitC 0.01% VitE 0.01% β-carotene	Kaplan-Meier, visual inspection for jaundice	Hardly any effect of Pro, VitC, Leu/Ile/Val in diet/tap water Jaundice onset 4 days later and mortality ↓ by enforced administration of Pro (80%), VitC (65%) No effect of VitE, β-carotene High Trx (36 µg/h) delayed (6 d) and decreased jaundice
		Tap water:	0.2 M Pro 0.08 M Leu/Ile/Val		
		Trx via minipump			

Continued

TABLE 3 Therapeutic Antioxidant Strategies in LEC Rats—cont'd

Citation	Treatment/Drug	Regimen	Measured Variables	Outcome
Yamashita et al. (1996) [87]	Phenylbutyl nitrone (PBN, spin trap)	128 mg/kg (s.c., every second day, 13–30 weeks)	BW, mortality, GPT, TBARS (liver) 8-OHdG, H&E	Delayed hepatitis 15.5 (Con) versus 19.9 weeks (PBN). Delayed mortality 15.3 (Con) versus 21.4 weeks (PBN) TBARS ↓
Shibata et al. (1999) [88]	Linolenic acid (LNA) Linoleic acid (LA)	±68% LA or 42% LNA (in diet for 10–16 weeks)	Lipid composition (serum and liver), AST, ALT, serum Cu	Reduced incidence hepatitis (75% → 50%). Delayed hepatitis (+33 d), LNA=LA, no effect on oxidative stress
Yamamoto et al. (2001) [89]	DL-α-Lipoic acid	10, 30, 100 mg/kg (p.o. for 5 days/week from week 8 to 12)	Food intake, BW, serum AST, ALT, BILI, total prot., lipid peroxidation, liver Cu/nonhem Fe (subfractions), HO-1/Cu/Zn-SOD/GPX/GSSG-R/catalase	Dose-dependent reduction of Liver damage GPX ↑ (80%), GSSG-R ↑ (20%–27%), Cu/Zn-SOD equal activity in liver subfractions Dose-dependent ↓ (33%/50%) only in mitochondrial Cu and Fe Lipid peroxidation ↓ (34%)
Watanabe et al. (2001) [90]	Lycopene TJ-9 (baicalin)	0.005% lycopene or 1% TJ-09 (in diet for 6–76 weeks)	Histology, Cu/Fe/Zn (the liver and serum), GST (IHC)	Unaltered Cu/Fe/Zn and GST levels No reduction in hepatocarcinogenesis (HCC area 54% ↑)
Kitade et al. (2002) [91]	Lycopene TJ-9 (baicalin)	0.005% lycopene or 1% TJ-09 (in diet for 6–76 weeks)	Histology, Cu/Fe/Zn in liver, BW, liver weight, MDA/OH-Pro liver	Liver weight ↑, fibrosis area ↓ (43%), slightly lower MDA/OH-Pro levels
Frank et al. (2003) [92]	Curcumin	0.5% curcumin (in diet for 4–100 weeks)	BW, Kaplan-Meier, liver Cu, tumor classification	Survival ↓ Liver Cu slightly ↑ at 12 weeks versus Con. Similar liver/kidney tumor incidence, metastases ↓
Du et al. (2004) [93]	PUFA depletion Linoleic acid (LA) Docosahexaenoic acid (DHA)	10% (in diet)	BW, fatty acid composition liver lipids, AST, ALT, liver examination, COX1/COX2/p53-mRNA expression	No difference in males in the treatment groups. Survival time in females 17% +20% longer in LA and DHA/LA. Treatment effect not due to enhanced/avoided lipid peroxidation but due to stimulated bile acid synthesis.
Kitamura et al. (2005) [94]	N-acetylcysteine (NAC) Quercetin (QC) Phytic acid (PA)	1% (in diet for 15–21 weeks)	Body and organ weights, serum, biochemistry, histology, Cu/Fe (liver and kidney by tissue staining), acrolein protein adducts	No effect of QC/PA NAC: Cu/Fe ↓, liver and kidney damage ↓↓, no protein adducts, effect rather due to metal chelation than ROS scavenging
Shibata et al. (2006) [95]	Fermented brown rice (FBRA)	5%/10% (in diet for 8–20 weeks)	Incidence hepatitis, Kaplan-Meier, Cu (liver and serum)	No effect on liver Cu, serum Cu slightly ↑, hepatitis incidence ↓ (96% → 85%), same age of hepatitis onset, slightly prolonged survival rate

TABLE 3 Therapeutic Antioxidant Strategies in LEC Rats—cont'd

Citation	Treatment/Drug	Regimen	Measured Variables	Outcome
Asanuma et al. (2007) [96]	Phenylbutyl nitrone derivative (LPBNSH, spin trap)	0.1–2.0 mg/kg (IP, every second day, for 9–14 weeks)	Kaplan-Meier, Cu/Fe (liver and serum), LDH, Hb, BW, AST, ALT, H&E, 8-OHdG, TBARS (liver), CPK/BUN	0.1/0.5 mg/kg doses delayed hepatitis onset (12.3 → 15 weeks) and prolonged jaundice (4.8 → 8 days) 8-OHdG index and TBARS ↓ High doses may be toxic, no difference in liver Cu
Katayama et al. (2014) [97]	Coffee "Nestle clear taste"	Coffee versus water for 6–32 weeks	BW, water/coffee consumption, GPT, Kaplan-Meier, Cu/Fe in liver, H&E, GST-P foci, markers of inflammation	Two weeks delayed hepatitis, survival ratio ↑ (0.57 vs. 0.33), inflammation-related cytokines in liver ↓, liver Fe and GST-P preneoplastic foci ↓

8-OHdG, 8-oxo-2′-deoxyguanosine; *ALT*, (GPT) alanine aminotransferase; *AST*, (GOT) aspartate aminotransferase; *BILI*, bilirubin; *BUN*, blood urea nitrogen; *BW* body weight; *CPK*, creatine phosphokinase; *GST*, glutathione S-transferase; *H&E*, hematoxylin-eosin; *Hb*, hemoglobin; *HCC*, hepatocellular carcinoma; *IHC*, immunohistochemistry; *Ile*, isoleucine; *i.p.*, intraperitoneal injection; *Leu*, leucine; *LDH*, lactate dehydrogenase; *MDA*, malondialdehyde; *OH-Pro*, hydoxyproline; *p.o.*, orally; *Pro*, proline; *Cu/Zn-SOD* superoxide dismutase; *TBARS*, thiobarbituric acid reactive substances; *Trx*, thioredoxin; *Val*, valine; *VitC*, ascorbic acid; *VitE*, α-tocopherol.

TABLE 4 Other Therapy Strategies Against Hepatitis in the LEC Rat

Citation	Treatment/Drug	Regimen	Measured Variables	Outcome
Kasai et al. (1992) [98]	Ovariectomy Orchiectomy Testosterone Estradiol	Operation at 4–5 weeks, hormone supplement (starting at 12 weeks, 2× per week, s.c.; end at 30 weeks)	GPT, Kaplan-Meier	Similar age of hepatitis onset. Survival rate decreased (females: 50% → 12.5%, males: 75% → 14.3%), testosterone ↑ survival rate >90%, estradiol delayed hepatitis
Yokoi et al. (1994) [99]	Cyclosporin (CsA) D-PA	5 mg/kg CsA (6× per week) 100 mg/kg D-PA (6–16 weeks, then 3× per week for 13 weeks)	Kaplan-Meier, BW, GOT, GPT, ALP, serum albumin, BILI, PDI (RIA)	D-PA prevented hepatitis, CsA did not prevent liver damage, but death rate ↓ at 20 weeks (50% → 14%) Immune response associated with death but not with hepatitis.
Yokoi et al. (1995) [100]	TJN-101	100 mg/kg (6× per week, 6–16 weeks, then 3× per week for 13 weeks)	Kaplan-Meier, BW, serum parameters (GOT, GPT, ALP), autoimmunity	Similar death rate, time to death 7–10 weeks delayed, gradual (and not fulminant) ↑ of liver damage, low autoimmunity
Kato et al. (1996) [101]	Iron deprivation	Iron-deficient diet for up to 65 weeks	Kaplan-Meier Liver: Cu/Fe, nonheme-iron, H&E Serum: Cu/Fe, LDH, Hb, TUNEL, ALT, LDH, BILI	No hepatitis, lower fibrosis, no carcinomas Equal hepatic copper, hepatic iron ↓ (two- to threefold)
Sugawara and Sugawara (1999) [102]	Copper/iron deprivation	0.1 or 10 mg Cu/kg and/or 1.5 or 150 mg Fe/kg, 6–11 weeks	AST, LDH, Cp, BUN, H&E, Cu/Fe (liver, kidney, spleen, and intestine), metallothionein/ferritin (intestine and liver)	A Cu-sufficient but Fe-deficient diet (Cu 10 mg/kg; Fe 1.5 mg/kg) showed jaundice at 75 days but not the other dietary combinations

Continued

TABLE 4 Other Therapy Strategies Against Hepatitis in the LEC Rat—cont'd

Citation	Treatment/Drug	Regimen	Measured Variables	Outcome
Sheline et al. (2002) [103]	Thiamine	1% in drinking water for 5–60 weeks	H&E, Cu	Extended life span in LEC (>64 weeks), less cellular vacuolization and degeneration of hepatocytes, no change in liver Cu
Chang et al. (2005) [104]	L-Carnitine	1 mg/mL (in tap water for 6–58 weeks)	AST/ALT/BILI, H&E, carnitine, FFA (liver/serum), TUNEL, GST-foci, mtDNA, ROS, HNE/8-OHdG liver	AST, ALT, BILI, free fatty acids significantly ↓ at 24 weeks. Mitochondrial function significantly ↑, mtDNA damage ↓, mitochondrial ROS ↓. Low liver HNE/8-OHdG
Park et al. (2005) [105]	Panax notoginseng saponins (PNS) water extract	1% in diet for 6–26 weeks (B-group) 0.005% Lycopene (C-group)	HCC/fibrosis, BW Liver: Cu/Fe/Zn, weight, MDA/OH-Pro	HCC area ↑ in B, C (16%→28%); fibrosis area ↓ in B, C similar liver Cu and BW; MDA ↓ (90%), liver weight ↑
Otsuka et al. (2006) [106]	D-Galactosamine	1× 300 mg/kg (s.c. at 14 weeks) 300 mg/kg (s.c., 1× per week for 6–18 weeks)	BW, Kaplan-Meier Liver: 8-OHdG/GPX/CAT, AST, ALT, LDH, H&E	Mortality ↓ (17%→0%), 8-OHdG ↓ (79%), GPX ↑ (36%), AST/ALT ↑ (25%/125%), liver regeneration
Tsubota et al. (2008) [107]	Bovine lactoferrin (iron-binding glycoprotein)	2% (in diet for 4–40 weeks)	Kaplan-Meier, GOT, GPT, BILI, H&E, liver/serum Cu, mitochondria, MDA, 8-OHdG, OGG1, methylation CpG	Survival rate ↑ at 20 weeks (80% vs 12.5% in Con.) BILI, AST, ALT ↓ Milder steatosis, intact mitochondria, MDA and 8-OHdG levels ↓, mtDNA mutations ↓, OGG-1 ↑

8-OHdG, 8-oxo-2′-deoxyguanosine; *ALP*, alkaline phosphatase; *ALT*, (GPT) alanine aminotransferase; *AST*, (GOT) aspartate aminotransferase; *BILI*, bilirubin; *BW*, body weight; *CAT*, catalase; *D-PA*, D-penicillamine; *FFA*, free fatty acids; *GPx*, glutathione peroxidase; *H&E*, hematoxylin-eosin; *Hb*, hemoglobin; *HCC*, hepatocellular carcinoma; *LDH*, lactate dehydrogenase; *MDA*, malondialdehyde; *OGG1*, 8-oxoguanine glycosylase; *OH-Pro*, hydoxyproline; *PDI*, protein disulfide isomerase; *RIA*, radioimmunoassay; *ROS*, reactive oxygen species; *s.c.*, subcutaneous injection.

Strategically, the reported therapeutic interventions can be grouped into three categories:

1. Approaches to either limit copper uptake, stimulate copper excretion, or detoxify copper by chelation (Table 2).
2. Antioxidant or radical scavenging therapies (Table 3).
3. Strategies to ameliorate liver functions or to support liver regeneration (Table 4).

With respect to the first group of interventions, copper chelation by either D-penicillamine (D-PA), trientine, tetrathiomolybdate (TTM), N-benzyl-D-glucamine dithiocarbamate (BGD), or methanobactin (MB) avoids hepatitis in LEC or LPP rats (Table 2). This unequivocally confirms that the prime liver toxic agent in WD animals is copper. In further agreement, hepatitis was avoided upon a copper-deficient diet [73], and positive therapeutic effects of zinc that may interfere with the hepatic copper load were reported [77, 79, 81, 82]. However, none of these treatments resulted in liver copper depletions down to wild-type levels, but reductions ranged from 11% to 75%. This underlines the liver's capacity to detoxify or neutralize a considerable copper amount. However, this also casts serious doubts on a pronounced presence of a Fenton-chemistry-based catalytic copper damage at such stages, as these animals are healthy, for example, with no obvious signs of lipid peroxidation, but still contain high copper quantities.

The second line of treatments aimed at the avoidance of oxidative stress. Dietary antioxidants like vitamin E, β-carotene, lipoic acid, lycopene, acetylcysteine, or spin traps like phenylbutyl nitrone have been administered (Table 3). The outcomes of these studies are much more ambiguous than the ones above that deal with copper reduction or neutralization. Several antioxidants were found to be of no effect with respect to the avoidance of hepatitis, and curcumin even had a worsening effect (Table 3). A repeated observation of antioxidant treatments was a delay in hepatitis onset. Vitamin E delayed hepatitis in male rats (but not in females) [85]; nitrone compounds delayed hepatitis onset for 3–4 weeks [87, 96]; and enforced administration of proline resulted in a hepatitis delay of 4 days [86], fatty acids (linolenic acid and linoleic acid), in a 33-day delay [88, 93], and a coffee-enriched diet, in a delay of 2 weeks [97]. Thus, a general antioxidant therapy has at best a hepatitis-delaying but not an avoiding effect. Interestingly, pronounced positive therapeutic effects were obtained by thiol-containing antioxidant therapies. Lipoic acid demonstrated dose-dependent reductions in liver damage [89], and acetylcysteine revealed a strongly lowered liver and kidney damage [94], whereas the positive effect of the latter drug has been suggested to potentially arise from metal chelation.

The last group of interventions is diverse in their proposed mechanisms of action (Table 4). Drugs that initiate liver regeneration (TJN-101 and D-galactosamine) delayed the time of death or death rate [100, 106]. Interestingly, iron deprivation or iron-binding compounds avoided hepatitis and increased the survival rate [101, 107]. In contrast, an opposing study demonstrated hepatitis in LEC rats upon feeding an iron-deficient but copper-sufficient diet [102]. It was concluded that iron deficiency enhances hepatic copper deposition possibly due to increased absorption of Cu from the GI tract [102]. While these conflicting results await clarification, they clearly indicate a link between iron and copper toxicity in WD animals. Finally, drugs that support mitochondrial bioenergetics improved mitochondrial function, reduced liver damage, and strongly extended the life span in LEC rats [103, 104].

CONCLUSIONS

1. *With respect to the occurrence of liver cancer after hepatitis and jaundice*: DNA adducts formed either directly by hydroxyl radicals or via reactive aldehydes originating from lipid peroxidation have amply been demonstrated in WD rats. This genotoxicity may explain a late developing liver cancer in LEC rats, which is, however, very rare in WD patients [108, 109].
2. *With respect to oxidative stress at the stage of overt hepatitis and jaundice*: Free (aqueous) copper ions catalyze the formation of hydroxyl radicals that cause damage to DNA, proteins, and lipids. An overwhelming dataset demonstrates that such damage does occur in WD patients and WD animals at the stage of hepatitis and jaundice, but apparently not or to a much lesser degree in earlier disease stages. Antioxidative therapies mostly delay but do not avoid hepatitis. Thus, in untreated WD animals, a Fenton-/Haber-Weiss-based copper toxicity is one/the major executioner of hepatocyte damage or death in hepatitis, but most likely not the driving toxic mechanism leading to this disease stage.
3. *With respect to metallothionein copper buffering in early disease stages*: WD animals that do already have massive and progressively increasing copper depositions stay nevertheless clinically healthy for months. This demonstrates an enormous capacity of hepatocytes to buffer and neutralize excess copper. The major cellular copper buffer is metallothionein (MT). However, MT has been reported to still act as an antioxidant at the onset of hepatitis in WD animals [52]. Thus, MT is not saturated by copper at this disease stage. Moreover, drugs with lower copper affinity than MT (e.g., D-PA, the oldest oral chelation treatment in WD) should not or only partially deplete this storage [78]. D-PA nevertheless is highly efficient in avoiding hepatitis in WD rats (Table 2). These findings therefore question the theory that simple copper overloading of MT—that would result in free copper ions to cause hydroxyl radical emergence—may be the trigger to initiate hepatitis. A liberation of copper from MT would be more plausible, if either damage to MT, an increased MT degradation and copper release in the lysosomal compartment, or a diminished MT expression would occur. Thus, hepatitis in WD animals is at best only in parts due to MT copper saturation.
4. *With respect to other cellular copper stores and biochemical impairments in early disease stages*: Cytosolic MT is not neutralizing all excess hepatic copper. Already at early stages of WD, cellular compartments like mitochondria and lysosomes present with high copper loads (e.g., Refs. [28, 78, 110]). These copper deposits are not fully silent, as mitochondrial structural alterations are among the first observed features in WD patient and animal hepatocytes [29, 61, 111]. Furthermore, enzymatic deficiencies, ER stress, lysosomal abnormalities, and progressive functional impairments in mitochondria have been reported in either WD patients [30, 34, 35, 39] or WD animals [28, 43, 78]. Thus, a progressive copper load increasingly impairs the biochemical and bioenergetic functionality of WD hepatocytes. This damaging effect may be prominently mediated via a copper attack on thiol residues in susceptible proteins, as thiol-containing drugs (e.g., lipoic acid) restored enzymatic activities and avoided liver damage in a dose-dependent manner [89]. This copper is less tightly bound than the one bound to MT and could therefore be the origin of free toxic copper upon overload. This would explain why a Fenton-chemistry-based copper toxicity is clearly detectable at later disease stages but not in the beginning of the disease. Moreover, it would present a rationale as to why most, if not all, therapies that reduce this more loosely bound copper are highly effective in jaundice avoidance, despite the fact that none of these therapies reduced the copper burden down to wild-type levels.

In summary, two major, not mutually exclusive mechanisms take place in copper-burdened WD hepatocytes. The first is a progressive filling and potential saturation of cellular copper storage places like MT, lysosomes, and mitochondria. The second is an increasing biochemical and bioenergetic inefficiency. Upon failure of either one, free copper may appear, causing a Fenton-/Haber-Weiss-based toxicity that seals the hepatocytes' fate.

ACKNOWLEDGMENTS

We would like to apologize to all our colleagues whose work has not been cited here due to space restrictions and would like to sincerely thank E. E. Rojo and all members of the AG Zischka for critical reading of the manuscript.

REFERENCES

[1] Knöpfel M, Solioz M. Characterization of a cytochrome b(558) ferric/cupric reductase from rabbit duodenal brush border membranes. Biochem Biophys Res Commun 2002;291(2):220–5.
[2] Kim BE, Nevitt T, Thiele DJ. Mechanisms for copper acquisition, distribution and regulation. Nat Chem Biol 2008;4(3):176–85.
[3] Kaim W, Rall J. Copper—a "modern" bioelement. Angew Chem Int Ed Engl 1996;35(1):43–60.
[4] Rubino JT, Franz KJ. Coordination chemistry of copper proteins: How nature handles a toxic cargo for essential function. J Inorg Biochem 2012;107(1): 129–43.
[5] Blockhuys S, et al. Defining the human copper proteome and analysis of its expression variation in cancers. Metallomics 2017;9(2):112–23.
[6] Vonk WIM, Wijmenga C, van de Sluis B. Relevance of animal models for understanding mammalian copper homeostasis. Am J Clin Nutr 2008;88(3): 840s–845s.
[7] Tanzi RE, et al. The Wilson disease gene is a copper transporting ATPase with homology to the Menkes disease gene. Nat Genet 1993;5(4):344–50.
[8] Bull PC, et al. The Wilson disease gene is a putative copper transporting P-type ATPase similar to the Menkes gene. Nat Genet 1993;5(4):327–37.
[9] Roberts EA, Sarkar B. Liver as a key organ in the supply, storage, and excretion of copper. Am J Clin Nutr 2008;88(3):851S–854S.
[10] Pfeiffer RF. Wilson's disease. Semin Neurol 2007;27(2):123–32.
[11] Cope-Yokoyama S, et al. Wilson disease: histopathological correlations with treatment on follow-up liver biopsies. World J Gastroenterol 2010; 16(12):1487–94.
[12] Litwin T, et al. Brain metal accumulation in Wilson's disease. J Neurol Sci 2013;329(1–2):55–8.
[13] Cumings JN. The copper and iron content of brain and liver in the normal and in hepato-lenticular degeneration. Brain 1948;71:410–5. Pt. 4.
[14] European Association for Study of Liver. EASL clinical practice guidelines: Wilson's disease. J Hepatol 2012;56(3):671–85.
[15] Gitlin JD. Wilson disease. Gastroenterology 2003;125(6):1868–77.
[16] Stohs SJ, Bagchi D. Oxidative mechanisms in the toxicity of metal ions. Free Radic Biol Med 1995;18(2):321–36.
[17] Krumova K, Cosa G. Overview of reactive oxygen species. In: Singlet oxygen: applications in biosciences and nanosciences. vol. 1. The Royal Society of Chemistry; 2016. p. 1–21 [Chapter 1].
[18] Girotti AW. Mechanisms of lipid peroxidation. J Free Radic Biol Med 1985;1(2):87–95.
[19] Halliwell B, Gutteridge JM. Role of free radicals and catalytic metal ions in human disease: an overview. Methods Enzymol 1990;186:1–85.
[20] Fritz KS, Petersen DR. An overview of the chemistry and biology of reactive aldehydes. Free Radic Biol Med 2013;59:85–91.
[21] Bellanti F, et al. Lipid oxidation products in the pathogenesis of non-alcoholic steatohepatitis. Free Radic Biol Med 2017;111:173–85.
[22] Stickel F. Alcoholic cirrhosis and hepatocellular carcinoma. Adv Exp Med Biol 2015;815:113–30.
[23] Rae TD, et al. Undetectable intracellular free copper: the requirement of a copper chaperone for superoxide dismutase. Science 1999; 284(5415):805–8.
[24] Lippard SJ. Free copper ions in the cell? Science 1999;284(5415):748–9.
[25] Field LS, Luk E, Culotta VC. Copper chaperones: personal escorts for metal ions. J Bioenerg Biomembr 2002;34(5):373–9.
[26] Banci L, et al. Affinity gradients drive copper to cellular destinations. Nature 2010;465(7298):645–8.
[27] Tapia L, et al. Metallothionein is crucial for safe intracellular copper storage and cell survival at normal and supra-physiological exposure levels. Biochem J 2004;378(Pt 2):617–24.
[28] Lichtmannegger J, et al. Methanobactin reverses acute liver failure in a rat model of Wilson disease. J Clin Invest 2016;126(7):2721–35.
[29] Zischka H, et al. Liver mitochondrial membrane crosslinking and destruction in a rat model of Wilson disease. J Clin Invest 2011;121(4):1508–18.
[30] Summer KH, Eisenburg J. Low content of hepatic reduced glutathione in patients with Wilson's disease. Biochem Med 1985;34(1):107–11.
[31] von Herbay A, et al. Low vitamin E content in plasma of patients with alcoholic liver disease, hemochromatosis and Wilson's disease. J Hepatol 1994;20(1):41–6.
[32] Ogihara H, et al. Plasma copper and antioxidant status in Wilson's disease. Pediatr Res 1995;37(2):219–26.
[33] Nair J, et al. Lipid peroxidation-induced etheno-DNA adducts in the liver of patients with the genetic metal storage disorders Wilson's disease and primary hemochromatosis. Cancer Epidemiol Biomarkers Prev 1998;7(5):435–40.
[34] Gu M, et al. Oxidative-phosphorylation defects in liver of patients with Wilson's disease. Lancet 2000;356(9228):469–74.
[35] Nagasaka H, et al. Relationship between oxidative stress and antioxidant systems in the liver of patients with Wilson disease: hepatic manifestation in Wilson disease as a consequence of augmented oxidative stress. Pediatr Res 2006;60(4):472–7.

[36] Bruha R, et al. Decreased serum antioxidant capacity in patients with Wilson disease is associated with neurological symptoms. J Inherit Metab Dis 2012;35(3):541–8.

[37] Kalita J, et al. A study of oxidative stress, cytokines and glutamate in Wilson disease and their asymptomatic siblings. J Neuroimmunol 2014;274(1–2):141–8.

[38] Kalita J, et al. Role of oxidative stress in the worsening of neurologic Wilson disease following chelating therapy. Neuromolecular Med 2015;17(4):364–72.

[39] Oe S, et al. Copper induces hepatocyte injury due to the endoplasmic reticulum stress in cultured cells and patients with Wilson disease. Exp Cell Res 2016;347(1):192–200.

[40] Yamada T, et al. Elevation of the level of lipid-peroxidation associated with hepatic-injury in Lec mutant rat. Res Commun Chem Pathol Pharmacol 1992;77(1):121–4.

[41] Yamamoto F, et al. Elevated level of 8-hydroxydeoxyguanosine in DNA of liver, kidneys, and brain of Long-Evans Cinnamon rats. Jpn J Cancer Res 1993;84(5):508–11.

[42] Stephenson GF, Chan HM, Cherian MG. Copper-metallothionein from the toxic milk mutant mouse enhances lipid peroxidation initiated by an organic hydroperoxide. Toxicol Appl Pharmacol 1994;125(1):90–6.

[43] Ohhira M, et al. Changes in free radical-metabolizing enzymes and lipid peroxides in the liver of Long-Evans with Cinnamon-like coat color rats. J Gastroenterol 1995;30(5):619–23.

[44] Nair J, et al. Copper-dependent formation of miscoding etheno-DNA adducts in the liver of Long Evans Cinnamon (LEC) rats developing hereditary hepatitis and hepatocellular carcinoma. Cancer Res 1996;56(6):1267–71.

[45] Yamamoto Y, et al. Oxidative stress in LEC rats evaluated by plasma antioxidants and free fatty acids. J Trace Elem Exp Med 1997;10(2):129–34.

[46] Suzuki KT, et al. Production of ascorbate and hydroxyl radicals in the liver of LEC rats in relation to hepatitis. Res Commun Mol Pathol Pharmacol 1997;96(2):137–46.

[47] Nakamura M, et al. Metal-induced hydroxyl radical generation by Cu(+)-metallothioneins from LEC rat liver. Biochem Biophys Res Commun 1997;231(3):549–52.

[48] Rui M, Suzuki KT. Copper in plasma reflects its status and subsequent toxicity in the liver of LEC rats. Res Commun Mol Pathol Pharmacol 1997;98(3):335–46.

[49] Chung FL, et al. Endogenous formation and significance of 1,N2-propanodeoxyguanosine adducts. Mutat Res 1999;424(1–2):71–81.

[50] Yamamoto H, et al. Mechanism of enhanced lipid peroxidation in the liver of Long-Evans Cinnamon (LEC) rats. Arch Toxicol 1999;73(8–9):457–64.

[51] Yamamoto H, et al. In vivo evidence for accelerated generation of hydroxyl radicals in liver of Long-Evans Cinnamon (LEC) rats with acute hepatitis. Free Radic Biol Med 2001;30(5):547–54.

[52] Shishido N, et al. Cu-metallothioneins (Cu(I)8-MTs) in LEC rat livers 13 weeks after birth still act as antioxidants. Arch Biochem Biophys 2001;387(2):216–22.

[53] Klein D, et al. Gene expression in the liver of Long-Evans Cinnamon rats during the development of hepatitis. Arch Toxicol 2003;77(10):568–75.

[54] Choudhury S, et al. Evidence of alterations in base excision repair of oxidative DNA damage during spontaneous hepatocarcinogenesis in Long Evans Cinnamon rats. Cancer Res 2003;63(22):7704–7.

[55] Samuele A, et al. Oxidative stress and pro-apoptotic conditions in a rodent model of Wilson's disease. Biochim Biophys Acta 2005;1741(3):325–30.

[56] Nair J, et al. Apoptosis and age-dependant induction of nuclear and mitochondrial etheno-DNA adducts in Long-Evans Cinnamon (LEC) rats: enhanced DNA damage by dietary curcumin upon copper accumulation. Carcinogenesis 2005;26(7):1307–15.

[57] Yasuda J, et al. Reactive oxygen species modify oligosaccharides of glycoproteins in vivo: a study of a spontaneous acute hepatitis model rat (LEC rat). Biochem Biophys Res Commun 2006;342(1):127–34.

[58] Jia G, et al. Aldo-keto reductase 1 family B7 is the gene induced in response to oxidative stress in the livers of Long-Evans Cinnamon rats. Int J Oncol 2006;29(4):829–38.

[59] Huster D, et al. Consequences of copper accumulation in the livers of the Atp7b-/- (Wilson disease gene) knockout mice. Am J Pathol 2006;168(2):423–34.

[60] Marquez-Quinones A, et al. Proteasome activity deregulation in LEC rat hepatitis: following the insights of transcriptomic analysis. OMICS 2007;11(4):367–84.

[61] Roberts EA, Robinson BH, Yang S. Mitochondrial structure and function in the untreated Jackson toxic milk (tx-j) mouse, a model for Wilson disease. Mol Genet Metab 2008;93(1):54–65.

[62] Marquez-Quinones A, et al. HNE-protein adducts formation in different pre-carcinogenic stages of hepatitis in LEC rats. Free Radic Res 2010;44(2):119–27.

[63] Wang J, et al. Quantification of oxidative DNA lesions in tissues of Long-Evans Cinnamon rats by capillary high-performance liquid chromatography-tandem mass spectrometry coupled with stable isotope-dilution method. Anal Chem 2011;83(6):2201–9.

[64] Karmahapatra SK, et al. Redox regulation of apurinic/apyrimidinic endonuclease 1 activity in Long-Evans Cinnamon rats during spontaneous hepatitis. Mol Cell Biochem 2014;388(1–2):185–93.

[65] Yu Y, et al. Comprehensive assessment of oxidatively induced modifications of DNA in a rat model of human Wilson's disease. Mol Cell Proteomics 2016;15(3):810–7.

[66] Kumar V, et al. Relationship of antioxidant and oxidative stress markers in different organs following copper toxicity in a rat model. Toxicol Appl Pharmacol 2016;293:37–43.

[67] Zischka H, Lichtmannegger J. Pathological mitochondrial copper overload in livers of Wilson's disease patients and related animal models. Ann N Y Acad Sci 2014;1315:6–15.

[68] Burkhead JL, Gray LW, Lutsenko S. Systems biology approach to Wilson's disease. Biometals 2011;24(3):455–66.
[69] Medici V, Huster D. Animal models of Wilson disease. Handb Clin Neurol 2017;142:57–70.
[70] Togashi Y, et al. D-penicillamine prevents the development of hepatitis in Long-Evans Cinnamon rats with abnormal copper metabolism. Hepatology 1992;15(1):82–7.
[71] Jong-Hon K, et al. Prevention of spontaneous hepatocellular carcinoma in Long-Evans Cinnamon rats with hereditary hepatitis by the administration of D-penicillamine. Hepatology 1993;18(3):614–20.
[72] Suzuki KT, et al. Selective removal of copper bound to metallothionein in the liver of LEC rats by tetrathiomolybdate. Toxicology 1993;83(1–3):149–58.
[73] Sugawara N, Sugawara C. A copper deficient diet prevents hepatic copper accumulation and dysfunction in Long-Evans Cinnamon (LEC) rats with an abnormal copper metabolism and hereditary hepatitis. Arch Toxicol 1994;69(2):137–40.
[74] Sone K, et al. Inhibition of hereditary hepatitis and liver tumor development in Long-Evans Cinnamon rats by the copper-chelating agent trientine dihydrochloride. Hepatology 1996;23(4):764–70.
[75] Sugawara N, Lai YR, Sugawara C. Therapeutic effects of tetrathiomolybdate on hepatic dysfunction occurring naturally in Long-Evans Cinnamon (LEC) rats: a bona fide animal model for Wilson's disease. Res Commun Mol Pathol Pharmacol 1999;103(2):177–87.
[76] Sugawara N, et al. The effect of subcutaneous tetrathiomolybdate administration on copper and iron metabolism, including their regional redistribution in the brain, in the Long-Evans Cinnamon rat, a bona fide animal model for Wilson's disease. Pharmacol Toxicol 1999;84(5):211–7.
[77] Sugawara N, Katakura M, Sugawara C. Preventive effect of zinc compounds, polaprezinc and zinc acetate against the onset of hepatitis in Long-Evans Cinnamon rat. Res Commun Mol Pathol Pharmacol 1999;103(2):167–76.
[78] Klein D, et al. Dissolution of copper-rich granules in hepatic lysosomes by D-penicillamine prevents the development of fulminant hepatitis in Long-Evans Cinnamon rats. J Hepatol 2000;32(2):193–201.
[79] Santon A, et al. Relationship between metallothionine and zinc in the protection against DNA damage in zinc-treated Long-Evans Cinnamon rat liver. Eur J Histochem 2004;48(3):317–20.
[80] Klein D, et al. Tetrathiomolybdate in the treatment of acute hepatitis in an animal model for Wilson disease. J Hepatol 2004;40(3):409–16.
[81] Medici V, et al. Efficacy of zinc supplementation in preventing acute hepatitis in Long-Evans Cinnamon rats. Liver Int 2005;25(4):888–95.
[82] Gonzalez BP, et al. Zinc supplementation decreases hepatic copper accumulation in LEC rat: a model of Wilson's disease. Biol Trace Elem Res 2005;105(1–3):117–34.
[83] Shimada H, et al. Protection from spontaneous hepatocellular damage by N-benzyl-D-glucamine dithiocarbamate in Long-Evans Cinnamon rats, an animal model of Wilson's disease. Toxicol Appl Pharmacol 2005;202(1):59–67.
[84] Summer KH, et al. The biogenic methanobactin is an effective chelator for copper in a rat model for Wilson disease. J Trace Elem Med Biol 2011;25(1):36–41.
[85] Yamazaki K, et al. Effects of dietary vitamin E on clinical course and plasma glutamic oxaloacetic transaminase and glutamic pyruvic transaminase activities in hereditary hepatitis of LEC rats. Lab Anim Sci 1993;43(1):61–7.
[86] Hawkins RL, et al. Proline, ascorbic acid, or thioredoxin affect jaundice and mortality in Long Evans Cinnamon rats. Pharmacol Biochem Behav 1995;52(3):509–15.
[87] Yamashita T, et al. The effects of alpha-phenyl-tert-butyl nitrone (PBN) on copper-induced rat fulminant hepatitis with jaundice. Free Radic Biol Med 1996;21(6):755–61.
[88] Shibata T, et al. Unsaturated fatty acid feeding prevents the development of acute hepatitis in Long-Evans Cinnamon (LEC) rats. Anticancer Res 1999;19(6B):5169–74.
[89] Yamamoto H, et al. The antioxidant effect of DL-alpha-lipoic acid on copper-induced acute hepatitis in Long-Evans Cinnamon (LEC) rats. Free Radic Res 2001;34(1):69–80.
[90] Watanabe S, et al. Effects of lycopene and Sho-saiko-to on hepatocarcinogenesis in a rat model of spontaneous liver cancer. Nutr Cancer 2001;39(1):96–101.
[91] Kitade Y, et al. Inhibition of liver fibrosis in LEC rats by a carotenoid, lycopene, or a herbal medicine, Sho-saiko-to. Hepatol Res 2002;22(3):196–205.
[92] Frank N, et al. No prevention of liver and kidney tumors in Long-Evans Cinnamon rats by dietary curcumin, but inhibition at other sites and of metastases. Mutat Res 2003;523-524:127–35.
[93] Du C, et al. Dietary polyunsaturated fatty acids suppress acute hepatitis, alter gene expression and prolong survival of female Long-Evans Cinnamon rats, a model of Wilson disease. J Nutr Biochem 2004;15(5):273–80.
[94] Kitamura Y, et al. Effects of N-acetylcysteine, quercetin, and phytic acid on spontaneous hepatic and renal lesions in LEC rats. Toxicol Pathol 2005;33(5):584–92.
[95] Shibata T, et al. Inhibitory effects of fermented brown rice and rice bran on the development of acute hepatitis in Long-Evans Cinnamon rats. Oncol Rep 2006;15(4):869–74.
[96] Asanuma T, et al. A new amphiphilic derivative, N-{[4-(lactobionamido)methyl]benzylidene}-1,1-dimethyl-2-(octylsulfanyl)ethylamin e N-oxide, has a protective effect against copper-induced fulminant hepatitis in Long-Evans Cinnamon rats at an extremely low concentration compared with its original form alpha-phenyl-N-(tert-butyl) nitrone. Chem Biodivers 2007;4(9):2253–67.
[97] Katayama M, et al. Coffee consumption delays the hepatitis and suppresses the inflammation related gene expression in the Long-Evans Cinnamon rat. Clin Nutr 2014;33(2):302–10.
[98] Kasai N, et al. Effects of sex hormones on fulminant hepatitis in LEC rats: a model of Wilson's disease. Lab Anim Sci 1992;42(4):363–8.
[99] Yokoi T, et al. Effects of cyclosporin-A and D-penicillamine on the development of hepatitis and the production of antibody to protein disulfide isomerase in LEC rats. Res Commun Mol Pathol Pharmacol 1994;85(1):73–81.

[100] Yokoi T, et al. Occurrence of autoimmune antibodies to liver microsomal proteins associated with lethal hepatitis in LEC rats: effects of TJN-101 ((+)-(6S,7S,R-biar)- 5,6,7,8-tetrahydro-1,2,3,12-tetramethoxy-6,7-dimethyl-10,11- methylenedioxy-6-dibenzo[a,c]cyclooctenol) on the development of hepatitis and the autoantibodies. Toxicol Lett 1995;76(1):33–8.

[101] Kato J, et al. Hepatic iron deprivation prevents spontaneous development of fulminant hepatitis and liver cancer in Long-Evans Cinnamon rats. J Clin Invest 1996;98(4):923–9.

[102] Sugawara N, Sugawara C. An iron-deficient diet stimulates the onset of the hepatitis due to hepatic copper deposition in the Long-Evans Cinnamon (LEC) rat. Arch Toxicol 1999;73(7):353–8.

[103] Sheline CT, et al. Cofactors of mitochondrial enzymes attenuate copper-induced death in vitro and in vivo. Ann Neurol 2002;52(2):195–204.

[104] Chang B, et al. L-Carnitine inhibits hepatocarcinogenesis via protection of mitochondria. Int J Cancer 2005;113(5):719–29.

[105] Park WH, Lee SK, Kim CH. A Korean herbal medicine, Panax notoginseng, prevents liver fibrosis and hepatic microvascular dysfunction in rats. Life Sci 2005;76(15):1675–90.

[106] Otsuka T, et al. Prevention of lethal hepatic injury in Long-Evans Cinnamon (LEC) rats by D-galactosamine hydrochloride. J Med Invest 2006;53(1–2):81–6.

[107] Tsubota A, et al. Bovine lactoferrin potently inhibits liver mitochondrial 8-OHdG levels and retrieves hepatic OGG1 activities in Long-Evans Cinnamon rats. J Hepatol 2008;48(3):486–93.

[108] Pfeiffenberger J, et al. Hepatobiliary malignancies in Wilson disease. Liver Int 2015;35(5):1615–22.

[109] van Meer S, et al. No increased risk of hepatocellular carcinoma in cirrhosis due to Wilson disease during long-term follow-up. J Gastroenterol Hepatol 2015;30(3):535–9.

[110] Klein D, et al. Association of copper to metallothionein in hepatic lysosomes of Long-Evans Cinnamon (LEC) rats during the development of hepatitis [se e comments]. Eur J Clin Invest 1998;28(4):302–10.

[111] Sternlieb I. Mitochondrial and fatty changes in hepatocytes of patients with Wilson's disease. Gastroenterology 1968;55(3):354–67.

Part III

Epidemiology

Chapter 7

Epidemiology of Wilson Disease

Thomas Damgaard Sandahl and Peter Ott
Department of Hepatology and Gastroenterology, Aarhus University Hospital, Aarhus, Denmark

INTRODUCTION

This chapter will address classical epidemiological aspects of Wilson disease such as its incidence and prevalence as well as a discussion of the impact of using clinical and genetic methodologies to assess this question.

Owing to the rarity of Wilson disease, most reports of the clinical phenotype of the disease are small and prone to selection bias. The second part of this chapter will therefore include a comprehensive review of the major papers on clinical presentation and phenotype of Wilson disease in an attempt to obtain a global perspective.

Incidence and Prevalence of Wilson Disease

A prevalence of 1:30,000–1:100,000 is often cited and was proposed in 1984 [1]. At Hardy-Weinberg equilibrium, this would correspond to a frequency of disease-causing mutated alleles of approximately 1:180–1:300 and a carrier frequency of 1:90–1:150. However, these estimates were based on limited data [2] and deserve reevaluation.

Diagnostic criteria for Wilson disease have been consistent over a long period [3, 4], while prevalence estimates—based on clinical diagnosis—have increased with time along with diagnostic awareness. But still, they may be biased by an unknown number of undiagnosed cases. To overcome this, population genetics has been employed. At the present stage, clinically and genetically based estimates of prevalence differ to a variable degree. To illustrate this, we will initially describe the results from studies of the Wilson disease population in Sardinia.

THE SARDINIAN CASE

Sardinia is a high-prevalence area for Wilson disease, and different methods have been used to estimate the actual prevalence of Wilson disease on the island. These studies illustrate how the chosen methodology affects the prevalence estimate.

Crude prevalence. Number of known patients in given population. The first Sardinian epidemiological study from 1985 included all 68 patients diagnosed from 1902 to 1983 [5]. From the patients alive on certain dates in 1951, 1961, 1971, and 1981, the authors calculated the prevalences to be 1:58,000, 1:44,000, 1:44,000, and 1:34,000, respectively. The increase in prevalence was likely attributable to a combination of improved diagnostic awareness over time and poor survival of Wilson disease patients in the earlier decades.

Incidence related to number of births. The Sardinian researchers also looked at the number of diagnoses per decade that was steadily increasing [5]. From 1971 to 1981, 16 cases of Wilson disease were diagnosed. With 266,944 births in that period, the incidence relative to births was estimated to be $16/266,944 = 1:16,684$. Assuming normal survival, this will equal the prevalence in the population. That suggested a large number of undiagnosed cases and stimulated diagnostic activity. In accordance, a later study from 2013 [6] included 192 patients corresponding to a prevalence of 1:8700. Thus, increased interest in Wilson disease increased the crude prevalence from 1:58,000 in 1951 to 1:8700 in 2013.

Population based genetics. In the Sardinian Wilson disease population, one mutation (−441/−427del) constitutes 61.7% of all affected alleles [7]. Screening of 5290 Sardinian newborns identified 122 heterozygotes (carriers) with this mutation, corresponding to a 1.15% allelic frequency of that mutation and $1.15\%/0.617 = 1.92\%$ for all Wilson disease-causing mutations in ATP7B. Assuming Hardy-Weinberg equilibrium, the incidence of Wilson disease should be $0.0192^2 = 1:2707$ of births [7]. Assuming normal life expectancy and full penetrance, this figure would also equal the prevalence in the population. A later study using another genetically based methodology reached almost the same result [6].

These reports are intriguing because of the contrast between the high prevalence predicted from the genetic studies (1:2700) [6, 7] and the estimates based on clinical diagnoses (1:16,000 and 1:8700) [5, 6]. The lack of confidence limits makes it hard to judge how different these estimates really are, one explanation would be that 50%–75% of Wilson disease patients in Sardinia were still undiagnosed after 30 years of enhanced diagnostic activity, which is hard to believe. Alternatively, the *penetrance* of Wilson disease-causing mutations in Sardinia is in fact much lower than the generally expected 100%.

PREVALENCE IN OTHER POPULATIONS

The Sardinian case illustrates the implications of methodology on prevalence estimates. The estimates from other parts of the world given below will thus be described as related to the methods employed.

High prevalence areas. Small isolated populations with very high risk of Wilson disease have been identified. Among areas with the highest reported prevalences are a village in Crete (1:15 births) [8], Kalymnos, Greece (1:740) [7], and a remote mountain area in Romania (1:1130) [9]. Wilson disease patients in such areas often present with specific mutations [10–12].

Prevalence in larger areas. Table 1 summarizes a small number of studies where all diagnosed cases have been identified in larger population. These estimates range from 1:250,000 in Germany to 1:34,000 in Sardinia. They will obviously be hampered by undiagnosed cases, and the estimates tend to increase with year of publication presumably due to diagnostic awareness.

Two large register-based studies from France [26] and Taiwan [17] provide important information. The French study included 906 Wilson disease patients with a crude prevalence of 1:63,000. However, in the age group ranging from 20 to 29 years, the prevalence was 1:41,400, and in best health-care areas, such as around Paris and in Lorraine, the crude prevalence was 1:35,100. If a number of elderly patients died undiagnosed without treatment, the prevalence will be somewhat higher in the future. The Taiwan study [17] identified 495 patients; in 2005, 365 were alive corresponding to a prevalence of 1:61,500. However, since more than 50% were diagnosed in 2000–05, the crude prevalence in Taiwan will likely increase with time.

TABLE 1 Crude Prevalence Estimates Based on Diagnosed Cases of Wilson Disease Relative to the Population

Country/Area/Year	Population	Wilson Disease Cases	Prevalence	Reference
Germany/1975			1:370,000	[13]
East Germany/1979	16.9 MIO	126	1:216,000	[14]
Sardinia/1981	1.6 MIO	47	1:34,000	[5]
Park/Scotland/1991	5.1 MIO	21	1:250,000	[2]
Iceland/1995	330	8	1:41,500	[15]
Albania/1995	2.9 MIO	40	1:72,500	[16]
Taiwan/2005	22.5 MIO	365	1:61,500	[17]
Austria/2005	8.1 MIO	149	1:55,000	[18]
Latvia/2008	1.9 MIO	40	1:47,500	[19]
Serbia/2009	7.1 MIO	112	1:63,000	[20]
Denmark/2011	5.5 MIO	49	1:112,000	[21]
Czech Republic/2011	10.6 MIO	113	1:93,805	[22]
Ireland/2014	4.6 MIO	50	1:217,000	[23]
China/2014	153,370	9	1:17,000	[24]
Croatia/2016	4.3 MIO	155	1:28,000	[25][a]
France/2017	56.8 MIO	906	1:63,000	[26]

[a]Including personal communication with Dr. Mirjana Kalauz.

A large Chinese study [24] included screening for Kayser-Fleischer (KF) Rings in 153,370 Han Chinese individuals and identified nine patients with Wilson disease. If the sample was representative for the Han Chinese population, the prevalence was 1:17,000 at minimum, as KF rings are not always present in Wilson disease. However, with only nine cases, even modest selection bias could severely hamper this conclusion.

Saito [27] used a method that was based on population genetics without genetic testing. He used Dahlberg's formula according to which the frequency of disease-causing alleles can be calculated from the ratio of first-cousin marriages in the population and among parents to Wilson disease patients. Based on detailed data from 168 Wilson disease families, an allele frequency of 0.52%–0.60% was calculated corresponding to a prevalence of 1:33,000 in a population without first-cousin marriages and 1:20,000 in a population with 4.6% first-cousin marriages as was the case in Japan at that time. Because first-cousin marriages were declining, he also predicted a declining prevalence of Wilson disease in Japan [27]. Saito's study is widely quoted and difficult to reproduce because the used methodology only works in populations where first-cousin marriages are common and known. Later studies have claimed that a wrong use of Dahlberg's formula overestimated the prevalence [2].

Incidence related to number of births. Table 2 lists studies that estimated the prevalence of Wilson disease from the number of diagnosed cases relative to the number of births. Estimates are only available from Europe and tend to be somewhat higher than crude prevalence estimates, suggestive of increased diagnostic awareness in later decades. They tend to center around 1:40,000–1:50,000 with a higher prevalence in Sardinia of 1:16,700.

BIOCHEMICAL SCREENING FOR WILSON DISEASE

Attempts to screen for Wilson disease by biochemical methods are listed in Table 3. These studies employed detection of ceruloplasmin in either whole dried blood ([33–36]) or urine [37, 38]. Screening of newborns was useless [33], while screening of children 0.5–15 years resulted in estimates of 1:1400–1:8084 (Table 3). These estimates were most likely far too high due to sampling biases because all [34] or a significant proportion [33, 35] of participants were patients in pediatric departments where Wilson disease cases would tend to concentrate. In addition, sample sizes were too low with only one or two cases identified. Screening of 48,819 school children in Japan by urine ceruloplasmin [37] may be the most unbiased sample, but with only two cases, the confidence limits were large 1:24,000 (1:20,100–1:201,600). The same holds true for a smaller study in 3-year-olds [38]. Even large, these studies were too small and too biased to provide useful estimates of the prevalence of Wilson disease. Recent evaluations concluded that a relevant test for screening for Wilson disease is not available at present [39].

POPULATION BASED GENETIC METHODS

The genetic methods take advantage of the ability of current technology to study a large number of individuals. The methods aim to assess the frequency q of disease-causing mutations in the ATP7B gene pool and estimate the prevalence of Wilson disease to be q^2. This assumes Hardy-Weinberg equilibrium (no consanguinity), normal life expectancy of Wilson disease patients, and 100% penetrance of the disease-causing mutations. Consanguinity is relevant in many

TABLE 2 Prevalence Estimates Based on the Number of Presenting Cases Relative to the Number of Births

Country/Area/Period	Cases Diagnosed	No. of Births	Prevalence Estimate	Reference
Switzerland 1946–55	19		1:44,800	[28]
Eastern Germany 1949–77	123	4.2 MIO/28 years	1:34,400	[29]
Western Germany 1962			1:86,000	[13]
Sardinia 1971–81	16	266,944	1:16,700	[5]
Ireland 1980–89	12	637	1:53,000	[30]
Denmark 1990–2009	28	70,000/year	1:49,500	[31]
Czech Republic 1995–2008	76	95,000/year	1:16,500	[22]
Poland 1996–2016	156	385,000/year	1:49,000	[32]

TABLE 3 Screening for Wilson Disease by Measurement of Ceruloplasmin in Blood or Urine

Year	Country	No	WD	Prevalence	Reference
Blood ceruloplasmin					
1999	Japan	126,810	0	0	[33]
1999	Japan	24,165	3	1:8084	[33]
1999	Japan	2789	2	1:1400	[34]
2002	South Korea	3667	1	1:3667	[35]
2006	The United States	1398	0	0	[36]
Urine holoceruloplasmin					
2002	Japan	48,819	2	1:24,400	[37]
Urine ceruloplasmin					
2008	Japan	11,362	1	1:11,362	[38]

TABLE 4 Predicted Prevalence of Wilson Disease as Function of the Frequency of Disease-Causing Alleles, q, and Percentage of First-Cousin Marriages in a Theoretical Population

		\multicolumn{5}{c}{Percentage of First-Cousin Marriages}				
q	Carrier Frequency	0	0.05	0.1	0.25	0.5
1:300	1:151	1:90,000	1:47,761	1:32,506	1:16,599	1:9143
1:200	1:101	1:40,000	1:25,447	1:18,659	1:10,364	1:5953
1:100	1:51	1:10,000	1:7940	1:6584	1:4354	1:2783
1:50	1:26	1:2500	1:2266	1:2073	1:1649	1:1231

Full penetrance and normal life expectancy are assumed, and other sources of consanguinity are neglected.

countries and may increase the prevalence dramatically when q is 1:100 or lower (Table 4). First-cousin parents have been reported in 32%–54% of Wilson disease patients [40, 41] illustrating the importance of this. Today, the life expectancy in Wilson disease is close to normal. The 100% penetrance assumption will be discussed further below.

In Sweden, the H1069Q or the T977M mutations constituted 44% of disease-causing alleles in 24 unrelated Swedish Wilson disease patients [42]. Among 2460 blood donors, these alleles together constituted 0.175% of ATP7B alleles [43]. From that, a frequency of disease-causing alleles of 0.175%/0.44 = 0.39% can be calculated corresponding to a prevalence of homozygotes or compound heterozygotes of 1:63,175. The confidence interval must be large because only nine mutated alleles were detected in the donor pool.

In a similar study from the United States [44], DNAs from 2601 newborns were examined for the H1069Q mutation that accounts for 1/3 of mutated alleles in US Caucasian Wilson disease patients. Seven heterozygotes were found in 2456 successfully analyzed samples, leading to a prevalence estimate of 1:50,000, again with large confidence limits (1:18,000–1:700,000).

A study from the United Kingdom [45] apparently challenged this conclusion. The British reference center for genetic testing for Wilson disease sequenced ATP7B from blood sampled from 1000 newborns (1K sample) supplied by a regional screening program. They identified 24 alleles with single-nucleotide variants predicted by a computer program to cause dysfunction of ATP7B. From this observation, they calculated a prevalence of Wilson disease of 1:7026. No confidence limits were provided, but we calculate 95% CI, 1:1800–1:15,000. However, only 12 of the alleles were known to cause Wilson disease, and based on that, we estimate a prevalence of 1:28,000 (95% CI, 1:9100–1:104,000). Because of the wide confidence limits, it is uncertain whether this estimate is really different from US estimate of 1:50,000 [44].

The UK study raises some questions. Half of the single-nucleotide variants that were predicted by the computer program to cause dysfunction of ATP7B have never been identified in Wilson disease patients. The study clearly demonstrated that the clinical "penetrance" of the computer predicted disease-causing variants is not 100%. Taking this into account, the UK and US studies are not clearly different from those based on clinical diagnosis. The UK study also included more limited analysis of 5376 samples (the 5K sample) in which they examined only exons 8, 14, and 18 and identified 23 alleles with possibly disease-causing mutations. Since 51% of known disease-causing alleles in the United Kingdom originated from these exons, we calculate a frequency of disease-causing alleles in that sample to be $(23/0.51)/(2 \times 5376)$ or 0.0042 and an estimated prevalence of 1:57,000, again in good agreement with the US study [44].

Four studies from South Korea using the same methodology [46–49] and based on almost identical mutations reported prevalence estimates ranging from 1:3000 to 1:31,000. This illustrates the uncertainty of these methods. Still, it could be argued that the largest study including 14,835 samples from newborns and a prevalence estimate of 1:7500 [46] suggest a higher prevalence in South Korea than in Europe. The same may be true for other East Asian populations where the R778L mutation is the most prevalent [50].

WHAT TO LEARN FROM DIFFERENT STUDIES OF PREVALENCE?

In large studies, the crude prevalence was around 1:40,000–60,000 in most countries. These estimates will still be biased by undiagnosed patients, at least in the past. Estimates based on the number of diagnosed cases relative to the number of births are only available from Europe and tend to center around 1:40,000–1:50,000, except for high-prevalence areas such as Sardinia and Czech Republic with estimates around 1:16,000. Genetically based studies provided somewhat conflicting result. In the Caucasian population, prevalence estimates ranged from 1:55,000 in the US study [44] to 1:28,000–1:57,000 in the UK study [45]. East Asian genetic studies reported varying results but could suggest a higher prevalence around 1:15,000. However, these studies were severely underpowered.

Reports from the best studied area, Sardinia, suggest that the genetic prevalence estimate may be four times higher than the actual prevalence of clinical Wilson disease. This raises the question whether the penetrance is really 100% in this specific population and in other populations. A few patients have been described that apparently never developed symptomatic disease [51, 52]. In a Polish case report, a 54-year-old woman was examined because her four siblings had Wilson disease and neurological symptoms [52]. She had KF ring and abnormal ^{64}Cu uptake into ceruloplasmin and was homozygous for the H1069Q mutation. She refused treatment and never developed symptoms. ALT was never higher than 64 IU/L, and MRI of the brain was normal at the final examination 30 years later at the age of 84. Because asymptomatic siblings to Wilson disease patients routinely receive treatment, we do not know how common this situation is. So-called asymptomatic siblings will often have some type of phenotype, such as fibrosis at liver biopsy, and we cannot rule out that it was also the case with the 84-year-old lady described above. Also, we do not know her intake of copper and zinc that may have protected her. However, this case raises the question of penetrance. If penetrance is not 100%, the diagnostic criteria will have to be revised [3].

From these data, the original prevalence estimate from 1984 of 1:30,000–1:50,000 [1] still appears to be valid for the Caucasian population except for high-prevalence areas such as Sardinia. Genetic studies in Caucasian populations are currently not in disagreement with these estimates. With a disease-causing allele frequency around 1:200, the rate of first-cousin parenthood will significantly affect the prevalence locally. However scarce, the data could suggest a somewhat higher prevalence in East Asia, maybe as high as 1:7500–1:15,000. For most other populations, useful data are missing. The question if the penetrance is lower than 100% will be an important focus for future research.

CLINICAL PRESENTATION
Age at Presentation

Wilson disease is usually diagnosed in the age range 5–35 years [53], but the full range is rather broad, and Wilson disease should be considered at any age if symptoms are suggestive. Thus, the two youngest symptomatic cases reported were 2- [54] and 3-year-old children [55]. The oldest reported cases are a 72-year-old woman and her 74-year-old brother [56]. In larger cohorts, 8%–12% of Wilson disease patients is diagnosed after the age of 35 [57, 58], and in a large European cohort, 4% of patients with neurological Wilson disease were diagnosed after the age of 40 [59].

On average, age of onset is later in neurological than in hepatic Wilson disease patients. Thus, in the Polish study, mean age of onset was 28.2 ± 8 years in neurological and 23.7 ± 8 in hepatic Wilson disease patients [60]. Increased diagnostic awareness should reduce the age of onset, and accordingly, in a German cohort [61], the mean age of first symptom was 20.2 years in neurological patients and 15.5 years in hepatic.

Presenting Phenotype

Wilson disease patients have been described in most parts of the world, with the peculiar exception of Africa, where only a small Egyptian cohort [62] and two case reports from Nigeria [63, 64] are found. While Wilson disease is always caused by mutations in the ATP7B gene, the most common disease mutations differ across global regions [12, 65–67]. Nevertheless, the clinical presentations seem to be rather similar.

Studies reporting the clinical presentations are summarized in Table 5. Only few are population-based. The wide scatter most likely reflects sampling bias with more neurological cases seen in neurological departments and more hepatic cases

TABLE 5 Phenotype Distribution in Different Areas

Country	n	Neuropsychiatric	Chronic Liver Disease	Acute Liver Failure	Hemolysis	Asymptomatic Sibling	Reference
Europe total	1782	45%	41%	5%	1%	9%	
Poland	614	56%	27%			17%	[60]
Sweden	28	64%	25%	0%	4%	14%	[42]
Czech and Slovakia	227	33%	63%			4%	[68]
East Germany	72	64%	31%			6%	[69]
Denmark	49	35%	35%	20%	2%	8%	[21]
The United Kingdom	34	26%	35%	12%	6%	21%	[70]
West Germany	163	34%	54%	5%		7%	[61]
Austria	229	27%	56%	5%	0%	12%	[71]
Spain	15	33%	47%	20%	0%	0%	[72]
Ireland	50	50%	32%		0%	18%	[72]
Hungary	42	57%	43%		0%	0%	[30]
Czech Republic	117	47%	44%			9%	[22]
Serbia	142	61%	35%			4%	[20]
Middle East total	195	39%	49%			12%	
Egypt	77	19%	45%	0%		35%	[62]
Saudi Arabia	71	45%	55%				[73]
Pakistan	47	53%	47%				[74]
Americas		52.5%	33%			15%	
Canada	48	44%	27%			29%	[75]
Brazil	36	61%	39%				[76]
Asia total	346	58%	28%	0%	3%	11%	
Bangalore, India	282	78%	16%			5%	[41]
China	39	44%	49%		5%	3%	[77]
India	25	52%	20%	0%	4%	24%	[78]
World total		47%	39%	3%	1%	10%	

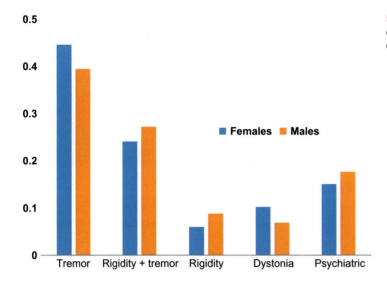

FIG. 1 The relative distribution of presenting symptoms in a cohort of 427 patients with neurological phenotype of Wilson disease (166 females/261 males) [60].

TABLE 6 Symptoms in Patients With Predominantly Hepatic Wilson Disease [53, 86, 92, 95–98]

Jaundice, anorexia, and vomiting	14%–44%
Ascites/edema	14%–50%
Variceal hemorrhage	3%–8%
Hemorrhagic diathesis	3%–10%
Hemolysis	5%–20%
Hepatosplenomegaly	15%–49%
Acute liver failure	7%–25%
Asymptomatic	5%–23%

Range of percentages is reported. Patient may present with more than one symptom.

reported from hepatological departments. As judged by Table 5, the distribution among symptomatic cases is approximately 50% with neurological/psychiatric disease, 40% with liver disease, and only few with hemolysis or other rare manifestations. There are no obvious global differences.

The dominant neurological symptoms are illustrated in Fig. 1 and include tremor, rigidity, and dystonia. Tremor was seen in 70%–75%. Around 25% of neurological patients present with psychiatric symptoms [60]. Patients with neurological presentation will often have elevated liver function tests and abnormal hepatic histology even if they have never had liver-related complaints. In one study, the liver biopsy was abnormal in 31 out of 34 patients with neurological presentation, and 14 had cirrhosis [18].

Patients with hepatic Wilson disease may present with a variety of symptoms as illustrated in Table 6.

Interestingly, the clinical presentation is age-dependent with more hepatic phenotypes in children. Thus, in larger series of children, hepatic presentation constituted 94% [32], 84% [79], 69% [80], and 67% [81]. It is likely that some of those children would have developed neurological symptoms and had their diagnosis been delayed. Consequently, a higher diagnostic awareness among pediatricians will change the picture toward the hepatological presentations.

Gender Differences

Males may be slightly more at risk to develop clinical Wilson disease as judged by larger series with male-female ratios of 1.28:1 ($n=494$) in Taiwan [17] and 1.13:1 ($n=617$) in Poland [60], while this was not the case in France (1.04:1, $n=906$) [26] or the European database (1.02:1, $n=1025$) [82]. Females are more often diagnosed with hepatic presentation than males. Thus, hepatic presentation was seen in 41% of female patients versus 25% of males in the Polish cohort [60] and in 75% of females versus 65% of males in the European database [82]. Females were diagnosed 2–3 years later than males and presented with higher 24h urine Cu [17, 26, 83]. Also, females were more likely to present with acute liver failure, 8F/0M in a German cohort [61] and 41F/12M in the European database [82]. It was speculated [60] that estrogen interference with brain dopamine and iron metabolism could be involved, but basically, these gender effects are poorly understood.

Diagnostic Delay

As pointed out by Walshe in 1992 [84], patients with Wilson disease often experience a substantial "diagnostic delay" from first sign of Wilson disease to diagnosis. Reports of this delay range from 10.5 to 25 months [2, 21, 60, 61, 84]. Individual patients had experienced delays up to 10 years [21, 84]. As the disease will progress during this period, the diagnostic delay is of clinical concern. Reduction of this delay requires the inclusion of Wilson disease in algorithms for the examination of abnormal liver function tests, Coombs-negative hemolysis, or relevant neuropsychiatric disease.

PROGNOSIS OF WILSON DISEASE

Randomized studies have never been performed, but before the introduction of D-penicillamine, median survival was 6–7 years [22], and the survival rate is now approaching that of the background population. Thus, in a large Austrian cohort of 225 patients diagnosed from 1961 to 2013 [71], the age- and sex-adjusted survival rates 10, 20, 30, and 40 years after diagnosis were 0.96 ± 0.01, 0.92 ± 0.02, 0.88 ± 0.04, and 0.83 ± 0.06, respectively. In comparison, Wilson disease patient survival rate was 0.94 relative to the background population in the Polish cohort [85], equal to the background population in the Czech Republic [22] and Germany [86], while a somewhat worse outcome was reported in Serbia [20]. The best predictor of liver transplantation or death was cirrhosis at the time of diagnosis [71]. The long-term survival in patients undergoing liver transplantation for Wilson disease is comparable with other indications. [87–94]. The indications are acute liver failure in around 1/3 of cases and chronic liver disease in 2/3. These data suggest that the current treatment of Wilson disease provides excellent survival even though a few patients may need liver transplantation.

ACKNOWLEDGMENT

The work was supported by a grant from The Memorial Foundation of Manufacturer Vilhelm Pedersen and wife.

REFERENCES

[1] Scheinberg IH, Sternlieb I. Wilson's disease. Philadelphia: WB Saunders; 1984.
[2] Park RH, et al. Wilson's disease in Scotland. Gut 1991;32(12):1541–5.
[3] Ferenci P, et al. Diagnosis and phenotypic classification of Wilson disease. Liver Int 2003;23(3):139–42.
[4] Scheinberg IH, Sternlieb I. Wilson's disease. Annu Rev Med 1965;16:119–34.
[5] Giagheddu A, et al. Epidemiologic study of hepatolenticular degeneration (Wilson's disease) in Sardinia (1902-1983). Acta Neurol Scand 1985;72(1):43–55.
[6] Gialluisi A, et al. The homozygosity index (HI) approach reveals high allele frequency for Wilson disease in the Sardinian population. Eur J Hum Genet 2013;21(11):1308–11.
[7] Zappu A, et al. High incidence and allelic homogeneity of Wilson disease in 2 isolated populations: a prerequisite for efficient disease prevention programs. J Pediatr Gastroenterol Nutr 2008;47(3):334–8.
[8] Dedoussis GV, et al. Wilson disease: high prevalence in a mountainous area of Crete. Ann Hum Genet 2005;69(Pt 3):268–74.
[9] Cocos R, et al. Genotype-phenotype correlations in a mountain population community with high prevalence of Wilson's disease: genetic and clinical homogeneity. PLoS One 2014;9(6).
[10] Hofer H, et al. Identification of a novel Wilson disease gene mutation frequent in Upper Austria: a genetic and clinical study. J Hum Genet 2012;57(9):564–7.
[11] Panichareon B, et al. Six novel ATP7B mutations in Thai patients with Wilson disease. Eur J Med Genet 2011;54(2):103–7.
[12] Ferenci P. Regional distribution of mutations of the ATP7B gene in patients with Wilson disease: impact on genetic testing. Hum Genet 2006;120(2):151–9.

[13] Przuntek H, Hoffmann E. Epidemiologic study of Wilson's disease in West Germany. Nervenarzt 1987;58(3):150–7.
[14] Bachmann H, Lossner J, Biesold D. Wilson's disease in the German Democratic Republic. I. Genetics and epidemiology. Z Gesamte Inn Med 1979;34(24):744–8.
[15] Thomas GR, et al. Wilson disease in Iceland: a clinical and genetic study. Am J Hum Genet 1995;56(5):1140–6.
[16] Adhami EJ, Cullufi P. Wilson's disease in Albania. Panminerva Med 1995;37(1):18–21.
[17] Lai CH, Tseng HF. Population-based epidemiologic study of Wilson's disease in Taiwan. Eur J Neurol 2010;17(6):830–3.
[18] Ferenci P, et al. Diagnostic value of quantitative hepatic copper determination in patients with Wilson's disease. Clin Gastroenterol Hepatol 2005;3(8):811–8.
[19] Krumina A, et al. From clinical and biochemical to molecular genetic diagnosis of Wilson disease in Latvia. Genetika 2008;44(10):1379–84.
[20] Svetel M, et al. Long-term outcome in Serbian patients with Wilson disease. Eur J Neurol 2009;16(7):852–7.
[21] Moller LB, et al. Clinical presentation and mutations in Danish patients with Wilson disease. Eur J Hum Genet 2011;19(9):935–41.
[22] Bruha R, et al. Long-term follow-up of Wilson disease: natural history, treatment, mutations analysis and phenotypic correlation. Liver Int 2011;31(1):83–91.
[23] Reilly M, Daly L, Hutchinson M. An epidemiological study of Wilson's disease in the Republic of Ireland. J Neurol Neurosurg Psychiatry 1993;56(3):298–300.
[24] Cheng N, et al. Wilson disease in the South Chinese Han population. Can J Neurol Sci 2014;41(3):363–7.
[25] Ljubic H, et al. ATP7B gene mutations in Croatian patients with Wilson disease. Genet Test Mol Biomarkers 2016;20(3):112–7.
[26] Poujois A, et al. Characteristics and prevalence of Wilson's disease: a 2013 observational population-based study in France. Clin Res Hepatol Gastroenterol 2018;42(1):57–63.
[27] Saito T. An expected decrease in incidence of Wilson's disease due to decrease in consanguinity. Jpn J Hum Genet 1985;30(3):249–53.
[28] Tschumi A, Colombo JP, Moser H. Wilson's disease in Switzerland. Clinical, genetic and biochemical studies. Schweiz Med Wochenschr 1973;103(4):140–5 [Conclusion].
[29] Bachmann H, et al. The epidemiology of Wilson's disease in the German Democratic Republic and current problems from the viewpoint of population genetics. Psychiatr Neurol Med Psychol (Leipz) 1979;31(7):393–400.
[30] O'Brien M, et al. Epidemiology of Wilson's disease in Ireland. Mov Disord 2014;29(12):1567–8.
[31] Moller LB, et al. Homozygosity for a gross partial gene deletion of the C-terminal end of ATP7B in a Wilson patient with hepatic and no neurological manifestations. Am J Med Genet A 2005;138(4):340–3.
[32] Naorniakowska M, et al. Clinical presentations of Wilson disease among Polish children. Dev Period Med 2016;20(3):216–21.
[33] Yamaguchi Y, et al. Mass screening for Wilson's disease: results and recommendations. Pediatr Int 1999;41(4):405–8.
[34] Ohura T, et al. Pilot study of screening for Wilson disease using dried blood spots obtained from children seen at outpatient clinics. J Inherit Metab Dis 1999;22(1):74–80.
[35] Hahn SH, et al. Pilot study of mass screening for Wilson's disease in Korea. Mol Genet Metab 2002;76(2):133–6.
[36] Kroll CA, et al. Retrospective determination of ceruloplasmin in newborn screening blood spots of patients with Wilson disease. Mol Genet Metab 2006;89(1–2):134–8.
[37] Owada M, et al. Mass screening for Wilson's disease by measuring urinary holoceruloplasmin. J Pediatr 2002;140(5):614–6.
[38] Nakayama K, et al. Early and presymptomatic detection of Wilson's disease at the mandatory 3-year-old medical health care examination in Hokkaido Prefecture with the use of a novel automated urinary ceruloplasmin assay. Mol Genet Metab 2008;94(3):363–7.
[39] Hahn SH. Population screening for Wilson's disease. Ann N Y Acad Sci 2014;1315:64–9.
[40] Saito T. An assessment of efficiency in potential screening for Wilson's disease. J Epidemiol Community Health 1981;35(4):274–80.
[41] Taly AB, et al. Wilson disease: description of 282 patients evaluated over 3 decades. Medicine (Baltimore) 2007;86(2):112–21.
[42] Waldenstrom E, et al. Efficient detection of mutations in Wilson disease by manifold sequencing. Genomics 1996;37(3):303–9.
[43] Olsson C, et al. Determination of the frequencies of ten allelic variants of the Wilson disease gene (ATP7B), in pooled DNA samples. Eur J Hum Genet 2000;8(12):933–8.
[44] Olivarez L, et al. Estimate of the frequency of Wilson's disease in the US Caucasian population: a mutation analysis approach. Ann Hum Genet 2001;65(Pt 5):459–63.
[45] Coffey AJ, et al. A genetic study of Wilson's disease in the United Kingdom. Brain 2013;136(Pt 5):1476–87.
[46] Jang JH, et al. Carrier frequency of Wilson's disease in the Korean population: a DNA-based approach. J Hum Genet 2017;62(9):815–8.
[47] Kim GH, et al. Estimation of Wilson's disease incidence and carrier frequency in the Korean population by screening ATP7B major mutations in newborn filter papers using the SYBR green intercalator method based on the amplification refractory mutation system. Genet Test 2008;12(3):395–9.
[48] Park HD, et al. Carrier frequency of the R778L, A874V, and N1270S mutations in the ATP7B gene in a Korean population. Clin Genet 2009;75(4):405–7.
[49] Song MJ, et al. Estimation of carrier frequencies of six autosomal-recessive Mendelian disorders in the Korean population. J Hum Genet 2012;57(2):139–44.
[50] Mak CM, et al. Mutational analysis of 65 Wilson disease patients in Hong Kong Chinese: identification of 17 novel mutations and its genetic heterogeneity. J Hum Genet 2008;53(1):55–63.
[51] Hefter H, et al. Late diagnosis of Wilson's disease in a case without onset of symptoms. Acta Neurol Scand 1995;91(4):302–5.
[52] Czlonkowska A, Rodo M, Gromadzka G. Late onset Wilson's disease: therapeutic implications. Mov Disord 2008;23(6):896–8.
[53] European Association for Study of Liver. EASL Clinical Practice Guidelines: Wilson's disease. J Hepatol 2012;56(3):671–85.
[54] Beyersdorff A, Findeisen A. Morbus Wilson: case report of a two-year-old child as first manifestation. Scand J Gastroenterol 2006;41(4):496–7.

[55] Wilson DC, et al. Severe hepatic Wilson's disease in preschool-aged children. J Pediatr 2000;137(5):719–22.
[56] Ala A, et al. Wilson disease in septuagenarian siblings: raising the bar for diagnosis. Hepatology 2005;41(3):668–70.
[57] Pilloni L, et al. Wilson's disease with late onset. Dig Liver Dis 2000;32(2):180.
[58] Reddy K. Late onset Wilson frequently overlooked. Gastroenterology 2006;131:.
[59] Ferenci P, et al. Late-onset Wilson's disease. Gastroenterology 2007;132(4):1294–8.
[60] Litwin T, Gromadzka G, Czlonkowska A. Gender differences in Wilson's disease. J Neurol Sci 2012;312(1–2):31–5.
[61] Merle U, et al. Clinical presentation, diagnosis and long-term outcome of Wilson's disease: a cohort study. Gut 2007;56(1):115–20.
[62] Abdel Ghaffar TY, et al. Phenotypic and genetic characterization of a cohort of pediatric Wilson disease patients. BMC Pediatr 2011;11:56.
[63] Longe AC, Glew RH, Omene JA. Wilson's disease. Report of a case in a Nigerian. Arch Neurol 1982;39(2):129–30.
[64] Esezobor CI, et al. Wilson disease in a Nigerian child: a case report. J Med Case Reports 2012;6:200.
[65] Gomes A, Dedoussis GV. Geographic distribution of ATP7B mutations in Wilson disease. Ann Hum Biol 2016;43(1):1–8.
[66] Rosencrantz R, Schilsky M. Wilson disease: pathogenesis and clinical considerations in diagnosis and treatment. Semin Liver Dis 2011;31(3):245–59.
[67] Ala A, et al. Wilson's disease. Lancet 2007;369(9559):397–408.
[68] Vrabelova S, et al. Mutation analysis of the ATP7B gene and genotype/phenotype correlation in 227 patients with Wilson disease. Mol Genet Metab 2005;86(1–2):277–85.
[69] Caca K, et al. High prevalence of the H1069Q mutation in East German patients with Wilson disease: rapid detection of mutations by limited sequencing and phenotype-genotype analysis. J Hepatol 2001;35(5):575–81.
[70] Nazer H, et al. Wilson's disease: clinical presentation and use of prognostic index. Gut 1986;27(11):1377–81.
[71] Beinhardt S, et al. Long-term outcomes of patients with Wilson disease in a large Austrian cohort. Clin Gastroenterol Hepatol 2014;12(4):683–9.
[72] Brage A, et al. Clinical and molecular characterization of Wilson disease in Spanish patients. Hepatol Res 2007;37(1):18–26.
[73] Firneisz G, et al. Common mutations of ATP7B in Wilson disease patients from Hungary. Am J Med Genet 2002;108(1):23–8.
[74] Parkash O, et al. Wilson's disease: experience at a tertiary care hospital. J Coll Physicians Surg Pak 2013;23(7):525–6.
[75] Moores A, et al. Wilson disease: Canadian perspectives on presentation and outcomes from an adult ambulatory setting. Can J Gastroenterol 2012;26(6):333–9.
[76] Bem RS, et al. Wilson's disease in southern Brazil: a 40-year follow-up study. Clinics (Sao Paulo) 2011;66(3):411–6.
[77] Gu YH, et al. Mutation spectrum and polymorphisms in ATP7B identified on direct sequencing of all exons in Chinese Han and Hui ethnic patients with Wilson's disease. Clin Genet 2003;64(6):479–84.
[78] Kalra V, Khurana D, Mittal R. Wilson's disease—early onset and lessons from a pediatric cohort in India. Indian Pediatr 2000;37:595–601.
[79] Asadi Pooya AA, Eslami NS, Haghighat M. Wilson disease in southern Iran. Turk J Gastroenterol 2005;16(2):71–4.
[80] Rukunuzzaman M, et al. Childhood Wilson disease: Bangladesh perspective. Mymensingh Med J 2017;26(2):406–13.
[81] Sintusek P, Chongsrisawat V, Poovorawan Y. Wilson's disease in Thai children between 2000 and 2012 at King Chulalongkorn Memorial Hospital. J Med Assoc Thai 2016;99(2):182–7.
[82] Ferenci P, Merle U, Czlonkowska A, Brůha R. Impact of gender on the clinical presentation of Wilson disease. J Hepatol 2010;52:S23–41.
[83] Litwin T, Gromadzka G, Czlonkowska A. Neurological presentation of Wilson's disease in a patient after liver transplantation. Mov Disord 2008;23(5):743–6.
[84] Walshe JM, Yealland M. Wilson's disease: the problem of delayed diagnosis. J Neurol Neurosurg Psychiatry 1992;55(8):692–6.
[85] Czlonkowska A, et al. Wilson's disease-cause of mortality in 164 patients during 1992-2003 observation period. J Neurol 2005;252(6):698–703.
[86] Stremmel W, et al. Wilson disease: clinical presentation, treatment, and survival. Ann Intern Med 1991;115(9):720–6.
[87] Eghtesad B, et al. Liver transplantation for Wilson's disease: a single-center experience. Liver Transpl Surg 1999;5(6):467–74.
[88] Emre S, et al. Orthotopic liver transplantation for Wilson's disease: a single-center experience. Transplantation 2001;72(7):1232–6.
[89] Haberal M, et al. Liver transplantation for Wilson's cirrhosis: one center's experience. Transplant Proc 1999;31(8):3160–1.
[90] Martin AP, et al. A single-center experience with liver transplantation for Wilson's disease. Clin Transplant 2008;22(2):216–21.
[91] Pabon V, et al. Long-term results of liver transplantation for Wilson's disease. Gastroenterol Clin Biol 2008;32(4):378–81.
[92] Schilsky ML, Scheinberg IH, Sternlieb I. Liver transplantation for Wilson's disease: indications and outcome. Hepatology 1994;19(3):583–7.
[93] Sevmis S, et al. Liver transplantation for Wilson's disease. Transplant Proc 2008;40(1):228–30.
[94] Weiss KH, et al. Outcome and development of symptoms after orthotopic liver transplantation for Wilson disease. Clin Transplant 2013;27(6):914–22.
[95] Walshe JM. Wilson's disease presenting with features of hepatic dysfunction: a clinical analysis of eighty-seven patients. Q J Med 1989;70(263):253–63.
[96] Scott J, et al. Wilson's disease, presenting as chronic active hepatitis. Gastroenterology 1978;74(4):645–51.
[97] Steindl P, et al. Wilson's disease in patients presenting with liver disease: a diagnostic challenge. Gastroenterology 1997;113(1):212–8.
[98] Lin L, et al. Hepatic manifestations in Wilson's disease: report of 110 cases. Hepatogastroenterology 2015;62(139):657–60.

Part IV

Diagnosis

Chapter 8

The Diagnostic Approach to Wilson Disease

Michelle Angela Camarata[*,†,‡] and Aftab Ala[†,‡]
*Department of Surgery, Section of Transplant and Immunology, Yale School of Medicine, New Haven, CT, United States, †Department of Gastroenterology and Hepatology, Royal Surrey County Hospital, Guilford, United Kingdom, ‡Department of Clinical and Experimental Medicine, University of Surrey, Guilford, United Kingdom

ESTABLISHING A DIAGNOSIS OF WILSON DISEASE

Apart from genetic testing, there is no single diagnostic test to establish a diagnosis of Wilson disease. The diagnosis of WD should be considered in any person between 3 and 40 years of age that presents with unexplained hepatic, neurological, or psychiatric disease [1]. There are rare cases that have been diagnosed later in life even up to the eighth decade of life [2]. It is particularly important to exclude the diagnosis in children or young adults who present with elevated liver enzymes, jaundice or other manifestations of liver disease, unusual extrapyramidal or cerebellar motor disorders, psychiatric disease, unexplained hemolysis, and with or without a family history of liver or neurological disease. Failure to recognize the diagnosis will lead to disease progression with devastating consequences, including need for liver transplant or preventable death.

In the majority of cases, a combination of clinical and laboratory features is required to make a diagnosis. Clinical presentations vary widely, and the diagnosis is often not clear-cut.

THE LEIPZIG SCORE AND ITS UTILITY IN DIAGNOSTIC WORK UP OF A WILSON DIAGNOSIS

While in most cases the diagnosis can be made by a combination of clinical and biochemical testing, none of the available tests are specific for the diagnosis of WD. Clinical symptoms may be absent or nonspecific in a large proportion of patients. A diagnostic algorithm based on all available tests was proposed by the Working Party at the eighth International Meeting on WD, Leipzig 2001. The Leipzig score was developed to guide clinicians as to whether further testing is needed or whether the diagnosis is established [3]; this might be a weakness of the scoring system as it was expert-driven and not data-driven. Following the Leipzig meeting, the scoring system was supported by AASLD and EASL clinical guidelines [4]. A score of 4 or above is required to establish the diagnosis of WD; if the score is below 4, then an alternative diagnosis needs to be considered. The Leipzig scoring system has now been validated in both adult and pediatric patients with WD [5]. Table 1 demonstrates the investigations that comprise the Leipzig scoring system and the allocated scores for each test result.

DIFFERENTIAL DIAGNOSIS

Wilson disease should always enter into the differential diagnosis of all patients with unexplained chronic hepatitis and cryptogenic cirrhosis. In acute hepatitis, diagnostic suspicion of WD should be raised in cases presenting with rapid onset of jaundice and hemolytic anemia [6]. In relatively advanced and progressive cases, presentation may involve predominantly behavioral, psychological, or psychiatric manifestations. There are instances also when the disease initially presents with neuropsychiatric symptoms that masquerade as antisocial problems, particularly in the young adolescent, which should heighten the index of suspicion [7].

Wilson Disease. https://doi.org/10.1016/B978-0-12-811077-5.00008-6
© 2019 Elsevier Inc. All rights reserved.

TABLE 1 Leipzig Scoring for Wilson Disease

	Allocated Score per Item
Typical clinical signs and symptoms	
Kayser-Fleischer Rings	
Present	2
Absent	0
Neurologic symptoms[a]	
Severe	2
Mild	1
Absent	0
Serum ceruloplasmin	
Normal (>0.2 g/L)	0
0.1–0.2 g/L	1
<0.1 g/L	2
Coombs-negative hemolytic anemia	
Present	1
Absent	0
Other tests	
Liver copper (in the absence of cholestasis)	
>5 times ULN (>4 µmol/g)	2
0.8–4 µmol/g	1
Normal (<0.8 µmol/g)	−1
Rhodamine-positive granules[b]	1
Urinary copper (in the absence of acute hepatitis)	
Normal	0
1–2 times ULN	1
>2 times ULN	2
Normal but five times ULN after penicillamine	2
Mutation analysis	
On both chromosomes detected	4
On one chromosome detected	1
No mutations detected	0

Evaluation Based on Total Leipzig Score	
Total Score	Evaluation
≥4	Diagnosis established
3	Diagnosis possible, more tests needed
≤2	Diagnosis very unlikely

[a]Or typical abnormalities at brain magnetic resonance imaging.
[b]If no quantitative liver copper available.

CLINICAL FEATURES SUGGESTIVE OF WILSON DISEASE (WD) DIAGNOSIS (Fig. 1)

Patients with symptomatic WD most frequently present with liver disease and/or neuropsychiatric symptoms. Those detected by family screening are often asymptomatic (also termed *presymptomatic*) [8]. Failure to initiate appropriate therapy for WD or the disruption of ongoing treatment ultimately will result in progression of disease to hepatic insufficiency, neuropsychiatric decline, and ultimately hepatic failure and death without transplantation [9]. Fig. 1 summarizes the variety of clinical manifestations that can be present in patients with WD.

FIG. 1 Clinical manifestations suggestive of Wilson disease.

Hepatic
- Persistently elevated serum aminotransferases
- Chronic hepatitis
- Cirrhosis (decompensated or compensated)
- Fulminant hepatic failure with jaundice (± haemolytic anaemia)

Neurological
- Tremor
- Choreiform movements
- Parkinsonism or akinetic rigid syndrome—i.e., partial parkinsonism
- Gait disturbances
- Dysarthria
- Pseudo bulbar palsy
- Rigid dystonia
- Seizures
- Migraine and headaches
- Insomnia
- Dysphagia

Ophthalmic
- K-F rings
- Sunflower cataracts

Psychiatric
- • Depression
- Neuroses
- Personality changes
- • Psychosis
- Difficulty with concentration and memory

Other systems (rare)
- Renal abnormalities: aminoaciduria and nephrolithiasis
  - Cardiomyopathy ± arrhythmias

Liver Disease

The onset of liver disease may precede the onset of neurological symptoms by as much as 10 years [10]. Most patients with neurological symptoms will have some degree of liver damage at presentation. Younger patients may be asymptomatic when identified by family screening or through serial evaluations of isolated liver test abnormalities [10]. In patients with cirrhosis and hepatic insufficiency, jaundice, ascites, edema, or other stigmata of chronic liver disease, including hepatic encephalopathy, may be observed. When untreated, the liver disease progresses to cirrhosis with hepatic insufficiency, liver failure, and death. Some patients present with an acute hepatitis with an associated nonimmune-related hemolytic anemia [6]. These patients are most often in their second decade of life. Wilson disease should enter into the diagnosis of any patient with acute hepatitis without any markers of recent infection. Among patients presenting with Wilsonian acute hepatitis, the female-to-male ratio is almost 2:1 in some series and even higher in others [11]. Patients diagnosed with WD who have a history of jaundice may have previously experienced an episode of hemolysis. Without the lifesaving intervention of liver transplantation, acute liver failure (ALF) due to Wilson disease is frequently fatal [12].

Neurological/Neuropsychiatric Disease

Patients presenting with neuropsychiatric symptoms are frequently diagnosed at a later age than those with hepatic symptoms [13]. There may be significant liver disease present at the time of diagnosis in those with predominantly neuropsychiatric symptoms. However, hepatic histology is not generally available for these patients because the diagnosis is often established by detecting a decreased ceruloplasmin level and presence of K-F rings on slit-lamp examination. Common neurological signs and symptoms include motor abnormalities with parkinsonian characteristics of dystonia, hypertonia and rigidity, chorea or athetosis, tremors, and dysarthria. Disabling muscle spasms can result in contractures. Speech difficulties may be present including dysarthria and dysphonia. Impaired swallowing and dysphagia can predispose patients to having a significant aspiration risk. Rarely, patients can present with a polyneuropathy or dysautonomia [14]. Magnetic resonance imaging or computed tomography of the brain may be useful in delineating changes in the basal ganglia and other brain regions, including the pons [15].

Kayser-Fleischer Rings

It is essential that all patients in whom WD is suspected undergo a slit-lamp examination, performed by an experienced ophthalmologist, for the detection of Kayser-Fleischer (K-F) rings. K-F rings are most apparent at the periphery of the cornea with a golden-brown appearance. They are caused by the granular deposition of copper on the inner surface of the cornea in Descemet's membrane [16]. The upper pole is affected first. Although sometimes visible to the naked eye, slit-lamp examination is often necessary to confirm the presence or absence of K-F rings. Rings indistinguishable from K-F rings have also been seen in other forms of chronic liver diseases, especially long-standing cholestasis and cryptogenic cirrhosis [17].

K-F rings are present in patients with neurological disease, with only rare exceptions. Although useful diagnostically when present, they may be absent particularly in younger patients with hepatic manifestations only. Another finding that may suggest WD is the presence of sunflower cataracts, also best observed by slit-lamp examination. These ophthalmologic features are reversible with medical therapy or after liver transplantation [18]. The reappearance of either of these eye changes in a medically treated patient indicates active disease and suggests noncompliance or failure of therapy.

Other Manifestations

Very occasionally, WD can affect other organ systems. Changes induced by copper toxicity in the kidneys include nephrocalcinosis, hematuria, and aminoaciduria [19]. Involvement of the musculoskeletal system may result in symptoms of arthritis, arthralgias, and premature osteoarthrosis [20]. When copper accumulates in the myocardium, cardiomyopathy and arrhythmias may result [21].

Acute Hepatitis and Wilson Disease

Approximately 5% of patients with WD present with acute liver failure [22]. In this setting, copper parameters, which are normally used to support a diagnosis of WD, are less reliable. The diagnosis of WD in the acute setting of hepatitis also deserves particular attention because of multiple novel characteristics. Firstly, the acute hepatitis is associated with hemolytic anemia with unconjugated hyperbilirubinemia and markedly elevated serum and urinary levels of copper; it is specifically not immunologically mediated. Secondly, these patients are mostly females in the second decade of life, and K-F rings may not yet be observed. An interesting observation has been the relatively low levels of alkaline phosphatase (ALP) in these patients. This specific feature has led to the finding that a ratio of ALP-bilirubin <2 may be of diagnostic value in acute hepatitis-related WD as initially proposed by Berman et al. In some patients with ALF due to Wilson disease this ratio was above 2 [23]. However, when an alkaline phosphatase-bilirubin ratio of <4 and aspartate aminotransferase (AST)-alanine aminotransferase (ALT) ratio of >2.2 are used in combination, a nearly 100% identification of WD can be made in patients with acute liver failure. The combination of clinical symptoms and "conventional" WD diagnostic parameters (serum or urinary copper and ceruloplasmin) appears to be less sensitive and specific in the acute setting but probably remains important for the diagnosis. In cases of uncertainty, the diagnosis should be characterized by liver biopsy if possible or after liver transplantation with explant analysis of hepatic copper and mutational analysis to enable screening of asymptomatic siblings.

INVESTIGATIONS

Ceruloplasmin

A test to determine the serum concentration of serum ceruloplasmin is routinely available in all clinical laboratories. It is desirable that ceruloplasmin be measured by means of enzymatic methods, to determine oxidase activity, because this best reflects the copper content of the protein. However, immunologic assays are most often utilized by commercial laboratories [24].

The normal range for serum ceruloplasmin is ~20–50 mg/dL, but the lower range may vary by laboratory. About 95% of homozygotes with WD have values of <20 mg/dL. Up to 5% of all homozygotes and up to 15%–50% of persons with liver disease may have normal levels, which is defined as concentrations above 20 mg/dL [25].

Serum ceruloplasmin is typically decreased in patients with neurological WD but may be in the normal range in some patients with hepatic WD. Normal levels are sometimes present in WD patients with active liver disease, probably as a consequence of an acute-phase response in the liver or as a result of estrogen supplementation. WD patients rarely have serum concentrations exceeding 30 mg/dL. The predictive value of ceruloplasmin in the diagnosis of WD increases in patients with levels of ceruloplasmin that are well below the lower limit of normal [26].

Low serum concentrations of ceruloplasmin may also be observed in hypoproteinemic states, such as protein-calorie malnutrition, nephrotic syndrome, protein-losing enteropathy, and other forms of severe decompensated liver disease, and in up to 20% of asymptomatic heterozygous carriers of the WD gene [27]. Rarer causes of serum ceruloplasmin deficiency include hereditary aceruloplasminemia and Menkes' disease [28].

Nonceruloplasmin Bound Copper (NCC)

The total concentration of copper in plasma or serum represents ceruloplasmin-bound copper plus nonceruloplasmin-bound ("free") copper, which is bound mainly to albumin, peptides, or amino acids. Because the former is reduced in proportion to the degree of hypoceruloplasminemia, the total copper in a patient with WD may be low in the face of a typically raised free copper concentration. The nonceruloplasmin-bound copper is the difference between the serum copper concentration in micrograms per deciliter and three times the serum ceruloplasmin concentration in milligrams per deciliter in the plasma ceruloplasmin copper concentration. The value for total plasma copper ranges from 80 to 120 μg/dL, and the value for NCC is usually 10% of the total value (~8–12 μg/dL).

Levels of NCC are typically above 25 μg/dL in symptomatic WD patients before treatment. Using commercial assays, the calculated level of NCC may be paradoxically low or negative, and this is likely due to use of the immunologic assay for ceruloplasmin that may overestimate the concentration of the protein and thereby the amount of the copper in the protein [29]. Further refinement of the assay is needed so that it may be useful for all patients. An alternative way of determining the "free" copper is to measure the exchangeable copper, that is, copper that is not removed from serum proteins by the chelator EDTA [30], and further testing is needed to determine the clinical utility of this assay.

Urinary Excretion of Copper

The copper excreted in urine is a reflection of the NCC. The amount of copper excreted in the urine in a 24 h period is useful in the diagnosis and monitoring of WD. The rate of excretion may exceed 100 μg/24 h in symptomatic patients. In patients presenting with chronic liver disease, the urinary copper level is elevated above normal levels but may not reach diagnostic levels of >100 μg/24 h. false-positive increases in urinary copper level may be seen in the face of significant proteinuria and urinary loss of ceruloplasmin and rarely in other liver diseases where copper storage is increased or in acute liver failure. A provocative test for urinary copper excretion with the use of the chelating agent penicillamine has been studied in children [31] but may be no better than changing the threshold of urinary copper excretion to 40 μg of copper in 24 h. Contamination of the sample and incomplete collections can also affect the result. Results can therefore be difficult to interpret unless strict precautions are taken. When urinary excretion of copper is tested, it is crucial that a metal-free container be used and that the adequacy of the collection be monitored by correlation with volume excreted or with creatinine excretion. Heterozygotes may also have a higher copper excretion than controls, but these rarely exceed the upper end of normal [32].

Concentration of Copper in the Liver

The main pathological hallmark of WD is copper accumulation in the hepatic parenchyma. Thus, biopsy and histological examination form another diagnostic modality that can support the diagnosis. The normal copper content of liver is <55 μg/g dry weight. Accurate analysis needs an adequate sample of the liver (at least 1 cm of a 1.6 mm diameter core) as particularly in the later stages of the disease copper may not be deposited evenly and thus sampling errors may occur. Measurement of copper content in the liver is the most important diagnostic test in patients in whom other data are suggestive but not diagnostic of disease. A hepatic copper concentration >250 μg/g dry weight is often found in WD and with some caveats remains the best biochemical test for the disease [33]. The threshold of 250 μg/g dry weight has been challenged recently as demonstrated by Yang et al. [34]. They prospectively evaluated the diagnostic accuracy of hepatic copper content as determined using the entire core of a liver biopsy core. They found that for the diagnosis of WD, the most useful cutoff value is 209 μg/g dry weight.

Diagnosis is sometimes only considered retrospectively, after liver biopsy has been done. Under these circumstances, liver biopsy specimens can be retrieved from paraffin blocks for quantitative copper measurement. However, the sensitivity and specificity of this test have never been clearly established.

Liver copper quantitation is an important indicator of disease; however, a value below 250 μg/g dry weight does not exclude the possibility of WD. Similar to ceruloplasmin values, a lowering of the threshold for hepatic copper content below 250 μg/g dry weight increases sensitivity for diagnosis but decreases specificity. In addition, specimens with extensive fibrosis and few parenchymal cells can provide copper concentrations that are falsely low [34]. Furthermore, greatly increased hepatic copper concentrations can be seen in long-term cholestasis. Therefore, the results of the hepatic copper concentration estimation should be taken in the context of the histological, clinical, and biochemical data.

Liver Histology

A liver biopsy is only required if the clinical and other investigations do not establish WD or if there is a clinical suspicion of an alternative diagnosis. However, it can provide useful information for staging and exclusion of other causes of liver disease.

In the early stages of the disease, diffuse cytoplasmic copper accumulation can be seen only by special immunohistochemical stains for copper detection, which are not routinely available. This early accumulation of copper is associated with macrosteatosis, microsteatosis, and glycogenated nuclei that are features that can be seen in various other disorders, for example, nonalcoholic steatohepatitis [35]. The ultrastructural abnormalities seen range from enlargement and separation of the mitochondrial inner and outer membranes, with widening of the intercristal spaces, to increases in the density and granularity of the matrix or the occurrence of large vacuoles. In the absence of cholestasis, these changes are regarded as pathognomonic of WD [36]. Ultrastructural analysis might be useful for helping to distinguish between heterozygous carriers and patients [37].

The detection of copper in hepatocytes by routine histochemical evaluation is highly variable. Particularly, in the early stages of the disease, copper is mainly in the cytoplasm bound to metallothionein and is not histochemically detectable. The amount of copper varies from nodule to nodule in cirrhosis and may vary from cell to cell in precirrhosis.

The initial stages of WD progress to an intermediate stage, which is characterized by periportal inflammation, mononuclear cellular infiltrates, erosion of the limiting plate, lobular necrosis, and bridging fibrosis, and these features are indistinguishable from those of autoimmune hepatitis. Mallory bodies can be seen in up to 50% of biopsy specimens. Cirrhosis almost invariably follows with either a micronodular or a mixed macronodular-micronodular histological pattern. In patients with fulminant hepatic failure, parenchymal apoptosis, necrosis, and collapse might predominate. It's interesting to note that half of patients at the time of diagnosis have cirrhosis but some older patients with WD do not have cirrhosis or even signs of liver disease.

Neuroradiology

Advances in neuroimaging have improved our understanding of the pathophysiology of WD. Structural brain MRI in patients with the disease has shown widespread lesions in the putamen, globus pallidus, caudate, thalamus, midbrain, pons, and cerebellum as well as cortical atrophy and white matter changes. In general, these lesions show high signal intensity on T2-weighted images and low intensity on T1 scan [38]. Although MRI changes are present in many WD patients and even patients without neurological symptoms, these changes tend to be more severe and widespread in patients with neurological WD. Proton-density MRI sequences seem to be especially sensitive in showing the extent of the neuropathology [39].

Molecular Genetic Studies

The identification of the ATP7B gene for WD has enabled the molecular genetic diagnosis of this disorder. There are now numerous disease-specific mutations of the gene described; however, the most common mutation is present in only 15%–30% of most populations [40]. Most patients are compound heterozygotes, possessing different mutations on each allele of *ATP7B*. Molecular analysis for the presence of disease-specific mutations is now possible and highly effective. Advances in DNA sequencing and screening technology enable the detection of mutations in almost all patients. The ability to firmly establish the diagnosis by this methodology depends on distinguishing disease-specific mutations from polymorphisms of the gene and is at times limited by the fact that some of the noncoding regions of the gene, which may also affect gene expression, are not analyzed.

Family members of an affected person may also be screened by haplotype analysis; however, this method is outdated given the more widespread availability of direct DNA sequencing for *ATP7B* mutations. Haplotype analysis involves analyzing the patterns of polymorphisms of the DNA in the region surrounding the WD gene to determine whether mutant regions present in the affected patient have also been inherited by asymptomatic family members [41]. This still may continue to have utility in testing families where no or only a single mutation of *ATP7B* was detected by direct sequencing.

Future advances in DNA analysis should make possible screening for disease-specific mutations for WD in an even more cost-effective manner, so that this test may someday be first-line. Indeed, someday de novo population screening may prove practical. At present, genetic testing should be used for family screening and used in parallel with standard clinical and biochemical testing as discussed above.

Family Screening

First-degree relatives must be screened to prevent disease progression with potentially devastating consequences of fulminant hepatic failure and death [4]. The probability of finding a homozygote in siblings is 25% and in the children is roughly 0.5%. Genetic testing for mutations of ATP7B is now the first-line test for siblings where two mutations were identified in the proband. In others, liver function tests, serum copper and ceruloplasmin concentration, and urinary copper analysis are performed. If necessary, investigations should be extended to test for K-F rings. Twenty-four-hour urinary copper might be difficult to interpret in WD heterozygotes [4]. The diagnosis could remain contentious when individuals without K-F rings have a low ceruloplasmin concentration. These individuals might need a liver biopsy for hepatic copper quantification to eliminate the diagnosis.

Haplotype analysis of markers around the *ATP7B* gene on chromosome 13 has been used in families to establish whether siblings of affected individuals have inherited the same pair of chromosomes [25]. This approach might be beneficial when it has not been possible to detect both mutations in the index case by mutation analysis.

REFERENCES

[1] Wilson SAK. Progressive lenticular degeneration: a familial nervous disease associated with cirrhosis of the liver. Brain 1912;34:295–509.
[2] Ala A, Borjigin J, Rochwarger A, Schilsky M. Wilson disease in septuagenarian siblings: raising the bar for diagnosis. Hepatology 2005;41(3):668–70.
[3] Ferenci P, Caca K, Loudianos G, Mieli-Vergani G, Tanner S, Sternlieb I, Schilsky M, Cox D, Berr F. Diagnosis and phenotypic classification of Wilson disease. Liver Int 2003;23(3):139–42.
[4] European Association for Study of Liver. EASL clinical practice guidelines: Wilson's disease. J Hepatol 2012 Mar;56(3):671–85.
[5] Xuan A, Bookman I, Cox DW, Heathcote J. Three atypical cases of Wilson disease: assessment of the Leipzig scoring system in making a diagnosis. J Hepatol 2007;47(3):428–33.
[6] Walshe JM. The acute haemolytic syndrome in Wilson's disease—a review of 22 patients. QJM 2013;106(11):1003–8.
[7] Azova S, Rice T, Garcia-Delgar B, Coffey BJ. New-onset psychosis in an adolescent with Wilson's disease. J Child Adolesc Psychopharmacol 2016;26(3):301–4.
[8] Socha P, Janczyk W, Dhawan A, Baumann U, D'Antiga L, Tanner S, Iorio R, Vajro P, Houwen R, Fischler B, Dezsofi A, Hadzic N, Hierro L, Jahnel J, McLin V, Nobili V, Smets F, Verkade HJ, Debray D. Wilson's disease in children: a position paper by the Hepatology Committee of the European Society for paediatric gastroenterology, hepatology and nutrition. J Pediatr Gastroenterol Nutr 2018;66(2):334–44.
[9] Ala A, Walker AP, Ashkan K, Dooley JS, Schilsky ML. Wilson's disease. Lancet 2007;369(9559):397–408.
[10] Lorincz MT. Neurologic Wilson's disease. Ann N Y Acad Sci 2010;1184:173–87.
[11] Litwin T, Gromadzka G, Członkowska A. Gender differences in Wilson's disease. J Neurol Sci 2012;312(1–2):31–5.

[12] Schilsky ML, Scheinberg IH, Sternlieb I. Liver transplantation for Wilson's disease: indications and outcome. Hepatology 1994;19(3):583–7.
[13] Steindl P, Ferenci P, Dienes HP, Grimm G, Pabinger I, Madl C, Maier-Dobersberger T, Herneth A, Dragosics B, Meryn S, Knoflach P, Granditsch G, Gangl A. Wilson's disease in patients presenting with liver disease: a diagnostic challenge. Gastroenterology 1997;113(1):212–8.
[14] Meenakshi-Sundaram S, Taly AB, Kamath V, Arunodaya GR, Rao S, Swamy HS. Autonomic dysfunction in Wilson's disease—a clinical and electrophysiological study. Clin Auton Res 2002;12(3):185–9.
[15] Pulai S, Biswas A, Roy A, Guin DS, Pandit A, Gangopadhyay G, Ghorai PK, Sarkhel S, Senapati AK. Clinical features, MRI brain, and MRS abnormalities of drug-naïve neurologic Wilson's disease. Neurol India 2014;62(2):153–8.
[16] Sridhar MS, Pineda R. Anterior segment optical coherence tomography to look for Kayser-Fleischer rings. Pract Neurol 2017;17(3):222–3.
[17] Nagral A, Jhaveri A, Nalawade S, Momaya N, Chakkarwar V, Malde P. Kayser-Fleischer rings or bile pigment rings? Indian J Gastroenterol 2015; 34(5):410–2.
[18] Srinivas K, Sinha S, Taly AB, Prashanth LK, Arunodaya GR, Janardhana Reddy YC, Khanna S. Dominant psychiatric manifestations in Wilson's disease: a diagnostic and therapeutic challenge!. J Neurol Sci 2008;266(1–2):104–8.
[19] Dzieżyc K, Litwin T, Członkowska A. Other organ involvement and clinical aspects of Wilson disease. Handb Clin Neurol 2017;142:157–69.
[20] Ye S, Dai T, Leng B, Tang L, Jin L, Cao L. Genotype and clinical course in 2 Chinese Han siblings with Wilson disease presenting with isolated disabling premature osteoarthritis: a case report. Medicine 2017;96(47).
[21] Bajaj BK, Wadhwa A, Singh R, Gupta S. Cardiac arrhythmia in Wilson's disease: an oversighted and overlooked entity. J Neurosci Rural Pract 2016;7(4):587–9.
[22] Schilsky ML. Liver transplantation for Wilson's disease. Ann N Y Acad Sci 2014;1315:45–9.
[23] Emre S, Atillasoy EO, Ozdemir S, Schilsky M, Rathna Varma CV, Thung SN, Sternlieb I, Guy SR, Sheiner PA, Schwartz ME, Miller CM. Orthotopic liver transplantation for Wilson's disease: a single-center experience. Transplantation 2001;72(7):1232–6.
[24] Merle U, Eisenbach C, Weiss KH, Tuma S, Stremmel W. Serum ceruloplasmin oxidase activity is a sensitive and highly specific diagnostic marker for Wilson's disease. J Hepatol 2009;51(5):925–30.
[25] Arianfar F, Fardaei M. Linkage analysis based on four microsatellite markers to screen for unknown mutation in families with Wilson disease. Clin Lab 2016;62(8):1541–6.
[26] Nicastro E, Ranucci G, Vajro P, Vegnente A, Iorio R. Re-evaluation of the diagnostic criteria for Wilson disease in children with mild liver disease. Hepatology 2010;52(6):1948–56.
[27] Ogimoto M, Anzai K, Takenoshita H, Kogawa K, Akehi Y, Yoshida R, Nakano M, Yoshida K, Ono J. Criteria for early identification of aceruloplasminemia. Intern Med 2011;50(13):1415–8.
[28] Squitti R, Siotto M, Cassetta E, El Idrissi IG, Colabufo NA. Measurements of serum non-ceruloplasmin copper by a direct fluorescent method specific to Cu(II). Clin Chem Lab Med 2017;55(9):1360–7.
[29] Poujois A, Trocello JM, Djebrani-Oussedik N, Poupon J, Collet C, Girardot-Tinant N, Sobesky R, Habès D, Debray D, Vanlemmens C, Fluchère F, Ory-Magne F, Labreuche J, Preda C, Woimant F. Exchangeable copper: a reflection of the neurological severity in Wilson's disease. Eur J Neurol 2017;24(1):154–60.
[30] Cieśla J, Koczańska M, Bieganowski A. An interaction of rhamnolipids with Cu(2+) ions. Molecules 2018;23(2).
[31] Schilsky ML. Non-invasive testing for Wilson disease: revisiting the D-penicillamine challenge test. J Hepatol 2007;47:172–3.
[32] Członkowska A, Rodo M, Wierzchowska-Ciok A, Smolinski L, Litwin T. Accuracy of the radioactive copper incorporation test in the diagnosis of Wilson disease. Liver Int 2018; [Epub ahead of print].
[33] Ferenci P, Steindl-Munda P, Vogel W, Jessner W, Gschwantler M, Stauber R, Datz C, Hackl F, Wrba F, Bauer P, Lorenz O. Diagnostic value of quantitative hepatic copper determination in patients with Wilson's disease. Clin Gastroenterol Hepatol 2005;3(8):811–8.
[34] Yang X, Tang XP, Zhang YH, Luo KZ, Jiang YF, Luo HY, Lei JH, Wang WL, Li MM, Chen HC, Deng SL, Lai LY, Liang J, Zhang M, Tian Y, Xu Y. Prospective evaluation of the diagnostic accuracy of hepatic copper content, as determined using the entire core of a liver biopsy sample. Hepatology 2015;62(6):1731–41.
[35] Sternlieb I. Mitochondrial and fatty changes in hepatocytes of patients with Wilson's disease. Gastroenterology 1968;55:354–67.
[36] Pilloni L, Lecca S, Van Eycken P, et al. Value of histochemical stains for copper in the diagnosis of Wilson's disease. Histopathology 1998;33:28–33.
[37] Alt E, Sternlieb I, Goldfischer S. The cytopathology of metal overload. Int Rev Exp Pathol 1990;31:165–88 [Review].
[38] Dusek P, Litwin T, Czlonkowska A. Wilson disease and other neurodegenerations with metal accumulations. Neurol Clin 2015 Feb;33(1):175–204.
[39] da Costa Mdo D, Spitz M, Bacheschi LA, Leite CC, Lucato LT, Barbosa ER. Wilson's disease: two treatment modalities. Correlations to pretreatment and posttreatment brain MRI. Neuroradiology 2009;51(10):627–33.
[40] Shah AB, Chernov I, Zhang HT, Ross BM, Das K, Lutsenko S, Parano E, Pavone L, Evgrafov O, Ivanova-Smolenskaya IA, Annerén G, Westermark K, Urrutia FH, Penchaszadeh GK, Sternlieb I, Scheinberg IH, Gilliam TC, Petrukhin K. Identification and analysis of mutations in the Wilson disease gene (ATP7B): population frequencies, genotype-phenotype correlation, and functional analyses. Am J Hum Genet 1997; 61(2):317–28.
[41] Schilsky ML, Ala A. Genetic testing for Wilson disease: availability and utility. Curr Gastroenterol Rep 2010;12(1):57–61.

Chapter 9

The Genetics of Wilson Disease

Michelle Angela Camarata[*,†,‡] and Si Houn Hahn[§]

[*]Department of Gastroenterology and Hepatology, Royal Surrey County Hospital, Guilford, United Kingdom, [†]Department of Clinical and Experimental Medicine, University of Surrey, Guilford, United Kingdom, [‡]Department of Surgery, Section of Transplant and Immunology, Yale School of Medicine, New Haven, CT, United States, [§]Division of Genetic Medicine, Department of Pediatrics, University of Washington School of Medicine, Seattle Children's Hospital, Seattle, WA, United States

Wilson disease (WD) is an inherited autosomal recessive condition. It is unusual in that autosomal recessive conditions are not usually present in consecutive generations, but this may occur in populations with a particularly high carrier frequency such as in WD [1]. Our group and others have reported on the occurrence of WD in two or more successive generations within the same family [2, 3]. For this reason, WD should not be excluded on the sole basis of a family history consistent with an autosomal dominant inheritance pattern. Furthermore, some studies have also found rare patients with atypical inheritance patterns such as the presence of three concurrent mutations in a single patient or segmental uniparental disomy [4]. Uniparental disomy occurs when both homologues of a chromosome originate from a single parent. It may therefore be important for clinicians to consider genotyping asymptomatic parents or obtaining full sequencing of *ATP7B* to confirm pathogenic variants.

ATP7B GENE AND ATPASE

WD is caused by mutations in the *ATP7B* gene, which encodes for a copper-transporting P-type ATPase [5] that transports copper into bile and incorporates it into ceruloplasmin. The gene was identified by positional cloning using yeast artificial chromosomes at 13q14.3 in 1993 [6, 7].

Currently, *ATP7B* is the only gene known to cause WD [7, 8]. Mutations in the *ATP7B* gene have been found in almost all exons. There have, however, been reports of individuals with a diagnosis of WD based on clinical and biochemical tests without two *ATP7B* mutations. The possibility of a second causative gene was raised, but no other causative genes have been identified to date [4, 9–13].

Dietary copper is absorbed in the digestive tract, bound to circulating albumin, and transported to the liver for metabolism and excretion [14]. The uptake of copper occurs via the copper transporter 1 (CTR1) on the basolateral side of hepatocytes. Antioxidant protein 1 (ATOX1) delivers copper to ATP7B, by copper-dependent protein-protein interactions [15]. Within hepatocytes, ATP7B performs two important functions in the trans-Golgi network (TGN) and in cytoplasmic vesicles. In the TGN, ATP7B incorporates six copper atoms into a molecule of apoceruloplasmin to form ceruloplasmin, which is then secreted into the plasma. Ceruloplasmin carries 90% of the copper present in the plasma. Mutations in ATP7B cause a reduction in conversion of apoceruloplasmin into ceruloplasmin, resulting in low levels in most patients due to the reduced circulating half-life of apoceruloplasmin relative to the protein with its complement of copper. In the cytoplasm, ATP7B incorporates excess copper into vesicles that are excreted via exocytosis across the apical canalicular membrane into bile [7, 8, 16, 17]. An increase in copper concentration has been shown to increase trafficking of the ATPase from the TGN to the cytoplasmic vesicles [18]. Defects in the function of the ATP7B transporter lead to copper accumulation particularly in the liver and brain and result in the clinical manifestations associated with WD.

MOLECULAR STRUCTURE OF *ATP7B*

The WD gene ATP7B is a large gene located on chromosome 13q14.3 and contains 20 introns and 21 exons, with a genomic length of 80 kb [7, 8]. The gene is synthesized in the endoplasmic reticulum and then relocated to the TGN within hepatocytes. The *ATP7B* gene is most highly expressed in the liver [8] but is also found in the brain, lung, kidney, mammary glands, and placenta [19, 20].

ATP7B (P-TYPE ATPASE) PROTEIN STRUCTURE

ATP7B belongs to class 1B (PIB) of the highly conserved P-type ATPase superfamily, which is responsible for the transport of copper and other heavy metals across cellular membranes [21]. The protein contains 1465 amino acids and the following functional domains: six copper-binding domains; a phosphatase domain (A-domain); phosphorylation domain (P-domain, amino acid residues 971–1035); nucleotide-binding domain (N-domain, amino acid residues 1240–1291); and M-domain, which is composed of eight hydrophobic transmembrane ion channels [16, 22, 23].

Each functional domain contains unique amino acid motifs such as SEHPL in the N-domain, DKTGT at the P-domain, and TGEA at the A-domain. The metal-binding domain (MBD) in the N-domain has six copper-binding sites, which each also contains the sequence motif GMXCXXC [24, 25]. These sites accept copper from the copper chaperone ATOX1 through protein-protein interactions. Similar to other ATPases, ATP7B transports ligand, in this instance copper, from the cytosol to cellular membranes using energy derived from ATP hydrolysis. The six N-terminal metal-binding sites are necessary for trafficking and are essential for copper transport. Studies have demonstrated that the MBDs 5 and 6 have stronger effects on the catalytic activation of ATP7B than MBDs 1–3 [22, 26].

Multiple steps are involved for the delivery of copper from the cytosol to cellular membranes. These include recognition of copper, delivery to the membrane portion of the transporter, and subsequent release on the other side of the membrane with binding and hydrolysis of ATP into ADP and Pi. Copper-induced changes, including conformational changes at the N-terminal copper-binding domains, have been suggested to be responsible for cellular trafficking [24]. Structural analysis of the N-terminus has revealed that after copper binds, both secondary and tertiary structural changes take place and that copper coordination induces phosphorylation of ATP7B that coincides with trafficking of the protein to the vesicular compartments [27, 28].

Following the binding of copper at the N-terminal MBD, it is transported across cellular membranes using ATP as an energy source. A number of residues have thought to be implicated in the binding of ATP in the N-Domain (H1069, G1099, G1101, 1102, G1149, and N115; [29]). Some of these residues are commonly mutated in WD. Following transport of copper across cellular membranes, free copper binds intracellularly to GG motifs on the MBDs, followed by transport onto the highly conserved Cys-Pro-Cys (CPC) sequence in MBD 6. This is one of the signature motifs characterizing ATPases involved in copper and other heavy metal transport that form the P1B subfamily [30]. Finally, dephosphorylation of acyl-phosphate at the A-domain discharges copper across the cellular membrane. Mutations causing altered copper homeostasis may occur at any of these points [31, 32].

VARIANTS IN THE *ATP7B* GENE

Most cases of WD are compound heterozygotes with a different mutation on each copy of the chromosome [33]. There are more than 600 pathogenic variants of the *ATP7B* gene; the majority affect the transmembrane region of ATP7B most commonly single-nucleotide missense and nonsense mutations, followed by insertions/deletions, and splice site mutations (Human Gene Mutation Database, accessed 29 April 2016 [34]). Other rare genetic mutations that have been reported include whole-exon deletions, promoter region mutations, three concurrent pathogenic variants, and monogenic disomy [4, 35]. In presymptomatic patients or in those with hepatic symptoms, mutations have mainly been found to be located in the M- and N-domains [13].

The histidine-to-glutamate substitution at position 1069 (p.H1069Q) in the histidine-containing SEHPL sequence in the N-domain of ATP7B is the most common cause of Wilson disease in northern Europeans. In the hepatocytes of patients homozygous for the p.H1069Q mutation, ATP7B was found in the endoplasmic reticulum instead of its normal location in the TGN 1, suggesting an abnormality in protein trafficking [36]. Insect models with the p.H1069Q mutation demonstrated reduced ATP-mediated catalytic phosphorylation but no significant protein misfolding. This suggests there may be a role for p.H1069Q in the orientation of the ATP7B catalytic site for ATP binding prior to hydrolysis [37, 38]. This mutation also leads to decreased heat stability [39].

Other common mutations in *ATP7B* include p.E1064A, p.R778L, p.G943S, and p.M769V. Mutations in p.E1064A also found in the SEHPL motif prevent ATP binding but do not result in protein misfolding, transport abnormalities, or thermal instability. The p.R778L mutation affects transmembrane transport of copper [40]. The p.G943S and p.M769V mutations result in defective copper metabolism but preserved ceruloplasmin levels that could explain the normal levels found in some patients [41].

A large number of WD-associated missense mutations including p.H1069Q and p.R778L cause significant reduction in protein levels through increased degradation [42–44]. Other mutations such as protein-truncating nonsense mutations (13% of known point mutations [45]) and frameshift mutations [46] are predicted to cause decay of mRNA [47, 48] or a severely

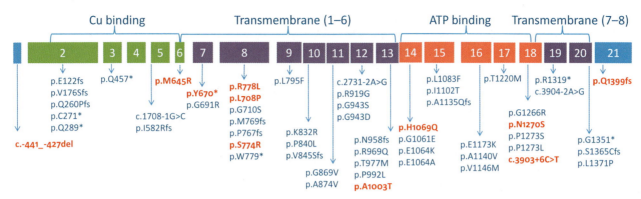

FIG. 1 A schematic representation of the ATP7B gene with common mutation sites including p.H1069Q, pR778L, and p.E1064K.

truncated protein, resulting in absent or diminished levels of protein. It is therefore expected that most patients with WD have absent or significantly reduced levels of ATP7B. Fig. 1 includes a schematic representation of the ATP7B gene with common mutations.

Certain genetic variants may affect the location of where ATP7B is targeted from the TGN to cytosolic vesicles. For instance, the p.Gly875Arg (c.2623A > G) variant results in a protein that is not stable and causes localization abnormalities in ATP7B. Studies have shown that the interaction of the A-domain with a copper-bound N-terminal stabilizes the protein for ER to Golgi transfer [49]. Under a low-copper environment, the p.Gly875Arg variant (in the A-domain) is sequestered in the endoplasmic reticulum. There is a loss of interaction between the A-domain and N-terminal that causes accumulation of ATP7B in the endoplasmic reticulum with increased intracellular copper. However, addition of exogenous copper to the cellular growth medium stabilizes the protein and the N-terminal/A-domain interaction, allowing it to complete its intended journey to the TGN allowing it to overcome its disease-causing phenotype. Theoretically, patients with this specific variant may be more sensitive to dietary copper deficiency. Copper levels influence the phenotype of this genetic variant, and this study demonstrates potential interaction between genetic and nongenetic factors for determining WD phenotype [49].

REGIONAL GENE FREQUENCY

The prevalence of WD varies geographically; similarly, the occurrence of specific mutations also varies between certain ethnic groups [50]. The most common mutation in Northern America and Central/Eastern Europe is the missense mutation p.His1069Gln on exon 14, with an allelic frequency ranging between 30% and 72% [51]. A few areas in the world have a particularly high prevalence of WD. In particular, Costa Rica, Sardinia, the Canary Islands, and Crete [51a, 52] have all been reported to have a high incidence. The incidence of WD in Costa Rica is the highest in the world at 4.9–6:100,000 inhabitants, with the majority having hepatic manifestations of the disease and >5% of patients presenting with fulminant liver failure. This high incidence is thought to be due to the founder effect, with 61% of patients sharing identical mutations at p.Asn1270Ser in *ATP7B* [52a]. In Sardinia, the screening of 5290 newborns for WD found an incidence of 1:2707 live births with a common mutation in $c. -441/-427$del in the promoter region, which may alter ATP7B expression. The prevalence of WD in the northeastern region of the Canary Islands in Spain was also found to be high with an incidence of approximately 1:2600 births due to the mutation, p.Leu708Pro [53]. A high incidence of WD of six out of 90 births has been also reported in the isolated mountains of Crete [54]. Table 1 lists other common mutations according to geographic location.

GENOTYPE-PHENOTYPE CORRELATION

Despite extensive investigation, direct genotype-phenotype correlation in WD has been difficult to establish [12, 33, 46, 64, 88]. The variety of mutations and compound heterozygous nature of WD complicates the process of characterizing each genetic variant and its clinical presentation. In addition, features such as age of onset and presenting symptoms may be hampered by delay in diagnosis and selection bias. The variability in phenotype in WD is likely due to a combination of genetic, metabolic, and environmental factors [89].

One of the most consistent genotype-phenotype correlations in WD is that mutations causing the loss of ATP7B function, mostly early stop mutations and mutations in functionally important regions, result in severe, early-onset disease with a predominantly hepatic presentation. Studies in mouse models such as the toxic milk (tx) mouse and the Jackson tx

TABLE 1 Common Mutations According to Geographic Location

Common Mutations	Region	Frequency	Reference
p.His1069Gln or H1069Q on exon 14	Northern America and Central/Eastern Europe	Poland (72%), Germany (63%), Bulgaria (59%), Czech Republic (57%), Yugoslavia (49%), Russia (49%), Hungary, (43%), United States (40%), Serbia (38%), Sweden (38%), Romania (38%), Brazil (37%), Greece (35%), Austria (34%), The Netherlands (33%), United Kingdom (19%), Iran, (19%), Denmark (18%), Turkey (17.4%), Italy (17.5%), France (15%)	[4, 46, 50–52, 55–70]
p.Met645Arg	Spain	27%	[71]
p.Tyr670Stop	Iceland	100%	[72, 73]
p.Arg778Leu or R778L on exon 8	Eastern Asia (i.e., China, Taiwan, Korea, Japan, and Thailand), North India	14%–49%	[74, 74a]
p.Asn958fs	Japan	17.95%	[75, 76]
p.Ala1003Val (11%), pCys271* (11%), p.Pro768Leu (9%), pArg969Glun (9%)	South India	11%	[77, 78]
pCys271* (16%), p.Gly1061Glu (11%)	East India	16%	[79]
pCys271* (20%), p.Glu122fs (11%)	West India	20%	[80, 81]
p.Gln1399Arg (32%) and p.Ser774Arg (16%)	Saudi Arabia	32%	[82]
p.Ala1003Thr	Lebanon	44.7%	[33]
IVS18+6T>C (42.2%), pAla1140Val (40.6%), p.Lys832Arg (26.5%)	Egypt	42.2%	[83, 84]
p.Asn1270Ser	Costa Rica	61%	[85]
c. −441/−427del in the promoter region	Sardinia	60.5%	[86]
p.Leu708Pro	Canary Islands	64%	[53]
p.Ala1135GlnfsX13	Venezuela	27%	[87]

mouse (tx^j) with fulminant hepatic disease harbor point mutations causing the loss of ATP7B function, but do not affect ATP7B synthesis [31, 90–92]. Point mutations in less important regions of the gene are generally associated with later onset and predominantly neurological or psychiatric presentation [64].

In general, individuals with protein-truncating mutations have earlier onset of disease by decreasing protein stability and quantity [45]. A milder phenotype has been linked with certain mutations with a partially preserved copper-transporting function [37, 40, 93].

Other specific mutations have been shown to have an association with age of onset and clinical presentation. The majority of individuals with the R778L mutation present with hepatic features and have an earlier disease onset [94]. Those with the H1069Q substitution have a largely neurological presentation with a later onset [66, 95]. Kayser-Fleischer Rings appear to be more common in H1069Q homozygous patients in Hungary at the time of diagnosis than in compound heterozygous individuals [62]. A study by Cheng et al. examined the genomic DNA from 1222 patients and 110 healthy controls using a rapid multiplex polymerase chain reaction (PCR) Mass Array method for detecting 100 mutant alleles of interest. In patients not found to have a mutation in the 110 alleles, PCR-Sanger sequencing was used. Findings demonstrated that pArg778Leu correlated with a younger age of onset and lower levels of ceruloplasmin and serum copper [96].

Other variants pArg919Gly and pThr935Met were associated with higher ceruloplasmin levels [96]. The variant pArg919Gly was also associated with neurological symptoms, whereas pThr935Met was associated with combined neurological and visceral manifestations [96]. The more severe the impact of the mutation was on ATP7B protein, the younger the age of onset and the lower the level of ceruloplasmin [96].

Other studies have examined the same mutation in homozygotes compared with compound heterozygotes to look for genotype-phenotype correlations. In a study of 76 members of a large, consanguineous Lebanese family, WD was demonstrated in nine subjects. All nine had the c.2299insC mutation, five were homozygous and four were compound heterozygous with p.Ala1003Thr. The study reported a correlation between c.2299insC and hepatic disease and between the p.Ala1003Thr mutation and neurological disease [33].

Other mutations that may affect the clinical phenotype of WD include *MTHFR* [67], *COMMD1* [97], *ATOX1* [98], *XIAP* [99], *PNPLA3* and hepatic steatosis [100], and *DMT1* [101], although none of these genes have been demonstrated to have significant diagnostic or predictive value.

Clinical features of WD are variable even in patients carrying the same homozygous mutation, individuals within the same family, and between monozygotic twins [102, 103]. While there are some reports that do suggest a correlation of clinical symptoms and biochemical results within families [104–106], others have shown significant differences in age of onset and presenting symptoms among siblings [107, 108] and family members with the same mutation [109]. In particular, different presentations in monozygotic twins indicate that there may be additional genetic influences and environmental factors that could affect the phenotype in WD.

CLINICAL MOLECULAR DIAGNOSIS

Direct Sanger sequencing of the *ATP7B* gene or molecular testing for familial mutations that were previously identified is the current gold standard for confirming a diagnosis of WD, particularly when supporting investigations are inconclusive or when the presentation is unusual [110]. Examples of such cases include a 3-year-old child with early-onset hepatic disease [111] or late-onset disease, in two siblings in their 70s—the oldest reported patients so far at time of diagnosis [79, 112, 113].

Due to the large size of the gene, sequencing the entire coding region is usually not feasible. In the past, most pathogenic variants in *ATP7B* were identified using a combination of PCR/RFLP, SSCP, DGGE, TTGE, DHPLC, and Sanger sequencing [46, 52, 71, 114, 115]. This complex approach is hampered by the low detection rate and high turnaround time. Identification of prevalent mutations in a given population facilitates mutation-based molecular diagnosis. However, although localized clusters of specific mutations have been described, a screening approach taking into account these regional variants may be affected by ethnic diversity and inaccurate information provided with samples. Direct sequencing of the *ATP7B* gene has become the preferred standard and provides the highest yield in clinical molecular diagnosis. Sequencing involves the use of a multiplex polymerase chain reaction (PCR) to amplify all 21 exons and splice sites of *ATP7B*, including promoter regions. Although large deletions or duplications cannot be detected with the conventional Sanger sequencing method, the probability of these being present in WD appears low [116]. The multiplex ligation-dependent probe amplification (MLPA) test should be considered in cases where clinical suspicious is still high with one pathogenic variant detected. Individuals with only one copy of a mutation should be carefully reviewed in the context of other biochemical and clinical findings. Molecular genetic testing using direct mutation analysis is very effective in identifying new patients and presymptomatic siblings [117].

As WD is an inherited autosomal recessive disorder, screening asymptomatic family members is important to prevent future complications in affected individuals. Once homozygous or compound heterozygous mutations in *ATP7B* have been established in the index patient, mutation detection facilitates family screening. In family members in whom clinical and biochemical features are uncertain, the demonstration of either heterozygous (carrier) or wild-type gene sequence prevents unnecessary treatment [47]. In asymptomatic family members, finding the same genotype confirms the diagnosis and allows for prompt treatment initiation to prevent disease progression.

If a diagnosis of WD is achieved on the basis of clinical and biochemical testing, but screening for *ATP7B* mutations is not available, family screening can be done by haplotype analysis of polymorphic markers flanking the disease gene [73, 79, 101]. Microsatellite or single-nucleotide polymorphisms (SNPs) in the *ATP7B* lateral wing are used for haplotyping, which is useful for screening relatives of patients with previously identified familial mutations. Although rare, the possibility of recombination events (typically between 0.5% and 5% of cases) still needs to be considered. The rate of recombination events is dependent on which flanking markers are studied. False positives may occur if haplotyping is used on patients with low-probability gene recombinations.

Prenatal diagnosis can be obtained by amniocentesis or chorionic villus sampling although this is not routinely recommended [118]. Disease-causing alleles of an affected family member have to be identified, or linkage established in the family before prenatal testing can be performed. Considering the treatability of WD when diagnosed early, the risks and benefits of prenatal diagnosis need to be strongly taken into account when deciding on prenatal testing.

Interpretation of variants of uncertain significance (VUS) has become a significant challenge affecting accurate interpretation and genetic counseling. Screening family members may help with the interpretation of VUS, but not all variants can be resolved with this approach. Functional analysis is often needed; however, no clinically applicable functional analysis method is currently available. A computational approach to predict the significance of these mutations is often helpful, but further development of a concrete model is required to demonstrate its use in facilitating clinical decision-making.

Genetic testing for WD remains complex and costly due to the variety of different mutations, the occurrence of regulatory mutations in noncoding sequence, and the relatively large size of the gene. A new high-throughput next-generation sequencing technology that has recently been developed can sequence 6 million base pairs of DNA per hour with accuracy >99%. In the future, specialized laboratories may be able to use these developments to sequence the entire genomic Wilson disease gene from patients, including not only the translated exons but also the important noncoding sequences that are not normally investigated, to detect all variants.

POPULATION SCREENING

Newborn screening aims to detect treatable congenital conditions that can affect a child's health and development. Recent tandem mass spectrometry (MS/MS) applications have markedly expanded the ability to screen for >50 metabolic diseases from a single dried blood spot. In addition to the original Wilson-Jungner classic screening criteria [119], the American College of Medical Genetics (ACMG) convened the Newborn Screening Expert Group to develop a standardized screening panel in 2006 [120]. WD is ideal for screening given the availability of effective treatment that can prevent disease progression if initiated at an early stage [121, 122]. Unfortunately, as of yet, no cost-effective tests for early screening have been developed for WD. Several small studies have been conducted using ceruloplasmin with limited outcomes [17, 123, 124]. A substantial number of newborns present with physiologically low ceruloplasmin; therefore alone, this test is not a sufficient screen in newborns. Testing for ceruloplasmin at around the age of 3 may be a more appropriate population screening method; however, mandatory health checkups with blood testing at this age are not universally available in the United States and globally.

Many treatable congenital disorders are caused by mutations that result in absent or diminished levels of proteins; thus, protein biomarkers have significant potential in the diagnosis/screening of congenital disorders. LC-MRM-MS has emerged as a robust technology that enables highly precise, specific, multiplex quantification of signature proteotypic peptides as stoichiometric surrogates of biomarker proteins. Currently, the use of peptide immunoaffinity enrichment is being studied [125, 126] to quantify ATP7B in dried blood spots. The signature peptide for ATP7B was either absent or reduced in WD patients providing compelling evidence that surrogate biomarker of ATP7B protein may be used as a novel platform to screen WD in dried blood spots [127]. These data may introduce a new method for screening for WD in newborns. Further clinical validation through a large-scale study will be required to determine the efficacy and applicability of the assay.

CONCLUSION

WD is a monogenic autosomal recessive disease due to pathogenic mutations in *ATP7B* resulting in abnormal copper homeostasis. While a combination of biochemical testing and clinical criteria may assist in the early diagnosis and treatment, genetic sequencing is an important confirmatory test particularly in cases of clinical uncertainty and in family screening. The current gold standard for WD diagnosis is direct sequencing of *ATP7B* or molecular testing for known familial mutations. *ATP7B* is the only identified gene known to cause WD; however, several mutation variants exist with significant geographic variation. Genotype-phenotype correlations have been reviewed although conclusions are not robust. Other influences including environmental factors and dietary copper intake may play an additional role in determining disease phenotype. Further developments in population and newborn screening are in need to detect WD at an early stage as prompt treatment initiation can prevent significant morbidity and mortality. Ceruloplasmin may be useful for screening after the age of three; however, it is insufficient to screen for WD in newborns due to its physiologically low level at this age group. Peptide immunoaffinity assays show promise for newborn screening in the future; however, further large-scale clinical studies are required to determine efficacy of these population-based screening methods for WD.

REFERENCES

[1] Wu F, et al. Wilson's disease: a comprehensive review of the molecular mechanisms. Int J Mol Sci 2015;16(3):6419–31.
[2] Bennett JT, Schwarz KB, Swanson PD, Hahn SH. An exceptional family with three consecutive generations affected by Wilson disease. JIMD Rep 2013;10:79–82.
[3] Park H, et al. Pseudo-dominant inheritance in Wilson's disease. Neurol Sci 2015;37(1):153–5.
[4] Coffey AJ, et al. A genetic study of Wilson's disease in the United Kingdom. Brain 2013;136(Pt 5):1476–87.
[5] Petrukhin KE, Lutsenko S. Characterization of the Wilson disease gene encoding a P-type copper transporting ATPase: genomic organization, alternative splicing and structure/function predictions. Hum Mol Genet 1994;3:1647–56.
[6] Bull PC, Cox DW. Wilson disease and Menkes disease: new handles on heavy metal transport. Trends Genet 1994;10(7):246–52.
[7] Tanzi RE, et al. The Wilson disease gene is a copper transporting ATPase with homology to the Menkes disease gene. Nat Genet 1993;5(4):344–50.
[8] Bull PC, et al. The Wilson disease gene is a putative copper transporting P-type ATPase similar to the Menkes gene. Nat Genet 1993;5(4):327–37.
[9] Kenney SM, Cox DW. Sequence variation database for the Wilson disease copper transporter, ATP7B. Hum Mutat 2007;28(12):1171–7.
[10] Lovicu M, et al. The canine copper toxicosis gene MURR1 is not implicated in the pathogenesis of Wilson disease. J Gastroenterol 2006;41(6):582–7.
[11] Mak CM, Lam C-W. Diagnosis of Wilson's disease: a comprehensive review. Crit Rev Clin Lab Sci 2008;45(3):263–90.
[12] Nicastro E, et al. Re-evaluation of the diagnostic criteria for Wilson disease in children with mild liver disease. Hepatology (Baltimore, MD) 2010;52(6):1948–56.
[13] Park S, et al. Identification of novel ATP7Bgene mutations and their functional roles in Korean patients with Wilson disease. Hum Mutat 2007;28(11):1108–13.
[14] Culotta V, Scott RA. Metals in cells. John Wiley & Sons; 2016.
[15] Walker JM, et al. The N-terminal metal-binding site 2 of the Wilson's disease protein plays a key role in the transfer of copper from Atox1. J Biol Chem 2004;279(15):15376–84.
[16] Cater MA, La Fontaine S, MERCER JFB. Copper binding to the N-terminal metal-binding sites or the CPC motif is not essential for copper-induced trafficking of the human Wilson protein (ATP7B). Biochem J 2007;401(1):143–53.
[17] Yamaguchi Y, et al. Mass screening for Wilson's disease: results and recommendations. Pediatr Int 1999;41(4):405–8.
[18] Scherfer M, Roelofsen H, Wolters H, Hofmann WJ, Müller M, Kuipers F, Stremmel W, Vonk RJ. Localization of the Wilson's disease protein in human liver. Gastroenterology 1999;117(6):1380–5.
[19] Michalczyk AA, Rieger J, Allen KJ, Mercer JF, Ackland ML. Defective localization of Wilson disease protein (ATp7B) in the mammary gland of the toxic milk mouse and the effects of copper supplementation. Biochem J 2000;352:565–71.
[20] Petrukhin K, Fischer SG. Mapping, cloning and genetic characterization of the region containing the Wilson disease gene. Nat Genet 1993;5(4):338–43.
[21] Gourdon P, et al. Crystal structure of a copper-transporting PIB-type ATPase. Nature 2011;475(7354):59–64.
[22] Cater MA, et al. Intracellular trafficking of the human Wilson protein: the role of the six N-terminal metal-binding sites. Biochem J 2004;380(3):805–13.
[23] Lenartowicz M, Krzeptowski W. Structure and function of ATP7A and ATP7B proteins—Cu-transporting ATPases. Postępy Biochem 2010;56(3):317–27.
[24] Fatemi N, Sarkar B. Molecular mechanism of copper transport in Wilson disease. Environ Health Perspect 2002;110(Suppl. 5):695–8.
[25] Sazinsky MH, et al. Structure of the ATP binding domain from the *Archaeoglobus fulgidus* Cu+-ATPase. J Biol Chem 2006;281(16):11161–6.
[26] Lutsenko S, et al. N-terminal domains of human copper-transporting adenosine triphosphatases (the Wilson's and Menkes disease proteins) bind copper selectively in vivo and in vitro with stoichiometry of one copper per metal-binding repeat. J Biol Chem 1997;272(30):18939–44.
[27] DiDonato M, Hsu HF, N. S. Copper induced conformational changes in the N-terminal domain of the Wilson disease copper-transporting ATPase. Biochemistry 2000;39(7):1890–6.
[28] Vanderwerf SM, Cooper MJ, Stetsenko IV, Lutsenko S. Copper specifically regulates intracellular phosphorylation of the Wilson's disease protein, a human copper-transporting ATPase. J Biol Chem 2001;276(39):36289–94.
[29] Dmitriev O, Tsivkovskii R, Abildgaard F, Morgan CT, Markley JL, Lutsenko S. Solution structure of the N-domain of WIlson disease protein: distinct nucleotide-binding environment and effects of disease mutations. Proc Natl Acad Sci U S A 2006;103(14):5302–7.
[30] Gupta A, Das S, Ray K. A glimpse into the regulation of the Wilson disease protein, ATP7B, sheds light on the complexity of mammalian apical trafficking pathways. Metallomics 2018;10:378–87.
[31] Huster D, et al. Consequences of copper accumulation in the livers of the Atp7b-/- (Wilson disease gene) knockout mice. Am J Pathol 2006;168(2):423–34.
[32] Schushan M, et al. A structural model of the copper ATPase ATP7B to facilitate analysis of Wilson disease-causing mutations and studies of the transport mechanism. Metallomics 2012;4(7):669–78.
[33] Usta J, et al. Phenotype-genotype correlation in Wilson disease in a large Lebanese family: association of c.2299insC with hepatic and of p. Ala1003Thr with neurologic phenotype. In: Dmitriev OY, editor. PLoS One 2014;9(11):e109727.
[34] Stenson PD, et al. The human gene mutation database: building a comprehensive mutation repository for clinical and molecular genetics, diagnostic testing and personalized genomic medicine. Hum Genet 2014;133(1):1–9.
[35] Bandmann O, Weiss KH, Kaler SG. Wilson's disease and other neurological copper disorders. Lancet Neurol 2015;14(1):103–13.

[36] Huster D, et al. Defective cellular localization of mutant ATP7B in Wilson's disease patients and hepatoma cell lines. Gastroenterology 2003; 124(2):335–45.
[37] Rodriguez-Granillo A, Sedlak E, Wittung-Stafshede P. Stability and ATP binding of the nucleotide-binding domain of the Wilson disease protein: effect of the common H1069Q mutation. J Mol Biol 2008;383(5):1097–111.
[38] Tsivkovskii R, Efremov RG, Lutsenko S. The role of the invariant His-1069 in folding and function of the Wilson's disease protein, the human copper-transporting ATPase ATP7B. J Biol Chem 2003.
[39] Ralle M, et al. Wilson disease at a single cell level: intracellular copper trafficking activates compartment-specific responses in hepatocytes. J Biol Chem 2010;285(40):30875–83.
[40] Dmitriev OY, et al. Difference in stability of the N-domain underlies distinct intracellular properties of the E1064A and H1069Q mutants of copper-transporting ATPase ATP7B. J Biol Chem 2011;286(18):16355–62.
[41] Okada T, et al. High prevalence of fulminant hepatic failure among patients with mutant alleles for truncation of ATP7B in Wilson's disease. Scand J Gastroenterol 2010;45(10):1232–7.
[42] de Bie P, et al. Distinct Wilson's disease mutations in ATP7B are associated with enhanced binding to COMMD1 and reduced stability of ATP7B. Gastroenterology 2007;133(4):1316–26.
[43] Payne AS, Kelly EJ, Gitlin JD. Functional expression of the Wilson disease protein reveals mislocalization and impaired copper-dependent trafficking of the common H1069Q mutation. Proc Natl Acad Sci U S A 1998;95(18):10854–9.
[44] van den Berghe PVE, et al. Reduced expression of ATP7B affected by Wilson disease-causing mutations is rescued by pharmacological folding chaperones 4-phenylbutyrate and curcumin. Hepatology (Baltimore, MD) 2009;50(6):1783–95.
[45] Merle U, et al. Truncating mutations in the Wilson disease gene ATP7B are associated with very low serum ceruloplasmin oxidase activity and an early onset of Wilson disease. BMC Gastroenterol 2010;10(1):8.
[46] Vrabelova S, et al. Mutation analysis of the ATP7B gene and genotype/phenotype correlation in 227 patients with Wilson disease. Mol Genet Metab 2005;86(1–2):277–85.
[47] Chang Y-F, Imam JS, Wilkinson MF. The nonsense-mediated decay RNA surveillance pathway. Annu Rev Biochem 2007;76(1):51–74.
[48] Mendell JT, et al. Nonsense surveillance regulates expression of diverse classes of mammalian transcripts and mutes genomic noise. Nat Genet 2004;36(10):1073–8.
[49] Gupta A, et al. Cellular copper levels determine the phenotype of the Arg875 variant of ATP7B/Wilson disease protein. Proc Natl Acad Sci U S A 2011;108(13):5390–5.
[50] Ferenci P. Regional distribution of mutations of the ATP7B gene in patients with Wilson disease: impact on genetic testing. Hum Genet 2006; 120(2):151–9.
[51] Caca K, et al. High prevalence of the ATP7B-H1069Q mutation in Wilson disease patients from Eastern Germany. J Hepatol 2000;32:134.
[51a] Zappu A, Magli O, Lepori MB, Dessì V, Diana S, Incollu S, Kanavakis E, Nicolaidou P, Manolaki N, Fretzayas A, De Virgiliis S, Cao A, Loudianos G. High incidence and allelic homogeneity of Wilson disease in 2 isolated populations: a prerequisite for efficient disease prevention programs. J Pediatr Gastroenterol Nutr 2008;47(3):334–8.
[52] Loudianos G, et al. Mutation analysis in patients of Mediterranean descent with Wilson disease: identification of 19 novel mutations. J Med Genet 1999;36(11):833–6.
[52a] Morera B, Barrantes R, Marin-Rojas R. The genetics of Wilson disease. Ann Hum Genet 2003;67:71–80.
[53] García Villarreal L, et al. High prevalence of the very rare Wilson disease gene mutation Leu708Pro in the Island of Gran Canaria (Canary Islands, Spain): a genetic and clinical study. Hepatology (Baltimore, MD) 2000;32(6):1329–36.
[54] Dedoussis GVZ, et al. Wilson disease: high prevalence in a mountainous area of Crete. Ann Hum Genet 2005;69(Pt 3):268–74.
[55] Todorov T, et al. Spectrum of mutations in the Wilson disease gene (ATP7B) in the Bulgarian population. Clin Genet 2005;68(5):474–6.
[56] Møller LB, et al. Clinical presentation and mutations in Danish patients with Wilson disease. Eur J Hum Genet 2011;19(9):935–41.
[57] Bost M, et al. Molecular analysis of Wilson patients: direct sequencing and MLPA analysis in the ATP7B gene and Atox1 and COMMD1 gene analysis. J Trace Elem Med Biol 2012;26(2–3):97–101.
[58] Deguti MM, et al. Wilson disease: novel mutations in the ATP7B gene and clinical correlation in Brazilian patients. Hum Mutat 2004;23(4):398.
[59] de Bem RS, et al. Wilson's disease in Southern Brazil: genotype-phenotype correlation and description of two novel mutations in ATP7B gene. Arq Neuropsiquiatr 2013;71(8):503–7.
[60] Machado AAC, et al. Neurological manifestations and ATP7B mutations in Wilson's disease. Parkinsonism Relat Disord 2008;14(3):246–9.
[61] Firneisz G, et al. Common mutations of ATP7B in Wilson disease patients from Hungary. Am J Med Genet 2002;108(1):23–8.
[62] Folhoffer A, et al. Novel mutations of the ATP7B gene among 109 Hungarian patients with Wilson's disease. Eur J Gastroenterol Hepatol 2007; 19(2):105–11.
[63] Kuppala D, et al. Wilson disease mutations in the American population: identification of five novel mutations in ATP7B. Open Hepatol J 2009; 1(1):1–4.
[64] Panagiotakaki E, et al. Genotype–phenotype correlations for a wide spectrum of mutations in the Wilson disease gene (ATP7B). Am J Med Genet A 2004;131A(2):168–73.
[65] Gomes A, Dedoussis GV. Geographic distribution of ATP7B mutations in Wilson disease. Ann Hum Biol 2016;43(1):1–8.
[66] Stapelbroek JM, et al. The H1069Q mutation in ATP7B is associated with late and neurologic presentation in Wilson disease: results of a meta-analysis. J Hepatol 2004;41(5):758–63.
[67] Gromadzka G, et al. Frameshift and nonsense mutations in the gene for ATPase7B are associated with severe impairment of copper metabolism and with an early clinical manifestation of Wilson's disease. Clin Genet 2005;68(6):524–32.

[68] Iacob R, et al. The His1069Gln mutation in the ATP7B gene in Romanian patients with Wilson's disease referred to a tertiary gastroenterology center. J Gastrointestin Liver Dis 2012;21(2):181–5.
[69] Ivanova-Smolenskaya I, et al. 5-29-06 Molecular analysis in Russian families with Wilson's disease. J Neurol Sci 1997;150:S314.
[70] Zali N, et al. Prevalence of ATP7B gene mutations in Iranian patients with Wilson disease. Hepat Mon 2011;11(11):890–4.
[71] Margarit E, et al. Mutation analysis of Wilson disease in the Spanish population—identification of a prevalent substitution and eight novel mutations in the ATP7B gene. Clin Genet 2005;68(1):61–8.
[72] Thomas GR, Jensson O, et al. Wilson disease in Iceland: a clinical and genetic study. Am J Hum Genet 1995;56(5):1140–6.
[73] Thomas GR, Roberts EA, et al. Haplotypes and mutations in Wilson disease. Am J Hum Genet 1995;56(6):1315–9.
[74] Wei Z, et al. Mutational characterization of ATP7B gene in 103 Wilson's disease patients from Southern China: identification of three novel mutations. Neuroreport 2014;25(14):1075–80.
[74a] Liu XQ, Zhang YF, Liu TT, Hsiao KJ, Zhang JM, Gu XF, Bao KR, Yu LH, Wang MX. Correlation of ATP7B genotype with phenotype in Chinese patients with Wilson disease. World J Gastroenterol 2004;10(4):590–3.
[75] Okada T, et al. Mutational analysis of ATP7B and genotype-phenotype correlation in Japanese with Wilson's disease. Hum Mutat 2000;15(5):454–62.
[76] Tatsumi Y, et al. Current state of Wilson disease patients in Central Japan. Intern Med (Tokyo, Japan) 2010;49(9):809–15.
[77] Kumar SS, et al. Genetics of Wilson's disease: a clinical perspective. Indian J Gastroenterol 2012;31(6):285–93.
[78] Santhosh S, et al. ATP7B mutations in families in a predominantly Southern Indian cohort of Wilson's disease patients. Indian J Gastroenterol 2006;25(6):277–82.
[79] Gupta A, et al. Molecular pathogenesis of Wilson disease: haplotype analysis, detection of prevalent mutations and genotype-phenotype correlation in Indian patients. Hum Genet 2005;118(1):49–57.
[80] Aggarwal A, Bhatt M. Update on Wilson disease. Int Rev Neurobiol 2013;110:313–48.
[81] Aggarwal A, et al. Wilson disease mutation pattern with genotype-phenotype correlations from Western India: confirmation of p.C271* as a common Indian mutation and identification of 14 novel mutations. Ann Hum Genet 2013;77(4):299–307.
[82] Al Jumah M, et al. A clinical and genetic study of 56 Saudi Wilson disease patients: identification of Saudi-specific mutations. Eur J Neurol 2004;11(2):121–4.
[83] Abdel Ghaffar TY, et al. Phenotypic and genetic characterization of a cohort of pediatric Wilson disease patients. BMC Pediatr 2011;11(1):56.
[84] Abdelghaffar TY, et al. Mutational analysis of ATP7B gene in Egyptian children with Wilson disease: 12 novel mutations. J Hum Genet 2008;53(8):681–7.
[85] Shah AB, et al. Identification and analysis of mutations in the Wilson disease gene (ATP7B): population frequencies, genotype-phenotype correlation, and functional analyses. Am J Hum Genet 1997;61(2):317–28.
[86] Figus A, et al. Molecular pathology and haplotype analysis of Wilson disease in Mediterranean populations. Am J Hum Genet 1995;57(6):1318–24.
[87] Paradisi I, De Freitas L, Arias S. Most frequent mutation c.3402delC (p.Ala1135GlnfsX13) among Wilson disease patients in Venezuela has a wide distribution and two old origins. Eur J Med Genet 2015;58(2):59–65.
[88] Cocoş R, et al. Genotype-phenotype correlations in a mountain population community with high prevalence of Wilson's disease: genetic and clinical homogeneity. In: Dermaut B, editor. PLoS One 2014;9(6):e98520.
[89] Leggio L, et al. Genotype–phenotype correlation of the Wilson disease ATP7B gene. Am J Med Genet A 2006;140A(8):933.
[90] Coronado V, Nanji M, Cox DW. The Jackson toxic milk mouse as a model for copper loading. Mamm Genome 2001.
[91] La Fontaine S, et al. Effect of the toxic milk mutation (tx) on the function and intracellular localization of Wnd, the murine homologue of the Wilson copper ATPase. Hum Mol Genet 2001;10(4):361–70.
[92] Theophilos MB, Cox DW, Mercer JF. The toxic milk mouse is a murine model of Wilson disease. Hum Mol Genet 1996;5(10):1619–24.
[93] Huster D, et al. Diverse functional properties of Wilson disease ATP7B variants. Gastroenterology 2012;142(4):947–56.e5.
[94] Wu Z-Y, et al. Molecular diagnosis and prophylactic therapy for presymptomatic Chinese patients with Wilson disease. Arch Neurol 2003;60(5):737–41.
[95] Kalita J, et al. R778L, H1069Q, and I1102T mutation study in neurologic Wilson disease. Neurol India 2010;58(4):627–30.
[96] Cheng N, et al. Spectrum of ATP7B mutations and genotype-phenotype correlation in large scale Chinese patients with Wilson disease. Clin Genet 2017;92:69–79.
[97] Weiss KH. Copper toxicosis gene MURR1 is not changed in Wilson disease patients with normal blood ceruloplasmin levels. World J Gastroenterol 2006;12(14):2239.
[98] Simon I. Analysis of the human Atox 1 homologue in Wilson patients. World J Gastroenterol 2008;14(15):2383.
[99] Weiss KH, et al. Genetic analysis of BIRC4/XIAP as a putative modifier gene of Wilson disease. J Inherit Metab Dis 2010;33(3):233–40.
[100] Stättermayer AF, et al. Genetic factors associated with histologic features of the liver and treatment outcome in chronic hepatitis C patients. Z Gastroenterol 2012;50(05):P51.
[101] Przybyłkowski A, Gromadzka G, Członkowska A. Polymorphisms of metal transporter genes DMT1 and ATP7A in Wilson's disease. J Trace Elem Med Biol 2014;28(1):8–12.
[102] Członkowska A, Gromadzka G, Chabik G. Monozygotic female twins discordant for phenotype of Wilson's disease. Mov Disord 2009;24(7):1066–9.
[103] Kegley KM, et al. Fulminant Wilson's disease requiring liver transplantation in one monozygotic twin despite identical genetic mutation. Am J Transplant 2010;10(5):1325–9.

[104] Chabik G, Litwin T, Członkowska A. Concordance rates of Wilson's disease phenotype among siblings. J Inherit Metab Dis 2014;37(1):131–5.
[105] Ferenci P, et al. Encephalopathy in Wilson disease: copper toxicity or liver failure? J Clin Exp Hepatol 2015;5:S88–95.
[106] Hofer H, et al. Identification of a novel Wilson disease gene mutation frequent in Upper Austria: a genetic and clinical study. J Hum Genet 2012; 57(9):564–7.
[107] Ala A, et al. Wilson's disease. Lancet (London, England) 2007;369(9559):397–408.
[108] Taly AB, et al. Wilson disease: description of 282 patients evaluated over 3 decades. Medicine 2007;86(2):112–21.
[109] Takeshita Y, et al. Two families with Wilson disease in which siblings showed different phenotypes. J Hum Genet 2002;47(10):0543–7.
[110] Caprai S, et al. Direct diagnosis of Wilson disease by molecular genetics. J Pediatr 2006;148(1):138–40.
[111] Wilson DC, et al. Severe hepatic Wilson's disease in preschool-aged children. J Pediatr 2000;137(5):719–22.
[112] Nanji MS, et al. Haplotype and mutation analysis in Japanese patients with Wilson disease. Am J Hum Genet 1997;60(6):1423–9.
[113] Weitzman E, et al. Late onset fulminant Wilson's disease: a case report and review of the literature. World J Gastroenterol: WJG 2014; 20(46):17656–60.
[114] Kim G-H, et al. Estimation of Wilson's disease incidence and carrier frequency in the Korean population by screening ATP7B major mutations in newborn filter papers using the SYBR green intercalator method based on the amplification refractory mutation system. Genet Test 2008; 12(3):395–9.
[115] Shimizu N, et al. Molecular analysis and diagnosis in Japanese patients with Wilson's disease. Pediatr Int 1999;41(4):409–13.
[116] Stenson PD, et al. The human gene mutation database (HGMD) and its exploitation in the fields of personalized genomics and molecular evolution. Hoboken, NJ: John Wiley & Sons, Inc.; 2012.
[117] Manolaki N, et al. Wilson disease in children: analysis of 57 cases. J Pediatr Gastroenterol Nutr 2009;48(1):72–7.
[118] Behari M, Pardasani V. Genetics of Wilsons disease. Parkinsonism Relat Disord 2010;16:639–44.
[119] Wilson JMG, Jungner G. Principles and practice of screening for disease; 1968. p. 34.
[120] American College of Medical Genetics Newborn Screening Expert Group. Newborn screening: toward a uniform screening panel and system—executive summary. Pediatrics 2006;117(5 Pt 2):S296–307.
[121] Hahn SH, et al. Pilot study of mass screening for Wilson's disease in Korea. Mol Genet Metab 2002;76(2):133–6.
[122] Roberts EA, Schilsky ML, American Association for Study of Liver Diseases (AASLD). Diagnosis and treatment of Wilson disease: an update. Hepatology (Baltimore, MD) 2008;47(6):2089–111.
[123] Owada M, et al. Mass screening for Wilson's disease by measuring urinary holoceruloplasmin. J Pediatr 2002;140(5):614–6.
[124] Schilsky ML, Shneider B. Population screening for Wilson's disease. J Pediatr 2002;140(5):499–501.
[125] Whiteaker JR, et al. A targeted proteomics-based pipeline for verification of biomarkers in plasma. Nat Biotechnol 2011;29(7):625–34.
[126] Whiteaker JR, et al. An automated and multiplexed method for high throughput peptide immunoaffinity enrichment and multiple reaction monitoring mass spectrometry-based quantification of protein biomarkers. Mol Cell Proteomics 2010;9(1):184–96.
[127] Jung S, Whiteaker JR, Zhao L, Yoo HW, Paulovich AG, Hahn SH. Quantification of ATP7B protein in dried blood spots by peptide immuno-SRM as a potential screen for Wilson's disease. J Proteome Res 2017;16(2):862–71.

Chapter 10

Biochemical Markers

Aurélia Poujois and France Woimant
French National Reference Centre for Wilson Disease, Neurology Department, Lariboisière Hospital, Assistance Publique-Hôpitaux de Paris, Paris, France

The diagnostic certainty of Wilson disease (WD) relies on the presence of two mutations in the ATP7B gene. However, even if the recent next-generation sequencing procedures speeded up the timeliness of results, genetic testing remains too long and may postpone the diagnosis and the instauration of treatment. Moreover, it may be noncontributory: in Lariboisière cohort (381 patients), the positivity rate of the molecular analysis is about 98%, so that 2% of patients with proved WD have only a single heterozygous mutation or even no mutation in the ATP7B gene. In everyday practice, the combination of clinical symptoms, radiological features, and specific anomalies of copper tests (with the association of low ceruloplasmin, low cupremia, and high cupruria) highly suggests the diagnosis and allows to start treatment before genetic confirmation of the disease. This chapter focuses on the usual copper tests evaluated as biochemical markers of the disease and highlights the recent contribution of exchangeable copper and relative exchangeable copper (REC) (ratio exchangeable copper/serum copper) in WD diagnosis.

TRADITIONAL BIOCHEMICAL MARKERS OF WILSON DISEASE

WD is a primary liver disease, and the biochemical liver tests—especially the serum aminotransferase activities—are generally abnormal [1], except at a very early age or during neurological form of the disease. Their contribution to diagnosis is limited as they are not specific of WD. Ceruloplasmin (Cp), total serum copper (Cu), and urinary copper (UCu) are the three traditional biochemical markers used for WD diagnosis.

Ceruloplasmin

Cp is the principal copper-carrying protein in man, binding 70%–95% of the circulating plasma copper. The protein is largely synthesized in hepatocytes as an inactive, unstable non-copper-bound form, called apoceruloplasmin (apoCp). Subsequently, the addition of six to eight atoms of copper per molecule produces the functional and more stable product holoceruloplasmin (holoCp) secreted into the plasma, which acts as a ferroxidase [2]. The functions of the holoprotein include roles in copper transport and iron metabolism, tissue angiogenesis, antioxidant defense, and coagulation [3]. Level of serum Cp may be measured enzymatically by its copper-dependent oxidase activity or by antibody-dependent assays such as radioimmunoassay, radial immunodiffusion, or nephelometry. The immunologic methods—currently used by automated clinical laboratories—are rapid and give an accurate measure of total Cp but don't distinguish between holoCp and apoCp and tend to overestimate enzymatically active Cp, as apoCp has no enzymatic activity [4–6]. The normal concentration of Cp measured by enzymatic test varies among laboratories but is around 0.20 g/L. (Table 1). Levels of serum Cp are physiologically very low in early infancy to the age of 6 months, peak at higher than adult levels in early childhood (around 0.3–0.5 g/L), and then settle to the adult range [7, 8]. Cp is generally lower in men than in women and is increased by estrogen, pregnancy, and contraceptive pill [4, 9–11]. Being an acute-phase response protein, Cp concentration can vary depending on the health status at the time of sample collections; it increases during inflammation, infections, and rheumatoid arthritis [12] and in patients with myocardial complications or cancer [13]. Further, it is subject to seasonal changes, lowest concentrations occurring in the winter months (November–February) [14].

A concentration below the reference range of 0.2 g/L is consistent with the diagnosis of Wilson disease, whereas a value <0.14 g/L is strongly consistent as this threshold found by Mak et al. allows diagnosing WD with 100% specificity and 93% sensitivity [15]. However, serum Cp alone is not sensitive for WD as low concentrations may be detected in approximately 20% of healthy heterozygous subjects for WD [16], in case of acute viral hepatitis, drug-induced liver

TABLE 1 Copper Balance in Normal and Untreated WD Subjects

	Normal	Suspicion of WD	Wilson Disease	False Negative	False Positive
Ceruloplasmin					
0.5 g/L	0.20–0.40	0.14–0.19	<0.14	**Increased value with** – estrogen, pregnancy, and contraceptive pill – inflammation – infections – rheumatoid arthritis – cancer **Normal value in** – 50% of patients with active WD liver disease with marked hepatic inflammation – 15%–36% of children with WD **Overestimation by immunologic assay**	**Decreased values in** – early infancy to the age of 6 months – winter seasons – healthy heterozygotes WD – acute viral hepatitis – drug-induced liver disease – alcoholic-induced liver disease – acquired copper deficiency – Menkes' disease – malabsorption – malnutrition – cachexia – renal protein loss – liver failure – aceruloplasminemia
Serum total copper					
0.5 µmol/L	14–21	11–13	<10	Idem as ceruloplasmin	Idem as ceruloplasmin
0.5 µg/L	890–1335	700–830	<635		
24 h urinary copper excretion					
0.5 µmol/24 h	<0.6	>0.6[a]	>1.6	**Normal** – if incorrect collection (incomplete, spot collection) – in 16%–23% of WD patients, especially children and asymptomatic siblings	**Increased in** – autoimmune hepatitis, – chronic active liver disease – cholestasis – acute hepatic failure of any origin with hepatocellular necrosis – healthy heterozygotes WD (intermediate levels)
0.5 µg/24 h	<40	>40[a]	>100		
Liver biopsy					
0.5 µmol/g dry tissue	0.2–0.9	0.65–3	>3.3	**Normal or intermediate values** (due to uneven copper distribution) in WD patients with – active liver diseases – regenerative nodules	**Increased levels in:** – cholestatic syndromes
0.5 µg/g dry tissue	13–57	40–190	209		
NCC					
0.5 µmol/L	1.6		>3.2		**Increased in:** – acute liver failure of any etiology – chronic cholestasis – copper intoxication Overestimation of ceruloplasmin by immunologic assay may result in negative values

TABLE 1 Copper Balance in Normal and Untreated WD Subjects—cont'd					
	Normal	Suspicion of WD	Wilson Disease	False Negative	False Positive
CuEXC					
0.5 µmol/L	0.62–1.15	>1.53	>2.08 in extrahepatic form		
0.5 µg/L	39–73	>97	>132 in extrahepatic form		
REC					
0.5 %	3–8.1		>15 in asymptomatic >18.5 in symptomatic patients		

Copper conversion: 1 µmol/L = 63.546 µg/L; *NCC*, nonceruloplasmin-bound copper calculated by the formula = (total serum copper (µmol/L) − 0.049) × ceruloplasmin (mg/L); *REC*, ratio exchangeable copper/total serum copper.
[a]24h urinary excretion in children and asymptomatic siblings.

disease, alcoholic-induced liver disease, acquired copper deficiency, Menkes' disease, malabsorption, malnutrition, cachexia, marked renal protein loss, liver failure, or aceruloplasminemia [17, 18]. Moreover, normal or near-normal concentrations may occur in about half of patients with active Wilson's liver disease with marked hepatic inflammation [19], in 15%–36% of children with WD [19], or in few patients with neurological presentations of WD [20] (Table 1).

The utility of routine Cp measurement for patients referred for the evaluation of liver anomalies has been assessed in different European and American cohorts [18, 21, 22]. In all, the positive predictive value was low (between 5.9% and 11%) in populations in whom 0.03%–0.19% had WD. This low positive predictive value is largely due to the low pretest probability of the diagnosis. Measuring serum Cp as a singular diagnostic test for WD or as part of the battery of unselected liver screening tests is inappropriate and low-yield. Considered in isolation and due to all these limitations, Cp cannot be regarded as a marker for Cu excess.

Serum Copper and Serum Nonceruloplasmin-Bound Copper (NCC)

Serum copper determination is usually made by either colorimetric or atomic absorption spectrometry methods, two techniques that correlate satisfactorily [23]. Normal ranges of the total serum copper (which includes copper incorporated in Cp) are around 14–21 µmol/L (Table 1).

Although a disease of copper overload, the total serum copper in WD is usually decreased in proportion to the decreased Cp in the circulation [5]. In WD patients, serum copper level is usually under 14 µmol/L. However, a low total serum copper determination is insufficient to set the diagnosis of WD as in patients with severe liver injury, serum copper may be within the normal range, independent of whether serum Cp levels are elevated or low. Moreover, during acute liver failure, levels of serum copper may be markedly elevated due to the sudden release of the metal from liver tissue stores. Normal or elevated serum copper levels in the face of decreased levels of Cp indicate an increase in the concentration of copper not bound to Cp in the blood [8] (Table 1).

The serum nonceruloplasmin-bound copper (NCC) has been proposed as a diagnostic test or monitoring tool for WD as it should estimate the toxic unbound (or "free") copper pool [24, 25]. The upper limit of normal for NCC is quoted in the literature [26] as 1.6 µmol/L. In most untreated WD patients, it is elevated above 3.2 µmol/L, but its specificity is low as the NCC concentration may be elevated in acute liver failure of any etiology, in chronic cholestasis, and in cases of copper intoxication (Table 1). Moreover, its calculation (NCC = (total serum copper (µmol/L) − 0.049) × ceruloplasmin (mg/L)) and concept are still a matter of debate. The main limit of this calculation is that no standardized reference method is yet available for ceruloplasmin. The immunologic methods cross-react with apoCp and have been accused of giving falsely high values. In WD, negative values for NCC are found in 10%–25% of patients where Cp are measured using an immunonephelometric method [24, 27], thus indicating fundamental flaws in either the NCC calculation or the NCC concept. This may be also due to the indirect way of the determination of Cp [28–30] especially in WD where total copper and Cp concentrations are low, sometimes below the detection limit of the analytic method. The measurement of Cp oxidase activity appears more appropriate and has been proposed to increase the detection threshold, but NCC calculation with this measurement has been rarely studied [31], and moreover, it is not performed routinely. Currently, EASL guidelines comment that NCC is not recommended for the diagnosis of Wilson disease [19].

Urinary Excretion of Copper

Urinary copper excretion is negligible in healthy individuals [32]. In WD, the ATP7B-mutated protein triggers the release of copper in the blood and then in the urinary tract, so the amount of copper excreted in the urine is useful for diagnosing WD. In untreated patients, the 24h urinary excretion of copper is thought to reflect the amount of nonceruloplasmin-bound copper in the circulation [31]. The reference limits for normal 24h excretion of copper vary among clinical laboratories, but many laboratories take 40 μg/24h (0.6 μmol/24h) as the upper limit of normal. The conventional level taken as diagnostic of untreated WD is >100 μg/24h (>1.6 μmol/24h) in symptomatic patients [32, 33]. Interpreting 24h urinary copper excretion may be difficult. Basal 24h urinary copper excretion may be below 100 μg/24h (<1.6 μmol/24h) at presentation in 16%–23% of patients, especially in children and asymptomatic siblings [34–36]. Some authors even propose a better threshold of 40 μg/24h (0.6 μmol/24h) for diagnosis in asymptomatic children [37, 38]. Another difficulty to interpret a high 24h urinary copper comes from overlap with findings in other types of liver disease (e.g., autoimmune hepatitis, chronic active liver disease, or cholestasis and in particular during acute hepatic failure of any origin). Heterozygotes may also have intermediate levels [35, 39] (Table 1).

For accurate determination of urinary copper excretion, two points should be carefully checked: The urine collection has to be complete (the exact urine volume and the total creatinine excretion per 24h shall be noticed; spot urine specimens should not be used as there is too much variability in the copper content), and copper contamination of the collection device should be avoided. The test is not validated in case of renal failure or impaired kidney function.

Measuring the 24h urinary copper excretion while giving D-penicillamine can be a useful provocative test. This test has only been standardized in a pediatric population in which 500 mg of D-penicillamine was administered orally at the beginning and again 12h later during the 24h urine collection, irrespective of body weight [40]. Compared with a spectrum of other liver diseases including autoimmune hepatitis, primary sclerosing cholangitis, and acute liver failure, a clear differentiation was found when the 24h urinary copper excretion was >1600 μg copper/24h (>25 μmol/24h). Reevaluation of the D-penicillamine challenge test in children found it valuable for the diagnosis of WD in patients with active liver disease (sensitivity 92%) but poor for excluding the diagnosis in asymptomatic siblings (sensitivity only 46%) [41]. As stated previously, recent data now suggest that using a lower threshold for urinary copper excretion (without D-penicillamine stimulation) of only 40 μg/24h (0.6 μmol/24h) increases sensitivity of the test in children and eliminates the need for the stimulation testing with D-penicillamine [42, 43]. This provocative test has been used in adults but is not standardized as many of the reported results of this test in adults used different dosages and timing for administration of D-penicillamine. Thus, this test is not recommended for the diagnosis of Wilson disease in adults [19].

Liver Biopsy With Copper Determination

Each isolated biochemical marker is insufficient to set the diagnosis, but the diagnosis is highly predictable if the copper pattern shows an association of a low Cp and Cu determination and an elevated basal 24h urinary copper. However, this classical triad may be present in 15% of heterozygote carrier as well. On the other hand, 3% of WD patients with confirmed biallelic disease-causing mutations have a normal copper balance (personal data, Lariboisière registry). In unexplained liver disease or in a few selected subjects with mild biological abnormalities and no or one mutation, the combination of 24h urinary copper, Cp, and Cu may be taken in default to make the diagnosis of WD. In these situations, liver biopsy with copper determination is recommended by American guidelines, but the decision to perform an invasive procedure in asymptomatic or paucisymptomatic subject is not always easy [8]. Moreover, the method of hepatic copper determination has not been standardized, and the threshold for diagnosis remains discussed [44–46]. Due to the inhomogeneous distribution of copper within the liver [47], the concentration may be falsely considered in the normal ranges in up to 18% of WD patients [44]. In children with cirrhosis, a single determination of liver copper content is not recommended because of its unreliability [48] (Table 1).

Biopsy specimens are obtained in routine fashion, without the need for copper-free solutions or instruments. The usual procedure for quantitative copper determination is to dry the 10–20 mg needle biopsy sample before placing it in a copper-free container. Paraffin-embedded specimens may also be analyzed for copper content but may be less reliable if the specimen is small. As copper is unevenly distributed in the liver, sample size is very important to correctly determine mean hepatic copper content and avoid false results: at least 1 cm of biopsy core length should be analyzed [46, 49]. In normal individuals, hepatic copper concentrations are between 15 and 55 μg/g (0.2–0.9 μmol/g) dry liver [50, 51]. A hepatic copper level ≥250 μg/g (4 μmol/g) dry tissue has historically been considered most valid for diagnosing WD [8, 19, 52] but without evaluation and validation in a sufficiently large group of patients with and without WD. In 2005, Ferenci et al. evaluated copper content in 114 liver biopsies of new untreated WD, in 219 patients with noncholestatic liver diseases, and in 26 without evidence of liver disease. 95/114 WD patients had a liver content >250 μg/g. By lowering the cutoff from 250 to 75 μg/g, the sensitivity of liver copper content to diagnose WD increased from 83.3% to 96.5% (91.3%–99.1%), but the specificity decreased from 98.6% to 95.4% [44]. A recent Chinese study determined prospectively dry-weight hepatic copper content in 178 WD patients and 513 patients with various liver diseases and evaluated which concentration thresholds for liver copper provide the highest diagnostic accuracy for WD. Two passes of liver biopsy were performed, and an entire core of liver biopsy sample was used for copper determination to decrease sampling variability. The sensitivity, specificity, positive predictive value, and negative predictive value of hepatic copper level for the diagnosis of WD in the absence of primary biliary cirrhosis and primary sclerosing cholangitis at the conventional cutoff value of 250 μg/g dry weight were 94.4%, 96.8%, 91.8%, and 97.8%, respectively. The most useful cutoff value was 209 μg/g (3.3 μmol/g) dry weight, with a sensitivity of 99.4% and a specificity of 96.1% [45] (Table 1).

Recently, an XRF spectroscopy-based approach on liver tissue was described as a promising diagnostic tool with major advantages [53]. This nondestructive method allows simultaneous quantitative analysis of multiple elements. It has a good reliability to evaluate the intrahepatic Cu level as a fair correlation was found between the detected Cu fluorescence intensity and the measured hepatic tissue Cu concentration. The versatility of XRF is underlined by the possibility to perform measurements on unfixed (fresh or frozen) or fixed tissues such as formalin-fixed paraffin-embedded samples. Authors showed that the intensity of copper related to iron and zinc significantly discriminated WD from other genetic or chronic liver diseases with 97.6% specificity and 100% sensitivity [53].

Because of the limits of the "golden triad" and the hepatic copper determination, new biological markers are investigated. Recent research efforts focused on a better determination of the labile fraction of serum copper (exchangeable copper).

EXCHANGEABLE COPPER AND REC: NEW SPECIFIC AND SENSIBLE BIOCHEMICAL MARKERS OF WILSON DISEASE

In 2009, a new validated assay for the direct determination of the labile fraction of copper complexed to albumin named exchangeable copper (CuEXC) was set up in Lariboisière Hospital (Paris) in healthy subjects and then evaluated as a diagnostic tool for WD [54, 55]. CuEXC is easily exchangeable in the presence of high-copper-affinity chelators such as EDTA and can be determined after the incubation of serum with EDTA during 1 h followed by ultrafiltration of the diluted serum. The analytic reliability of CuEXC determination was confirmed [54, 55]. The normal values range from 0.62 to 1.15 μmol/L (Table 1). The main advantage of this assay is that it doesn't depend on the dosage of Cp and offers an accurate view of the copper overload. Moreover, it can be successfully applied in routine framework to a large number of patients.

The CuEXC/total serum copper ratio (viz., REC as relative exchangeable copper) offers a promising alternative to assess the toxic fraction of copper in the blood of WD patients. El Balkhi et al. compared the REC to the usual tests used for WD diagnosis (total serum copper, urinary copper excretion, Cp, and NCC) in three different adult populations (WD, heterozygous, and normal subjects). The usual biological tests used for WD diagnosis yielded lower sensitivity and specificity compared to the REC. Regardless of sex, age, or the degree of underlying liver failure, in this study, REC appeared as an excellent biomarker for the diagnosis of WD offering 100% sensitivity and 100% specificity with a threshold >18.5%. REC was the only biomarker without false negatives or positives (Fig. 1) [55].

Another study conducted in the Long-Evans Cinnamon rat (LEC), a referenced WD animal model, confirmed the results [56]. The authors tested the validity of the new biomarker throughout the course of the liver disease, from a presymptomatic stage to the beginning of chronic liver failure. REC remained discriminating between LEC and normal rats and was not influenced by the presence of liver damage. The sensibility and specificity of REC was 100%.

The second proposed use of REC is familial screening in WD. A recent study focused on family screening in 127 asymptomatic subjects and compared REC with serum copper, Cp concentration, urinary copper excretion, and molecular analysis [57]. The subjects screened were either first- or second-degree relatives of the genetically confirmed index case and were classified into three groups according to molecular testing of the ATP7B gene (WD, heterozygous, or normal). REC

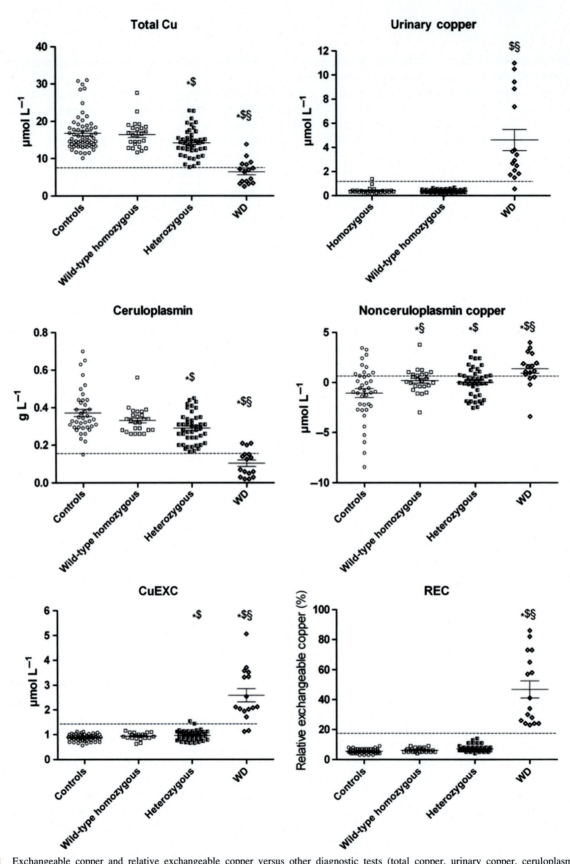

FIG. 1 Exchangeable copper and relative exchangeable copper versus other diagnostic tests (total copper, urinary copper, ceruloplasmin, and nonceruloplasmin-bound copper) to distinguish WD from other subjects. Populations studied were control subjects ($n=62$), wild-type homozygous ($n=25$), heterozygous ($n=45$), and patients with WD ($n=16$). For the diagnosis of WD, REC offers 100% sensitivity and 100% specificity with a threshold $>18.5\%$. *CuEXC*, exchangeable copper; *REC*, relative exchangeable copper; *, $, and § indicate statistical significant difference ($P<0.05$) versus controls, homozygous, and heterozygous, respectively. *Broken lines* represent the cutoffs obtained from ROC analysis. *(From El Balkhi S, Trocello JM, Poupon J, Chappuis P, Massicot F, Girardot-Tinant N, et al. Relative exchangeable copper: a new highly sensitive and highly specific biomarker for Wilson's disease diagnosis. Clin Chim Acta 2011;412(23–24):2254–60.)*

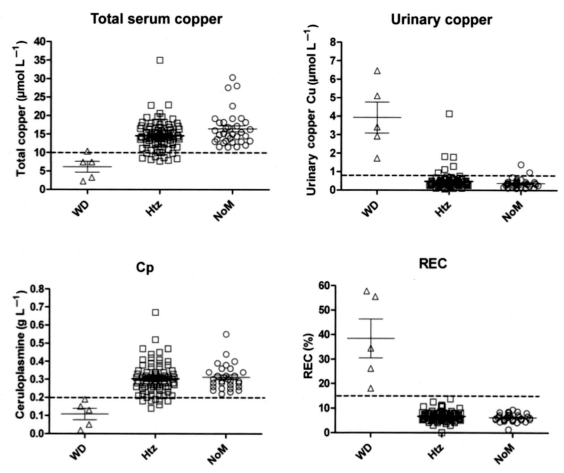

FIG. 2 Interest of REC in WD familial screening. Results of urinary copper, ceruloplasmin, total serum copper, and REC in WD patients, ATP7B heterozygous carriers, and subjects without ATP7B mutations. REC determination significantly discriminated subjects without Wilson disease (heterozygous and normal) from WD patients with a cutoff of 15%. *WD*, Wilson disease patients; *Htz*, ATP7B heterozygous carriers; *NoM*, subjects without ATP7B mutations; *Cp*, ceruloplasmin; *REC*, relative exchangeable copper; *dashes* correspond to cutoff of biological parameters in favor of WD. *(From Trocello JM, El Balkhi S, Woimant F, Girardot-Tinant N, Chappuis P, Lloyd C, et al. Relative exchangeable copper: a promising tool for family screening in Wilson disease. Mov Disord. 2014;29(4):558–62.)*

determination significantly discriminated subjects without Wilson disease (heterozygous and normal) from WD patients with a cutoff of 15% (Fig. 2). REC gives a new practical and fast answer to an appropriate diagnosis of the disease and appears to be a promising tool for family screening in WD, especially for heterozygous ATP7B carriers who could present with slight copper abnormalities.

REC determination in patients with other chronic/acute liver diseases (NASH, autoimmune, or infectious) has been also studied in adults and in pediatric populations. REC was below 14% in all non-Wilsonian patients, a result that supports the validity of the biomarker [58].

The severity of WD presentation is linked to the free copper accumulating not only in the liver but also in the brain. Defining at diagnosis, the diffusion of the copper overload is of great interest as the presence of extrahepatic lesions and especially neurological involvement render the prognosis more uncertain [33] and can require a progressive adjustment of doses or change of treatment [59, 60]. Exchangeable copper determination is helpful at diagnosis because initial CuEXC value is closely linked to the presence and the severity of the extrahepatic involvement. In a population of 48 newly diagnosed patients, CuEXC value was found significantly higher in patients with extrahepatic involvement, and a value over 2.08 μmol/L was able to determine the presence of extrahepatic lesions with satisfactory power (AUC=0.883, specificity 94.1%, and sensitivity 85.7%) [27]. This result suggests the presence of a clear threshold concentration of CuEXC above which targeted tissues such as the brain and eyes suffer from toxic exposure to copper, leading to neurological symptoms, corneal lesions, and brain MRI abnormalities (Fig. 3). Moreover, CuEXC at diagnosis provides information on the severity of extrahepatic involvement. CuEXC was positively correlated with the UWDRS neurological score (score evaluating

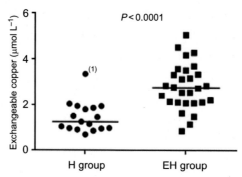

FIG. 3 Distribution of exchangeable copper values in WD with hepatic (H group) and extrahepatic symptoms (EH group). The normal reference range for exchangeable copper is 0.62–1.15 µmol/L. (1) corresponds to the only hepatic patient with hemolytic anemia. *(From Poujois A, Trocello JM, Djebrani-Oussedik N, Poupon J, Collet C, Girardot-Tinant N, et al. Exchangeable copper: a reflection of the neurological severity in Wilson's disease. Eur J Neurol. 2017;24(1):154–60.)*

disability and neurological symptoms), the severity of corneal copper deposits, and the dissemination of MRI cerebral lesions. But in the case of purely hepatic disease, CuEXC could not indicate the severity of liver damage, there being no correlation with the hepatic score [27]. Cases of acute liver failure may be different, however. Indeed, the only patient of the study with acute liver failure and Coombs-negative hemolytic anemia in the hepatic group had a high CuEXC. Moreover, in LEC rats, Schmitt et al. demonstrated that CuEXC levels were correlated with acute liver failure [56]. So, a high CuEXC in the context of fulminant or acute liver failure should be carefully interpreted.

CONCLUSION

A century after the initial description of the disease, biological markers of Wilson disease continue to evolve. The classical association of a low cupremia, low Cp, and high cupruria remains the gold standard, and the hepatic copper determination still has a specific place in certain situations. Since a few years, two new biochemical markers based on the direct determination of exchangeable copper had proved their value: (1) The relative exchangeable copper (REC) allows to confirm diagnosis of WD with very high sensibility and specificity, and (2) the exchangeable copper (CuEXC) determination reflects the diffusion and severity of extrahepatic involvement at diagnosis.

REFERENCES

[1] Clark JM, Brancati FL, Diehl AM. The prevalence and etiology of elevated aminotransferase levels in the United States. Am J Gastroenterol 2003;98(5):960–7.

[2] Bielli P, Calabrese L. Structure to function relationships in ceruloplasmin: a 'moonlighting' protein. Cell Mol Life Sci 2002;59(9):1413–27.

[3] Roeser HP, Lee GR, Nacht S, Cartwright GE. The role of ceruloplasmin in iron metabolism. J Clin Invest 1970;49(12):2408–17.

[4] Gibbs K, Walshe JM. A study of the caeruloplasmin concentrations found in 75 patients with Wilson's disease, their kinships and various control groups. Q J Med 1979;48(191):447–63.

[5] Walshe JM. Wilson's disease: the importance of measuring serum caeruloplasmin non-immunologically. Ann Clin Biochem 2003;40(Pt 2):115–21.

[6] Macintyre G, Gutfreund KS, Martin WR, Camicioli R, Cox DW. Value of an enzymatic assay for the determination of serum ceruloplasmin. J Lab Clin Med 2004;144(6):294–301.

[7] Montagna O, Grosso R, Santoro A, Mautone A. Plasma levels of the serum antioxidants (uric acid, ceruloplasmin, transferrin) in term and preterm neonates in the first week of life. Minerva Pediatr 1994;46(6):255–60.

[8] Roberts EA, Schilsky ML. Diagnosis and treatment of Wilson disease: an update. Hepatology 2008;47(6):2089–111.

[9] Demily C, Parant F, Cheillan D, Broussolle E, Pavec A, Guillaud O, et al. Screening of Wilson's disease in a psychiatric population: difficulties and pitfalls. A preliminary study. Ann Gen Psychiatry 2017;16:19.

[10] Ganaraja B, Pavithran P, Ghosh S. Effect of estrogen on plasma ceruloplasmin level in rats exposed to acute stress. Indian J Med Sci 2004;58(4):150–4.

[11] Méndez MA, Araya M, Olivares M, Pizarro F, González M. Sex and ceruloplasmin modulate the response to copper exposure in healthy individuals. Environ Health Perspect 2004;112(17):1654–7.

[12] Gruys E, Toussaint MJ, Niewold TA, Koopmans SJ. Acute phase reaction and acute phase proteins. J Zhejiang Univ Sci B 2005;6(11):1045–56.

[13] Senra Varela A, Lopez Saez JJ, Quintela Senra D. Serum ceruloplasmin as a diagnostic marker of cancer. Cancer Lett 1997;121(2):139–45.

[14] Kanikowska D, Grzymislawski M, Wiktorowicz K. Seasonal rhythms of "acute phase proteins" in humans. Chronobiol Int 2005;22(3):591–6.

[15] Mak CM, Lam CW, Tam S. Diagnostic accuracy of serum ceruloplasmin in Wilson disease: determination of sensitivity and specificity by ROC curve analysis among ATP7B-genotyped subjects. Clin Chem 2008;54(8):1356–62.

[16] Gromadzka G, Chabik G, Mendel T, Wierzchowska A, Rudnicka M, Czlonkowska A. Middle-aged heterozygous carriers of Wilson's disease do not present with significant phenotypic deviations related to copper metabolism. J Genet 2010;89(4):463–7.

[17] Miyajima H. Aceruloplasminemia. Neuropathology 2015;35(1):83–90.

[18] Cauza E, Maier-Dobersberger T, Polli C, Kaserer K, Kramer L, Ferenci P. Screening for Wilson's disease in patients with liver diseases by serum ceruloplasmin. J Hepatol 1997;27(2):358–62.

[19] European Association for Study of Liver. EASL clinical practice guidelines: Wilson's disease. J Hepatol 2012;56(3):671–85.

[20] Woimant F, Trocello JM. Disorders of heavy metals. Handb Clin Neurol 2014;120:851–64.

[21] Tapper EB, Rahni DO, Arnaout R, Lai M. The overuse of serum ceruloplasmin measurement. Am J Med 2013;126(10):926.e1–5.

[22] Kelly D, Crotty G, O'Mullane J, Stapleton M, Sweeney B, O'Sullivan SS. The clinical utility of a low serum ceruloplasmin measurement in the diagnosis of Wilson disease. Ir Med J 2016;109(1):341–3.

[23] Arnaud J, Chappuis P, Zawislak R, Houot O, Jaudon MC, Bienvenu F, et al. Comparison of serum copper determination by colorimetric and atomic absorption spectrometric methods in seven different laboratories. The S.F.B.C. (Société Française de Biologie Clinique) Trace Element Group. Clin Biochem 1993;26(1):43–9.

[24] Twomey PJ, Viljoen A, Reynolds TM, Wierzbicki AS. Non-ceruloplasmin-bound copper in routine clinical practice in different laboratories. J Trace Elem Med Biol 2008;22(1):50–3.

[25] Walshe JM. Monitoring copper in Wilson's disease. Adv Clin Chem 2010;50:151–63.

[26] Gaffney D, Fell GS, O'Reilly D. Wilson's disease: acute and presymptomatic laboratory diagnosis and monitoring. J Clin Pathol 2000;53(11):807–12.

[27] Poujois A, Trocello JM, Djebrani-Oussedik N, Poupon J, Collet C, Girardot-Tinant N, et al. Exchangeable copper: a reflection of the neurological severity in Wilson's disease. Eur J Neurol 2017;24(1):154–60.

[28] Twomey PJ, Viljoen A, House IM, Reynolds TM, Wierzbicki AS. Relationship between serum copper, ceruloplasmin, and non-ceruloplasmin-bound copper in routine clinical practice. Clin Chem 2005;51(8):1558–9.

[29] Gnanou JV, Thykadavil VG, Thuppil V. Pros and cons of immuno-chemical and enzymatic method in the diagnosis of Wilson's disease. Indian J Med Sci 2006;60(9):371–5.

[30] McMillin GA, Travis JJ, Hunt JW. Direct measurement of free copper in serum or plasma ultrafiltrate. Am J Clin Pathol 2009;131(2):160–5.

[31] Walshe JM. Serum 'free' copper in Wilson disease. QJM 2012;105(5):419–23.

[32] Tu JB, Blackwell RQ. Studies on levels of penicillamine-induced cupriuresis in heterozygotes of Wilson's disease. Metabolism 1967;16(6):507–13.

[33] Merle U, Schaefer M, Ferenci P, Stremmel W. Clinical presentation, diagnosis and long-term outcome of Wilson's disease: a cohort study. Gut 2007;56(1):115–20.

[34] Sanchez-Albisua I, Garde T, Hierro L, Camarena C, Frauca E, de la Vega A, et al. A high index of suspicion: the key to an early diagnosis of Wilson's disease in childhood. J Pediatr Gastroenterol Nutr 1999;28(2):186–90.

[35] Steindl P, Ferenci P, Dienes HP, Grimm G, Pabinger I, Madl C, et al. Wilson's disease in patients presenting with liver disease: a diagnostic challenge. Gastroenterology 1997;113(1):212–8.

[36] Giacchino R, Marazzi MG, Barabino A, Fasce L, Ciravegna B, Famularo L, et al. Syndromic variability of Wilson's disease in children. Clinical study of 44 cases. Ital J Gastroenterol Hepatol 1997;29(2):155–61.

[37] Gow PJ, Smallwood RA, Angus PW, Smith AL, Wall AJ, Sewell RB. Diagnosis of Wilson's disease: an experience over three decades. Gut 2000;46(3):415–9.

[38] Garcia-Villarreal L, Daniels S, Shaw SH, Cotton D, Galvin M, Geskes J, et al. High prevalence of the very rare Wilson disease gene mutation Leu708Pro in the Island of Gran Canaria (Canary Islands, Spain): a genetic and clinical study. Hepatology 2000;32(6):1329–36.

[39] Frommer DJ. Urinary copper excretion and hepatic copper concentrations in liver disease. Digestion 1981;21(4):169–78.

[40] Martins da Costa C, Baldwin D, Portmann B, Lolin Y, Mowat AP, Mieli-Vergani G. Value of urinary copper excretion after penicillamine challenge in the diagnosis of Wilson's disease. Hepatology 1992;15(4):609–15.

[41] Muller T, Koppikar S, Taylor RM, Carragher F, Schlenck B, Heinz-Erian P, et al. Reevaluation of the penicillamine challenge test in the diagnosis of Wilson's disease in children. J Hepatol 2007;47(2):270–6.

[42] Dhawan A, Taylor RM, Cheeseman P, De Silva P, Katsiyiannakis L, Mieli-Vergani G. Wilson's disease in children: 37-year experience and revised King's score for liver transplantation. Liver Transpl 2005;11(4):441–8.

[43] Nicastro E, Ranucci G, Vajro P, Vegnente A, Iorio R. Re-evaluation of the diagnostic criteria for Wilson disease in children with mild liver disease. Hepatology 2010;52(6):1948–56.

[44] Ferenci P, Steindl-Munda P, Vogel W, Jessner W, Gschwantler M, Stauber R, et al. Diagnostic value of quantitative hepatic copper determination in patients with Wilson's disease. Clin Gastroenterol Hepatol 2005;3(8):811–8.

[45] Roberts EA. "Not so rare" Wilson disease. Clin Res Hepatol Gastroenterol 2013;37(3):219–21.

[46] Yang X, Tang XP, Zhang YH, Luo KZ, Jiang YF, Luo HY, et al. Prospective evaluation of the diagnostic accuracy of hepatic copper content, as determined using the entire core of a liver biopsy sample. Hepatology 2015;62(6):1731–41.

[47] Faa G, Nurchi V, Demelia L, Ambu R, Parodo G, Congiu T, et al. Uneven hepatic copper distribution in Wilson's disease. J Hepatol 1995;22(3):303–8.

[48] Song YM, Chen MD. A single determination of liver copper concentration may misdiagnose Wilson's disease. Clin Biochem 2000;33(7):589–90.

[49] Ludwig J, Moyer TP, Rakela J. The liver biopsy diagnosis of Wilson's disease. Methods in pathology. Am J Clin Pathol 1994;102(4):443–6.
[50] Sternlieb I, Scheinberg IH. Prevention of Wilson's disease in asymptomatic patients. N Engl J Med 1968;278(7):352–9.
[51] Smallwood RA, Williams HA, Rosenoer VM, Sherlock S. Liver-copper levels in liver disease: studies using neutron activation analysis. Lancet 1968;2(7582):133.
[52] Wilson's BGJ. Disease: a clinician's guide to recognition, diagnosis, and management. Norwell, MA: Kluwer Academic; 2001.
[53] Kaščáková S, Kewish CM, Rouzière S, Schmitt F, Sobesky R, Poupon J, et al. Rapid and reliable diagnosis of Wilson disease using X-ray fluorescence. J Pathol Clin Res 2016;2(3):175–86.
[54] El Balkhi S, Poupon J, Trocello JM, Leyendecker A, Massicot F, Galliot-Guilley M, et al. Determination of ultrafiltrable and exchangeable copper in plasma: stability and reference values in healthy subjects. Anal Bioanal Chem 2009;394(5):1477–84.
[55] El Balkhi S, Trocello JM, Poupon J, Chappuis P, Massicot F, Girardot-Tinant N, et al. Relative exchangeable copper: a new highly sensitive and highly specific biomarker for Wilson's disease diagnosis. Clin Chim Acta 2011;412(23–24):2254–60.
[56] Schmitt F, Podevin G, Poupon J, Roux J, Legras P, Trocello JM, et al. Evolution of exchangeable copper and relative exchangeable copper through the course of Wilson's disease in the Long Evans Cinnamon rat. PLoS One 2013;8(12).
[57] Trocello JM, El Balkhi S, Woimant F, Girardot-Tinant N, Chappuis P, Lloyd C, et al. Relative exchangeable copper: a promising tool for family screening in Wilson disease. Mov Disord 2014;29(4):558–62.
[58] Guillaud O, Brunet A-S, Mallet I, Dumortier J, Pelosse M, Heissat S, et al. Relative exchangeable copper: a valuable tool for the diagnosis of Wilson disease. Liver Int 2017.
[59] Kalita J, Kumar V, Chandra S, Kumar B, Misra UK. Worsening of Wilson disease following penicillamine therapy. Eur Neurol 2014;71(3–4):126–31.
[60] Litwin T, Dziezyc K, Karlinski M, Chabik G, Czepiel W, Członkowska A. Early neurological worsening in patients with Wilson's disease. J Neurol Sci 2015;355(1–2):162–7.

Chapter 11

Diagnosis of Hepatic Wilson Disease

Palittiya Sintusek*,†, Eirini Kyrana† and Anil Dhawan†
*Division of Gastroenterology and Hepatology, King Chulalongkorn Memorial Hospital, Chulalongkorn University, Bangkok, Thailand, †Paediatric Liver, GI and Nutrition Centre, King's College Hospital, London, United Kingdom

INTRODUCTION

Wilson disease (WD) is a rare autosomal recessive disorder of copper metabolism with a variety of clinical and biochemical manifestations. WD should be considered in patients with unexplained liver impairment or neuropsychiatric abnormalities. Currently, there are several diagnostic parameters postulated for WD, but no single test has been specific enough. Molecular diagnosis with detection of *ATP7B* gene mutations is a significant development but always needs to be evaluated in the context of the clinical presentation and is a time-consuming method in cases with heterozygous expression of novel mutations. In severe cases of WD with acute liver failure, noninvasive methods of early diagnosis are needed to manage patients with WD as early as possible. Worldwide, since 2001, a diagnostic scoring system [1] has been used. Since 2012, the WD scoring system has been adopted into the diagnostic algorithm included in the European Association for the Study of the Liver (EASL) guidelines (Table 1). In this chapter, a critical appraisal will be conducted of each parameter used for the diagnosis of WD, together with other proposed diagnostic parameters.

PARAMETERS IN THE DIAGNOSTIC SCORING SYSTEM

Ceruloplasmin

Ceruloplasmin is a 132 kDa plasma glycoprotein that is synthesized mainly not only in the liver but also in the mammary gland [2], uterus [3], placenta [4], testis [4], lungs [5], brain [6], and monocytes [7]. Ceruloplasmin is considered to be an acute-phase protein; studies have demonstrated upstream transcription regulation of ceruloplasmin during inflammation [8] and hyperestrogenic states [9]. Several cytokines like IL-1 [10], IL-6 [10], TNF-α [11], LPS [11], and steroids can induce ceruloplasmin synthesis by hepatic cells. Apart from albumin and transcuprein [12], up to 95% of total circulating copper binds with ceruloplasmin [7]. The physiological function of ceruloplasmin is complex, and several functions have been proposed, including copper transport, antioxidant activity, and oxidation of ferrous ion (ferroxidase activity) and aromatic amines [13, 14]. The ATP7B protein incorporates six copper molecules into apoceruloplasmin protein (half-life 5 h) [15], to form holoceruloplasmin (half-life 5 days) [15], which includes the majority of ceruloplasmin in the circulation. A defect in the *ATP7B* gene reduces the synthesis of holoceruloplasmin, hence making a low serum ceruloplasmin level the most useful test in the diagnosis of WD. Fig. 1 summarizes the mechanism of copper transport in the human body.

Serum levels of ceruloplasmin can be measured by detecting copper-dependent oxidase activity toward specific substrates by enzymatic assays or by immunologic or antibody-dependent assays, including radioimmunoassay, radial immunodiffusion, or nephelometry, all of which are widely used. However, immunologic assays may overestimate serum ceruloplasmin, as this method measures both holoceruloplasmin and apoceruloplasmin. Normal values of serum ceruloplasmin depend on the age of the patient, and levels are very low in the first 6 months of life [16, 17], because the primary site of synthesis switches near birth from the lungs to the liver [5]. From birth, serum levels increase until they reach a peak in the pubertal period, after which time the levels reduce slightly to the adult level [18]. Normal serum values of ceruloplasmin [19] are shown in Table 2. There is no diurnal variation, no variation by sex, or no variation by weight for the serum ceruloplasmin value [20]. Existing guidelines [18, 21, 22] have recommended that a serum ceruloplasmin level of <0.2 g/L, when measured by immunologic methods, supports a diagnosis of WD. There are several conditions that can lead to false-positive or false-negative serum levels of ceruloplasmin (Table 3) [16–18, 23, 24]. As a general screening method or for suspected cases of WD, measurement of serum ceruloplasmin by immunologic assays is simple to perform and clinically

TABLE 1 Wilson Disease Diagnostic Score

Symptoms, Clinical Signs, and Laboratory Tests	Score
KF rings	
Present	2
Absent	0
Neuropsychiatric symptoms suggestive of WD (or typical brain MRI)	
Present	2
Absent	0
Coombs-negative hemolytic anemia (+high serum copper)	
Present	1
Absent	0
Urinary copper (in the absence of acute hepatitis)	
Normal	0
1–2 times ULN	1
>2 times ULN	2
Normal, but >5 times 1 day after challenge with 2 × 0.5 g of penicillamine	2
Liver copper quantitative	
Normal	−1
Up to 5 times ULN	1
>5 times ULN	2
Rhodanine-positive hepatocytes (only if quantitative copper is not available)	
Absent	0
Present	1
Serum ceruloplasmin (nephelometric assay: normal >20 mg/dL)	
Normal	0
10–20	1
<10	2
Mutation analysis	
Disease causing mutations on both chromosomes	4
Disease causing mutations on one chromosome	1
No disease causing mutation detected	0
Total score	
Assessment of the WD—diagnosis score	
4 or more: diagnosis of Wilson disease highly likely	
2–3: diagnosis of Wilson disease probable, do more investigations	
0–1: diagnosis of Wilson disease unlikely	

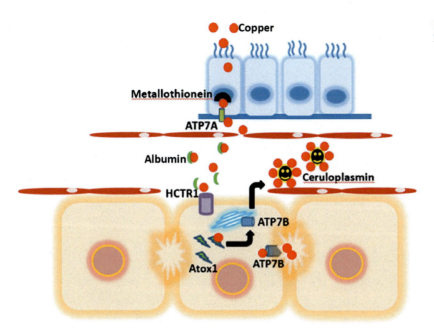

FIG. 1 Mechanism of copper transport in the human body.

TABLE 2 Normal Values of Serum Ceruloplasmin (mg/dL) in Various Age Groups by Immunologic Methods

Age and Gender	Number	Mean	Median	95% Reference Interval
Male				
6 months–2 years	149	26	26	16–37
3–5 years	148	26	26	18–37
6–8 years	149	30	30	20–45
9–11 years	149	30	29	20–44
12–14 years	148	27	26	17–42
15–17 years	147	25	24	17–42
Female				
6 months–2 years	122	26	25	18–37
3–5 years	150	26	25	18–36
6–8 years	149	29	28	19–40
9–11 years	144	30	29	20–43
12–14 years	147	29	28	20–41
15–17 years	146	29	27	20–51

Modified from Clifford SM, Bunker AM, Jacobsen JR, Roberts WL. Age and gender specific pediatric reference intervals for aldolase, amylase, ceruloplasmin, creatine kinase, pancreatic amylase, prealbumin, and uric acid. Clin Chim Acta 2011;412:788–90.

useful, but interpretation of the results should be made with caution, and the measurement of serum ceruloplasmin alone is not sufficient to diagnose or to exclude WD. The sensitivity and specificity of serum ceruloplasmin in the diagnosis of WD have been reported in several published studies but with varying results, depending on the participants included in each study; serum ceruloplasmin is reported to have a sensitivity of between 71% and 93% and a specificity of between 55% and 87% for the widely used immunologic assays. Table 4 [23–40] summarizes the diagnostic value of serum

TABLE 3 Conditions That Can Affect the Value of Serum Ceruloplasmin [16–18, 23, 24]

Low Serum Ceruloplasmin Level	High Serum Ceruloplasmin Level
- Hereditary aceruloplasminemia - Congenital copper deficiency (Menke's disease) - Acquired copper deficiency (Malabsorption syndrome) - Congenital disorder of glycosylation - 10%–20% of heterozygous carriers of WD - Renal or enteric protein loss (nephrotic syndrome, protein losing enteropathy, and sprue) - Sever hepatic insufficiency - Healthy infant (<6 months of age)	- Hyperestrogenic states (pill ingestion and pregnancy) - Steroid ingestion - Infection or inflammation states (60% of WD with fulminant hepatic failure) - 35% of hepatic WD - Immunologic methods for serum ceruloplasmin measurement

TABLE 4 Literature Reviews of the Diagnostic Value of Serum Ceruloplasmin for WD [23–40]

Study	Characteristics of Participants	Method	Sensitivity	Specificity	PPV	NPV	LR
Studies in non-WD							
Cauza et al.	2867 Hepatitis, aged 19–74 years	Immunologic	–	–	6%	–	–
Kelly et al.	1573 Patients with various diseases, aged 1–93 years	Immunologic	–	–	11%	–	–
Vieira et al.	50 Heterozygote mothers, mean age 58 years/heterozygote fathers mean age 62 years	Immunologic	–	–	–	–	–
Studies in WD							
Dhawan et al.	74 WD, aged 5.9–17.9 years	Immunologic	77%	–	–	–	–
Sallie et al.	18 FHF-WD, aged 6–23 years	Immunologic	89%	–	–	–	–
Manolaki et al.	57 WD, aged 4 months–18 years	Immunologic	88%	–	–	–	–
Steindl et al.	55 WD, aged 4–39 years	Immunologic	73%	–	–	–	–
Gow et al.	30 WD, aged 7–58 years	Immunologic	80%	–	–	–	–
	– 22 WD, aged 7–54 years		73%	–	–	–	–
	– 8 WD-FHF, aged 11–54 years		88%	–	–	–	–
Sanchez-Albisua et al.	26 WD, aged 4–16 years	Immunologic[a]	89%	–		–	
Lau et al.	37 WD, aged 9–44 years	Enzymatic	92%			–	
Sintusek et al.	21 WD, aged 10–16 years	Immunologic	90%	–		–	
Studies in both WD and non-WD							
Korman et al.	124 Non-WD-ALF and 16 WD-ALF, aged 14–78 years and 14–53 years						
	– Acute liver failure	Enzymatic	21%	84%	–	–	1
		Immunologic	56%	63%	–	–	2
	– Acute on chronic liver failure	Enzymatic	71%	97%	–	–	21
		Immunologic	71%	79%	–	–	3

TABLE 4 Literature Reviews of the Diagnostic Value of Serum Ceruloplasmin for WD [23–40]—cont'd

Study	Characteristics of Participants	Method	Sensitivity	Specificity	PPV	NPV	LR
Merle et al.	110 WD, 52 healthy, 51 various liver diseases (adult)						
– WD compared with healthy		Immunologic[b]	94%	79%	–	–	21.2
		Enzymatic[c]	94%	100%	–	–	–
– WD compared with various liver diseases		Immunologic[b]	94%	87%	–	–	13.1
		Enzymatic[c]	94%	97%	–	–	2.8
Mak et al.	57 WD, 17 family members, 25 mimic liver diseases, 690 healthy	Immunologic	98%	56%	48%	99%	2.2
Kim et al.	2834 Hepatitis, aged 2.3–7.9 years	Immunologic	93%	84%	–	–	5.9
Martin De Costa et al.	17 WD, 58 non-WD, aged 2–18 years	Immunologic	8%	94	–	–	–
Nicastro et al.	40 WD, 58 non-WD, aged 1.1–20.9 years (age matched)	Immunologic	95%	85%	–	–	–
Muller et al.	38 WD age 5–16 years, 60 non-WD aged 2.3–15 years	Immunologic	85%	89%	–	–	–
Sintusek et al.	43 WD, 278 non-WD (NAFLD and AIH) (children)	Immunologic	92%	78%	57%	97%	4.23

AIH, autoimmune hepatitis; *LR*, likelihood ratio; *NAFLD*, nonalcoholic fatty liver disease; *NPV*, negative predictive value; *PPV*, positive predictive value; *WD-FHF*, Wilson disease with fulminant hepatic failure.
[a]Used cutoff >15 mg/dL.
[b]Used cutoff >19 mg/dL.
[c]Used cutoff >50 U/L.

ceruloplasmin. There have been studies that have recommended the cutoff serum level of 0.14 and 0.16 g/L to increase the specificity of the test to 94.9% and 100%, respectively, while the sensitivity was 91.9% and 94.9%, respectively [23, 25]. However, retrospective data from King's College Hospital, London, United Kingdom [40a], found that the cutoff value of 0.16 g/L had sensitivity, specificity, positive predictive value (PPV), negative predictive value (NPV), and accuracy of 76.9%, 95.9%, 85.1%, 93.1%, and 91.4%, respectively. Therefore, large multicenter studies would be needed to validate new cutoff values for serum ceruloplasmin in the diagnosis of WD.

Serum Copper

The pathogenesis of WD is due to the accumulation of copper in many organs such as the liver, brain, and kidney and also in the blood. Serum copper does not reflect free copper and is not a helpful marker for the diagnosis of WD. The following equation demonstrates the status of copper in the serum [41]:

$$\text{Total copper (serum copper)} = \text{Free copper} + \text{Bound copper}$$

$$\text{Bound copper (}\mu\text{mol/L)} = 0.0472 \times \text{ceruloplasmin (mg/L)}$$

Free copper, or serum nonceruloplasmin-bound copper, is increased, while bound copper is decreased, resulting in variable values of serum copper. However, free copper calculation is not a reliable diagnostic test if serum ceruloplasmin is measured by an immunoassay method, as described previously.

Urinary Copper

Urinary copper measurement is one method to detect the amount of nonceruloplasmin-bound copper in the circulation, instead of using a serum free copper calculation. As spot urine copper content has a high variability, a 24 h urine collection is the standard method and is more accurate for copper measurement in the urine. However, urine collection may be very difficult in young children. While collecting urine for the diagnosis of WD, the container and the funnel to pour the urine, all must be made from copper-free material. A urine bag and Foley catheter are not appropriate methods due to their contamination with copper. For the same reason, the use of tap water should be avoided when cleaning any container used for urine collection. Disposable collection containers and devices are recommended. If the process of urine collection is accurate, the normal value of 24 h urinary copper is usually <40 μg/24 h. A diagnostic urinary copper level for WD is >100 μg/24 h [42]. However, studies have shown that in 16%–25% of patients with WD, their urinary copper was less than this cutoff level [31, 33, 43] especially in patients with WD who are asymptomatic. A study of 24 h urinary copper measurements in asymptomatic WD patients showed that the value of >40 μg/24 h can accurately detect this group of patients and should be the new threshold for WD diagnosis, especially in young children [39]. However, long-standing cholestatic liver conditions can be associated with high urinary levels of copper. Table 5 [18, 22] shows clinical conditions, other than WD, that are associated with high and low levels of urinary copper.

Urinary copper in 24 h excretion after D-penicillamine challenge test (PCT) might be useful especially in symptomatic WD. However, this test is described with a standard protocol only in children [38] where 500 mg D-penicillamine is administered orally followed with the same dose 12 h later. Urine after ingestion is collected for 24 h to measure the copper content. Urine copper >25 μmol copper/24 h is considered positive with high accuracy (sensitivity 76%–88.2% and specificity 93%–98.2% in symptomatic while only 46% in asymptomatic WD) [38, 40], while some centers use a fivefold increase in the copper excretion (Table 6) [27, 28, 30–32, 35, 38, 39, 42–44]. However, in adults and heterozygote carriers, the accuracy of this test is not well described, with only few studies demonstrating that PCT can discriminate WD patients from heterozygote carriers [45].

Liver Copper

Quantitative evaluation of copper in the liver is considered to be an invasive diagnostic method, as it involves liver biopsy, but is a specific diagnostic method for WD. Hepatic copper of >250 μg/g dry weight provides the best biochemical evidence for WD; hence, this cutoff value was included in the diagnostic score, while values <50 μg/g dry weight are considered to exclude a diagnosis of WD [46]. However, there have been several published studies with conflicting findings regarding hepatic copper, and it has been suggested that this cutoff of 250 μg/g dry weight is too high and other values have been suggested for the diagnosis of WD [42, 47, 48]. These studies are summarized in Table 7 [28, 39, 40, 47–49]. Liver conditions with long-standing cholestatic status and other diseases of abnormal copper metabolism can also have very high liver copper content, as shown in Table 8 [38, 46, 50]. Therefore, the other clinical findings must be compatible with the diagnosis of WD, and conditions that may mimic WD should be excluded.

The main limitation of liver copper measurement is the sampling error [51, 52]. The distribution of copper in the liver is not homogenous and can result in underestimation of copper content. There are several studies that have highlighted the sampling error in advanced liver disease, such as cirrhosis with nodules and fibrosis [38, 53, 54] as the centers of large regenerative nodules and tracts of fibrous tissue will both have lower copper concentration. However, some studies found no significant correlation of this factor [48, 49]. To overcome the sampling error, increasing the required liver biopsy size to at least 1 cm length of a 1.6 mm diameter of biopsy core [55, 56] or >1 mg liver tissue, with two biopsies from various areas [57] or

TABLE 5 Condition That Can Affect the Value of Urinary Copper [18, 22]

False-Positive Urinary Copper Value	False-Negative Urinary Copper Value
Chronic or severe liver diseases	*Conditions*
Primary biliary cirrhosis	Renal failure: oliguric phrase
Chronic active hepatitis	*Collection process*
Fulminant hepatic failure from any origin	Incomplete urine collection
Cholestasis syndromes (Alagille syndrome)	*Others*
Obstructive disease ex. Biliary atresia	Asymptomatic WD
Autoimmune hepatitis	
Sclerosing cholangitis	
Collecting process	
Container with copper contamination	
Others	
Heterozygote WD	

TABLE 6 Literature Reviews the Diagnostic Value of Urinary Copper Measurement From the Previous Study [27, 28, 30–32, 35, 38, 39, 42–44]

Study	Characteristics of Participants	Study	Cutoff Value µg Copper/24h	Basal or PCT	Sensitivity	Specificity	PPV	NPV
Studies in non-WD								
Vieira et al.	50 Heterozygote mothers, mean age 58 years/heterozygote fathers mean age 62 years	Cross-x	40	Basal	–	–	–	96%
			100	Basal	–	–	–	100
			1000	PCT	–	–	–	–
Tu and Blackwell et al.	10 Heterozygote parents, mean age 45 years and 10 WD, mean age 15 years	Cross-x	100	Basal	100%	90%	100%	95%
			1600	PCT	100%	100%	100%	100
Studies in WD								
Dhawan et al.	74 WD, aged 5.9–17.9 years	Retro	100	Basal	98%	–	–	–
		Retro	1600	PCT	95%	–	–	–
Steindl et al.	55 WD, aged 4–39 years	Retro	100	Basal	69%	–	–	–
Manolaki et al.	57 WD, aged 4 months–18 years	Retro	100	Basal	88%	–	–	–
Gow et al.	18 WD (not included 4WD with anuria), aged 7–58 years	Retro	100	Basal	88%	–	–	–
Merle et al.	163 WD, mean age 17.4 years	Retro	100	Basal	94%	–	–	–
Giacchino et al.	43 WD (children)	Retro	100	Basal	81.70%	–	–	–
Sintusek et al.	21 WD (children)	Retro	100	Basal	83%	–	–	–
Studies in both WD and non-WD								
Martin Da Costa et al.	17 WD, 58 non-WD, aged 2–18 years	Retro	60	Basal	100%	83%	–	–
			1600	PCT	88%	98%	–	–
Nicastro et al.	40 WD (mild liver disease), 58 non-WD, ages 1.1–20.9 years (age matched)	Retro	40	Basal	79%	88%	–	–
			100	Basal	66%	98%	–	–
			1600	PCT	12%	67%	–	–
Muller et al.	38 WD aged 5–16 years, 60 non-WD aged 2.3–15 years	Retro	100	Basal	86%	79%	–	–
			1600	PCT	76%	93%	–	–
Sintusek et al.	22 WD, 207 non-WD (NAFLD and AIH) (children)	Retro	40	Basal	97%	55%	35%	99%
			100	Basal	87%	87%	62%	96%
			1600	PCT	42%	99%	95%	82%

AIH, autoimmune hepatitis; *Basal*, 24h urine copper measurement; *Cross-x*, cross-sectional study; *NAFLD*, nonalcoholic fatty liver disease; *non-WD*, non-Wilson disease; *NPV*, negative predictive value; *PCT*, 24h urine copper post D-penicillamine challenge test; *PPV*, positive predictive value; *Retro*, retrospective study; *WD*, Wilson disease.

sampling from the deeper regions of the liver as copper concentrations will be higher than in the subcapsular regions [22], has been recommended. An observational study from King's College Hospital, London, United Kingdom, found there was no correlation between the size and weight of the liver and copper measurement for WD and control patients [58]. When genetic analysis is not available, in asymptomatic patients, liver copper content is a vital parameter to be considered in the next step of investigations. Furthermore, liver copper measurement is superior to histochemical staining of liver biopsies for copper, which

TABLE 7 Literature Reviews the Diagnostic Value of Liver Copper Measurement [28, 39, 40, 47–49]

Study	Characteristics of Participants	Study	Sampling Biopsy	Cutoff Value of Liver Copper Content	Sensitivity	Specificity	NPV	PPV
Ferenci et al.	149 WD, 219 noncholestatic liver disease, 26 without evidence of liver disease	Retro	Single	75	96.5	95.4	–	–
				250	83.3	98.6	98	–
Liggi et al.	35 WD with double-sampling biopsy and 30 WD with single-sampling biopsy	Retro	Single	250	80	–	–	–
				50–250	93	–	–	–
			Double	250	85.7	–	–	–
				50–250	97	–	–	–
Yang et al.	178 WD, 513 non-WD (exclude PBC and PSC)	Pros	Double	250	94.4	96.8	–	–
				209	99.4	96.1	91.8	97.8
Dhawan et al.	74 WD, aged 5.9–17.9 years	Retro	Single	250	82.40%	–	–	–
Nicastro et al.	30 WD, 24 non-WD, aged 1.1–20.9 years (age-matched)	Retro	Single	50	–	94	–	–
				75	93	–	–	–
				250	93	–	–	–
Muller et al.	22 WD aged 5–16 years, 26 non-WD aged 2.3–15 years	Retro	Single	250	91	92	–	–
Sintusek et al.	33 WD, 123 non-WD (NAFLD and AIH) (children)	Retro	Singer	50	95	60	51	96
				120	82	91	76	94
				250	76	98	94	90

AIH, autoimmune hepatitis; *NAFLD*, nonalcoholic fatty liver disease; *PBC*, primary biliary cirrhosis; *Pros*, prospective study; *PSC*, primary sclerosing cholangitis; *Retro*, retrospective study.

TABLE 8 Condition With High Value of Liver Copper Content [38, 46, 50]

Normal infant <6 months
Primary biliary cirrhosis
Heterozygote carriers of *ATP7B* mutation
Intrahepatic cholestasis as alcoholic steatohepatitis
Cholestasis syndromes as biliary atresia, paucity of intrahepatic bile ducts
Sclerosing cholangitis
Primary biliary cirrhosis
Autoimmune hepatitis
Idiopathic copper toxicosis syndromes such as Indian childhood cirrhosis

may be negative as copper deposited in the cytoplasm in the early stage of WD is undetectable. Beyond the size of the liver tissue, the technical process of the liver biopsy is important and has been well described [18] including the use of disposable suction or cutting needles. The liver biopsy must be placed in a copper-free container and sent for analysis of the copper content as soon as possible. The current guidelines from the American Association for the Study of Liver Diseases (AASLD) have included that paraffin-embedded specimens can also be analyzed for copper content [18]. The EASL guidelines are concerned that the result will be less reliable from fixed liver tissue rather than fresh tissue [22].

Routine histopathology of the liver in WD, without the use of special stains for copper, appears to have less diagnostic benefit with no specific or classic features. The histological findings in WD can mimic those of nonalcoholic fatty liver disease (NAFLD), autoimmune hepatitis (AIH), and MDR3 deficiency as they include mild steatosis, glycogenated nuclei, focal hepatocellular necrosis, and piecemeal necrosis [59]. However, it is useful to look for other diseases or additional liver pathology in the differential diagnosis [59]. Copper or copper-binding protein on histochemical staining with rhodamine or orcein, respectively, may be helpful, but copper may be unstained in the early stage of WD as copper in the cytoplasm may be only bound to metallothionein protein and cannot be detected. Previous studies found only 10%–25% of cases of WD had rhodamine-positive granules in hepatocytes [47, 59] because only the lysosomal copper deposition can be detected; in the later stage of WD, copper distributes to lysosomes and can stain histochemically. The technique of special staining might be different in each center, and it is necessary to evaluate and calibrate the methods with the control subjects. Special stains might give a clue to the diagnosis of WD in cases that have low copper content, due to liver sampling error. Particular liver diseases with long-standing cholestasis (primary biliary cirrhosis and sclerosing cholangitis) can also stain for hepatic copper, making it difficult to discriminate from WD; this is something that the clinician needs to keep in mind.

In summary, liver copper content is more specific for the diagnosis of WD when compared with histopathology or special stains for copper detection. Liver biopsy for histology and hepatic copper detection is an invasive investigation that is not suitable for screening for WD and should be reserved for situations where the diagnosis is unclear by simpler diagnostic approaches.

Eye Signs

Kayser-Fleischer (KF) Rings are due to the deposition of copper in Descemet's membrane and are the pathognomonic sign of WD that can be detected in almost all (up to 95%) patients with neurological symptoms [31, 32]. In contrast to KF rings, sunflower cataracts or the deposits of copper in the center of both the anterior and posterior lens are very rare [60]. Both eye signs can be easily detected by slit-lamp examination. In young and asymptomatic children with abnormal liver enzymes, KF rings are rarely found. Manolaki et al. [30] found that 21/42 (50%) children older than 8 years had KF rings and only 1/14 children younger than 8 years had KF rings. Sintusek et al. [35] found that 14/21 (66%) children had KF rings but only in children older than 7 years who were symptomatic at the time of evaluation. In adolescents and young adults, up to 50% of cases had KF rings with severe liver and/or neurological problems [28]. These eye signs can be rarely present in chronic cholestatic conditions. There are no vision problems with these eye signs, which can disappear following treatment for WD.

Coombs-Negative Hemolytic Anemia

Since 1934, there have been many WD cases reported that have been associated with nonimmune hemolytic anemia [61]. Coombs-negative hemolysis that accompanies hepatitis was the presenting manifestations in 5%–20% of WD patients [21, 62–64]. Most of these cases presented with acute liver failure, but some manifested firstly with acute (84%) or chronic intermittent hemolytic anemia before neurological or hepatic involvement [65]. An observational study reported by Walshe in 22 patients with WD [66] found a female predominance (female-male ratio was 15:7) with the average onset at 12.6 years. The pathogenesis of the hemolysis remains unclear. There does exist some evidence, both in humans [61, 67] and from preclinical studies [61], suggesting that excess copper that is suddenly released from the liver into the blood can lead to toxic effects on the erythrocyte membrane and induce hemolysis. When the copper release ceases, normally within a few days, the hemolytic process terminates, and the hematologic findings will return to normal within several weeks. There is still some debate regarding the mechanism of hemolysis, whether it is intravascular or extravascular. There have been several case reports [61, 67, 68] that have shown an increase in methemoglobin, haptoglobin, and renal injury without spherocytosis in the peripheral blood smear, supporting an intravascular cause. However, the improvement after splenectomy that was reported in the Walshe series suggested an extravascular origin. Therefore, all patients who present with nonimmune or Coombs-negative hemolysis should be considered for a diagnosis of WD.

Neurological and Psychological Symptoms and Signs

Neurological dysfunction can be the first manifestation of WD in 40%–60% of cases [69]. CNS damage in WD is limited to the motor system, whereas intellectual and sensory functions are spared. The most common symptoms are incoordination, tremor (both posture tremor and tremor at rest), dysarthria, dystonia, and mask-like facies [70, 71]. Unfortunately, the usual age at onset is 12–32 years [71], making it difficult to diagnose in young asymptomatic children. The neurological symptoms are rare in the prepubertal period [72]. The presence of KF rings is a suggestive diagnostic clue in WD, and from all who presented with neurological problems, only few did not have KF rings [73].

Neuroimaging is recommended in all suspected cases of WD with neurological symptoms. Structural brain magnetic resonance imaging (MRI) shows widespread pathology in the basal ganglion especially the putamen, the globus pallidus, the caudate area, and the thalamus reflecting the abnormal accumulation of copper. Other areas of the CNS can be involved, including the midbrain, pons, and the cerebellum as well as cortical atrophy [74].

Psychiatric manifestations have been reported in 20%–65% of individuals with WD, and in almost 50%, these symptoms were sufficiently severe to require psychiatric intervention [22, 75, 76], with diagnoses including adolescent-adjustment problems, bizarre behaviors, anxiety neurosis, mania, depression, psychosis, hysteria, schizoaffective disorder, and schizophrenia [71]. In cases with mild symptoms, such as personality changes, disturbances of mood, and poor school performance, a diagnosis of WD can get delayed for up to 12 years [42]. The psychological stress and the physical strains of WD are considerable, particularly if the diagnosis is delayed, and they can aggravate emotional lability or immature personality [71]. It is important for health-care providers to be alert and screen with a serum ceruloplasmin in such cases.

Genetic ATP7B Mutation Analysis

Mutation analysis is a reliable test for WD diagnosis. The only major concerns are the genetic heterogeneity and the presence of an uncommon mutational mechanism >600 mutations have now been identified. Coffey et al. [77] reported the first cases of WD associated with segmental uniparental isodisomy and the mutations in two consecutive generations in the United Kingdom. The variety of the genetic mutations can make it difficult to detect the abnormalities early. Hotspot detection might be useful for particular populations [78, 79] as *H1069Q* is the most common mutation in Western populations and *R778L* in Asian populations. However, most patients with WD are heterozygotes, and only a few have large deletions and might be described incorrectly as just having the one-point mutation on one chromosome. Recently, a large study on phenotype-genotype correlation was reported in 227 WD patients, which showed no significant correlation of each mutation [80]. More genetic association studies are required for the phenotype to be defined as accurately as possible. The genotype-phenotype correlation will be helpful to detect the mutations by using the microarray method [81], if the hotspot or the common specific mutation for each ethnic group has a negative result. However, in cases with novel mutations, direct sequencing is the best way to identify the mutation, but this takes time, usually several weeks. Also, it is never certain that the mutation detected reflects disease state. Functional analysis may be required for this. Therefore, in severe cases that require an urgent diagnosis, genetic analysis may not be available in time to make appropriate management decisions. Recently, Nemeth et al. [82] published a study on the clinical use of next-generation sequencing for WD that can rapidly detect the mutations, which may possibly prompt diagnosis for patients with WD with severe symptoms at presentation. This method was suggested for the diagnosis of WD in selected cases and considered to be useful, time-saving, reliable, and cost-effective.

PROPOSED PARAMETERS THAT FACILITATE WD DIAGNOSIS

As discussed above, there is no single parameter that is ideal for the diagnosis of WD. A scoring system [21] that is composed of several parameters has now been used for >13 years. The high diagnostic accuracy of this scoring system was reported by Koppikar and Dhawan, in 2005 [83]. However, there have been no prospective studies to confirm this scoring system, especially in children who are most likely to lack some of the important clinical diagnostic features of WD, including neurological manifestations and the KF rings. In a resource-limited setting where genetic analysis is not available or can be performed only in selected cases, other parameters, such as treatment response, have been used to confirm the diagnosis. In this section, we have tried to summarize other proposed parameters that may help in the diagnosis of WD.

AST to ALT Ratio

Previous and recent studies have supported that the AST to ALT ratio of >2.0 yielded a high diagnostic sensitivity and specificity for the diagnosis of WD in the setting of acute liver failure, in both children [84] and adults [36, 85, 86]. This parameter could be considered if diagnostic parameters such as urinary copper or liver copper are not possible due to acute renal failure or uncorrectable coagulopathy, but other signs and symptoms are suggestive of WD at the time of presentation. The AST may be higher than the ALT in WD-acute liver failure (WD-ALF) as a reflection of cytosolic or mitochondrial injury or as a reflection of established chronic liver disease [84]. As previously described, ALT in cirrhosis is usually lower than in other liver diseases. A low ALT can be a clue in the differential diagnosis of chronic liver disease.

Alkaline Phosphatase (ALP) to Total Bilirubin (TB) Ratio

Studies by Karman et al. and Berman et al. found that the ALP-to-TB ratio of <2.0 was useful for screening acute liver failure from WD, as ALP is usually low in the WD-ALF setting [36, 86], although this is in contrast to the study by Sallie et al. [29] Our observation of WD in children also found that the ALP-to-TB ratio of <2.0 was not useful in differentiating between WD-ALF and other causes of ALF [84] (Table 9). The reasons for these differences may be that the studies by Korman et al. and Berman et al. were adult studies rather than children, as ALP levels correlate with bone growth, and ALP

TABLE 9 Demographic Data and Investigative Results of WD-ALF, Indeterminate ALF, and ALF From Other Causes

Parameters	WD-ALF (N = 8)	Indeterminate ALF (N = 9)
A. Comparative data between WD-ALF and indeterminate ALF		
Age (year)	13.1 (7.8–16.0)	7.6 (1.0–12.2)*
Female-male (%female)	3:5 (37.5%)	1:8 (11%)
Zn (μmol/L)	5.80 (4.1–8.3)	9.8 (7.0–12.1)**
Copper (μmol/L)	11.7 (4.5–24.3)	18.1 (11.4–36.7)
Free copper (μmol/L)	6.00 (3.08–18.16)	6.41 (3.38–23.48)
ALP (IU/L)	140.5 (19–1120)	345 (159–758)
Low ALP for age (N)	5	0*
USG clue[a]	100%	0%**
TB (μmol/L)	80.5 (33–625)	282 (48–448)*
AST (IU/L)	124 (83–403)	2778 (2039–6637)**
ALT (IU/L)	57.5 (16–185)	2744 (1816–2857)*
AST to ALT	2.16 (1.34–8.25)	1.05 (0.74–1.53)**
AST to ALT > 2 (N)	3	0*
ALP to TB	2.26 (0.03–33.94)	1.48 (0.51–3.66)
ALP to TB < 2 (N)	4	6
B. Comparative data between WD-ALF and ALF from others		
Parameters	WD-ALF (N=15)	ALF from others (N=61)
Age (year)	13.2 (7.7–16.0)	6 (1.1–16.6)
Female-male (%female)	5:10 (33.3%)	24:37 (39.3%)
ALP (IU/L)	92 (19–1120)	255 (22–1286)*
TB (μmol/L)	57 (19–625)	150 (6–844)
AST (IU/L)	132 (39–403)	1421 (34–11,303)**
ALT (IU/L)	54 (11–190)	974 (8–7833)**
AST to ALT	2.83 (0.68–12.36)	1.37 (0.18–7.3)*
AST to ALT > 2 (N)	10	19*
ALP to TB	2.71 (0.03–33.94)	2.05 (0.10–37.5)
ALP to TB < 2 (N)	7	30*

*P-value <.05.
**P-value <.001.
[a]USG clues, USG finding of heterogeneous or nodular surface of liver.
ALP, alkaline phosphatase; ALT, alanine aminotransferase; AST, aspartate aminotransferase; TB, total bilirubin; WD-ALF, WD with acute liver failure presentation; WD-non-ALF, WD with nonacute liver failure presentation; Zn, zinc.

might be high during rapid bone growth in the preschool and adolescent period, making the ALP-to-TB ratio greater than expected. Therefore, the ALP-to-TB ratio may be inappropriate for use in the diagnosis of WD in children. An alternative analysis that corrects ALP levels for age could be a better parameter for screening WD-ALF in children.

Alkaline Phosphatase (ALP)

As the previous observational study and case report mentioned above show, a low ALP is observed in WD, especially in the acute liver failure setting [68, 84, 87] (Table 9) and especially in ALF with nonimmune hemolysis [68]. The reason why ALP is low in the WD-ALF setting with a hemolysis presentation is not well understood. One study tried to confirm the hypothesis that excess copper in serum and hemolysis causes low ALP but did not demonstrate this [68]. Reported causes of low ALP [68] include hypothyroidism, pernicious anemia, congenital hypophosphatasia, and zinc deficiency. However, very low ALP in WD is a transient phenomenon and may be associated with hypercupremia. A preclinical study has been reported of the association between copper and serum ALP [68] that found no significant difference between hypercupremia and low ALP. Another possibility is that zinc deficiency causes low ALP. As ALP is a metalloenzyme in which zinc is the important metal, zinc deficiency can potentially affect the enzyme activity [84, 88]. However, there relationships need clarification with further studies.

Others

As serum copper and calculated free copper cannot reflect the excess copper in the body, if serum ceruloplasmin is measured by immunoassay, other laboratories have postulated that relative exchangeable copper (REC) and radiocopper should be studied. REC was tested in patients who were newly diagnosed with WD with 100% sensitivity and specificity [89]. This biomarker is valuable to use and was validated in a large WD group, but is still not widely available. In another investigation, a radiocopper study [90–92] was used in patients with a suspected diagnosis of WD who had normal ceruloplasmin level. Radiocopper incorporation into this protein is significantly reduced compared with normal individuals, most heterozygotes and all homozygotes. However, this test is rarely used because of difficulties in obtaining the isotope.

REFERENCES

[1] Ferenci P, et al. Hepatic encephalopathy—definition, nomenclature, diagnosis, and quantification: final report of the working party at the 11th World Congresses of Gastroenterology, Vienna, 1998. Hepatology 2002;35:716–21.
[2] Jaeger JL, Shimizu N, Gitlin JD. Tissue-specific ceruloplasmin gene expression in the mammary gland. Biochem J 1991;280(Pt 3):671–7.
[3] Schilsky ML, Stockert RJ, Pollard JW. Caeruloplasmin biosynthesis by the human uterus. Biochem J 1992;288(Pt 2):657–61.
[4] Aldred AR, Grimes A, Schreiber G, Mercer JF. Rat ceruloplasmin. Molecular cloning and gene expression in liver, choroid plexus, yolk sac, placenta, and testis. J Biol Chem 1987;262:2875–8.
[5] Fleming RE, Gitlin JD. Primary structure of rat ceruloplasmin and analysis of tissue-specific gene expression during development. J Biol Chem 1990;265:7701–7.
[6] Klomp LW, Gitlin JD. Expression of the ceruloplasmin gene in the human retina and brain: implications for a pathogenic model in aceruloplasminemia. Hum Mol Genet 1996;5:1989–96.
[7] Mazumder B, Mukhopadhyay CK, Prok A, Cathcart MK, Fox PL. Induction of ceruloplasmin synthesis by IFN-gamma in human monocytic cells. J Immunol 1997;159:1938–44.
[8] Gitlin JD. Transcriptional regulation of ceruloplasmin gene expression during inflammation. J Biol Chem 1988;263:6281–7.
[9] German 3rd JL, Bearn AG. Effect of estrogens on copper metabolism in Wilson's disease. J Clin Invest 1961;40:445–53.
[10] Ramadori G, Van Damme J, Rieder H, Meyer zum Buschenfelde KH. Interleukin 6, the third mediator of acute-phase reaction, modulates hepatic protein synthesis in human and mouse. Comparison with interleukin 1 beta and tumor necrosis factor-alpha. Eur J Immunol 1988;18:1259–64.
[11] Fleming RE, Whitman IP, Gitlin JD. Induction of ceruloplasmin gene expression in rat lung during inflammation and hyperoxia. Am J Physiol 1991;260:L68–74.
[12] Linder MC. Ceruloplasmin and other copper binding components of blood plasma and their functions: an update. Metallomics 2016;8:887–905.
[13] Samokyszyn VM, Miller DM, Reif DW, Aust SD. Inhibition of superoxide and ferritin-dependent lipid peroxidation by ceruloplasmin. J Biol Chem 1989;264:21–6.
[14] Gutteridge JM. Antioxidant properties of caeruloplasmin towards iron- and copper-dependent oxygen radical formation. FEBS Lett 1983;157:37–40.
[15] Gitlin D, Janeway CA. Turnover of the copper and protein moieties of ceruloplasmin. Nature 1960;185:693.
[16] Schenker IF, Shemesh A, Polishuk WZ. Serum ceruloplasmin levels in normal pregnancy and in newborn infants. Harefuah 1971;81:362–3.
[17] Pojerova A, Tovarek J. Ceruloplasmin in early childhood. Acta Paediatr 1960;49:113–20.
[18] Roberts EA, Schilsky ML, American Association for Study of Liver Diseases (AASLD). Diagnosis and treatment of Wilson disease: an update. Hepatology 2008;47:2089–111.

[19] Clifford SM, Bunker AM, Jacobsen JR, Roberts WL. Age and gender specific pediatric reference intervals for aldolase, amylase, ceruloplasmin, creatine kinase, pancreatic amylase, prealbumin, and uric acid. Clin Chim Acta 2011;412:788–90.

[20] Cox DW. Factors influencing serum ceruloplasmin levels in normal individuals. J Lab Clin Med 1966;68:893–904.

[21] Ferenci P, et al. Diagnosis and phenotypic classification of Wilson disease. Liver Int 2003;23:139–42.

[22] European Association for the Study of the Liver. EASL Clinical Practice Guidelines: Wilson's disease. J Hepatol 2012;56:671–85.

[23] Mak CM, Lam CW, Tam S. Diagnostic accuracy of serum ceruloplasmin in Wilson disease: determination of sensitivity and specificity by ROC curve analysis among ATP7B-genotyped subjects. Clin Chem 2008;54:1356–62.

[24] Cauza E, et al. Screening for Wilson's disease in patients with liver diseases by serum ceruloplasmin. J Hepatol 1997;27:358–62.

[25] Kim JA, et al. Diagnostic value of ceruloplasmin in the diagnosis of pediatric Wilson's disease. Pediatr Gastroenterol Hepatol Nutr 2015;18:187–92.

[26] Kelly D, et al. The clinical utility of a low serum ceruloplasmin measurement in the diagnosis of Wilson disease. Ir Med J 2016;109:341–3.

[27] Vieira J, et al. Urinary copper excretion before and after oral intake of D-penicillamine in parents of patients with Wilson's disease. Dig Liver Dis 2012;44:323–7.

[28] Dhawan A, et al. Wilson's disease in children: 37-year experience and revised King's score for liver transplantation. Liver Transpl 2005;11:441–8.

[29] Sallie R, et al. Failure of simple biochemical indexes to reliably differentiate fulminant Wilson's disease from other causes of fulminant liver failure. Hepatology 1992;16:1206–11.

[30] Manolaki N, et al. Wilson disease in children: analysis of 57 cases. J Pediatr Gastroenterol Nutr 2009;48:72–7.

[31] Steindl P, et al. Wilson's disease in patients presenting with liver disease: a diagnostic challenge. Gastroenterology 1997;113:212–8.

[32] Gow PJ, et al. Diagnosis of Wilson's disease: an experience over three decades. Gut 2000;46:415–9.

[33] Sanchez-Albisua I, et al. A high index of suspicion: the key to an early diagnosis of Wilson's disease in childhood. J Pediatr Gastroenterol Nutr 1999;28:186–90.

[34] Lau JY, et al. Wilson's disease: 35 years' experience. Q J Med 1990;75:597–605.

[35] Sintusek P, Chongsrisawat V, Poovorawan Y. Wilson's disease in Thai children between 2000 and 2012 at King Chulalongkorn Memorial Hospital. J Med Assoc Thai 2016;99:182–7.

[36] Korman JD, et al. Screening for Wilson disease in acute liver failure: a comparison of currently available diagnostic tests. Hepatology 2008;48:1167–74.

[37] Merle U, Eisenbach C, Weiss KH, Tuma S, Stremmel W. Serum ceruloplasmin oxidase activity is a sensitive and highly specific diagnostic marker for Wilson's disease. J Hepatol 2009;51:925–30.

[38] Martins da Costa C, et al. Value of urinary copper excretion after penicillamine challenge in the diagnosis of Wilson's disease. Hepatology 1992;15:609–15.

[39] Nicastro E, Ranucci G, Vajro P, Vegnente A, Iorio R. Re-evaluation of the diagnostic criteria for Wilson disease in children with mild liver disease. Hepatology 2010;52:1948–56.

[40] Muller T, et al. Re-evaluation of the penicillamine challenge test in the diagnosis of Wilson's disease in children. J Hepatol 2007;47:270–6.

[40a] Sintusek P, Kyrana E, Dhawan A. Value of serum zinc in diagnosing and assessing severity of liver disease in children with Wilson disease. J Pediatr Gastroenterol Nutr 2018;67:377–82.

[41] Walshe JM, Clinical Investigations Standing Committee of the Association of Clinical Biochemists. Wilson's disease: the importance of measuring serum caeruloplasmin non-immunologically. Ann Clin Biochem 2003;40:115–21.

[42] Merle U, Schaefer M, Ferenci P, Stremmel W. Clinical presentation, diagnosis and long-term outcome of Wilson's disease: a cohort study. Gut 2007;56:115–20.

[43] Giacchino R, et al. Syndromic variability of Wilson's disease in children. Clinical study of 44 cases. Ital J Gastroenterol Hepatol 1997;29:155–61.

[44] Tu JB, Blackwell RQ. Studies on levels of penicillamine-induced cupriuresis in heterozygotes of Wilson's disease. Metabolism 1967;16:507–13.

[45] Frommer DJ. Urinary copper excretion and hepatic copper concentrations in liver disease. Digestion 1981;21:169–78.

[46] Smallwood RA, Williams HA, Rosenoer VM, Sherlock S. Liver-copper levels in liver disease: studies using neutron activation analysis. Lancet 1968;2:1310–3.

[47] Ferenci P, et al. Diagnostic value of quantitative hepatic copper determination in patients with Wilson's disease. Clin Gastroenterol Hepatol 2005;3:811–8.

[48] Yang X, et al. Prospective evaluation of the diagnostic accuracy of hepatic copper content, as determined using the entire core of a liver biopsy sample. Hepatology 2015;62:1731–41.

[49] Liggi M, et al. Uneven distribution of hepatic copper concentration and diagnostic value of double-sample biopsy in Wilson's disease. Scand J Gastroenterol 2013;48:1452–8.

[50] Tanner MS. Indian childhood cirrhosis and Tyrolean childhood cirrhosis. Disorders of a copper transport gene? Adv Exp Med Biol 1999;448:127–37.

[51] Goldfischer S, Sternlieb I. Changes in the distribution of hepatic copper in relation to the progression of Wilson's disease (hepatolenticular degeneration). Am J Pathol 1968;53:883–901.

[52] Maharaj B, et al. Sampling variability and its influence on the diagnostic yield of percutaneous needle biopsy of the liver. Lancet 1986;1:523–5.

[53] Nooijen JL, van den Hamer CJ, Houtman JP, Schalm SW. Possible errors in sampling percutaneous liver biopsies for determination of trace element status: application to patients with primary biliary cirrhosis. Clin Chim Acta 1981;113:335–8.

[54] Faa G, et al. Uneven hepatic copper distribution in Wilson's disease. J Hepatol 1995;22:303–8.

[55] Song YM, Chen MD. A single determination of liver copper concentration may misdiagnose Wilson's disease. Clin Biochem 2000;33:589–90.

[56] Ala A, Walker AP, Ashkan K, Dooley JS, Schilsky ML. Wilson's disease. Lancet 2007;369:397–408.

[57] McDonald JA, et al. Striking variability of hepatic copper levels in fulminant hepatic failure. J Gastroenterol Hepatol 1992;7:396–8.
[58] Sintusek P, Dhawan A. Liver copper estimation: does liver biopsy size really matter? Hepatology 2016;64:1381–2.
[59] Ludwig J, Moyer TP, Rakela J. The liver biopsy diagnosis of Wilson's disease. Methods in pathology. Am J Clin Pathol 1994;102:443–6.
[60] Cairns JE, Williams HP, Walshe JM. "Sunflower cataract" in Wilson's disease. Br Med J 1969;3:95–6.
[61] McIntyre N, Clink HM, Levi AJ, Cumings JN, Sherlock S. Hemolytic anemia in Wilson's disease. N Engl J Med 1967;276:439–44.
[62] Walshe JM. Wilson's disease presenting with features of hepatic dysfunction: a clinical analysis of eighty-seven patients. Q J Med 1989;70:253–63.
[63] Niederau C, Stremmel W, Strohmeyer G. Early detection of hereditary hemochromatosis and Wilson's disease. Z Gastroenterol Verh 1991;26:82–7.
[64] Schilsky ML, Scheinberg IH, Sternlieb I. Liver transplantation for Wilson's disease: indications and outcome. Hepatology 1994;19:583–7.
[65] El Raziky MS, Ali A, El Shahawy A, Hamdy MM. Acute hemolytic anemia as an initial presentation of Wilson disease in children. J Pediatr Hematol Oncol 2014;36:173–8.
[66] Walshe JM. The acute haemolytic syndrome in Wilson's disease—a review of 22 patients. QJM 2013;106:1003–8.
[67] Robitaille GA, Piscatelli RL, Majeski EJ, Gelehrter TD. Hemolytic anemia in Wilson's disease. A report of three cases with transient increase in hemoglobin A2. JAMA 1977;237:2402–3.
[68] Hoshino T, et al. Low serum alkaline phosphatase activity associated with severe Wilson's disease. Is the breakdown of alkaline phosphatase molecules caused by reactive oxygen species? Clin Chim Acta 1995;238:91–100.
[69] Walshe JM. Wilson's disease. The presenting symptoms. Arch Dis Child 1962;37:253–6.
[70] Oder W, et al. Neurological and neuropsychiatric spectrum of Wilson's disease: a prospective study of 45 cases. J Neurol 1991;238:281–7.
[71] Cartwright GE. Diagnosis of treatable Wilson's disease. N Engl J Med 1978;298:1347–50.
[72] Sternlieb I. Wilson's disease: indications for liver transplants. Hepatology 1984;4:15S–7S.
[73] Demirkiran M, Jankovic J, Lewis RA, Cox DW. Neurologic presentation of Wilson disease without Kayser-Fleischer rings. Neurology 1996;46:1040–3.
[74] Aisen AM, et al. Wilson disease of the brain: MR imaging. Radiology 1985;157:137–41.
[75] Akil M, Schwartz JA, Dutchak D, Yuzbasiyan-Gurkan V, Brewer GJ. The psychiatric presentations of Wilson's disease. J Neuropsychiatry Clin Neurosci 1991;3:377–82.
[76] Akil M, Brewer GJ. Psychiatric and behavioral abnormalities in Wilson's disease. Adv Neurol 1995;65:171–8.
[77] Coffey AJ, et al. A genetic study of Wilson's disease in the United Kingdom. Brain 2013;136:1476–87.
[78] Ferenci P. Phenotype-genotype correlations in patients with Wilson's disease. Ann N Y Acad Sci 2014;1315:1–5.
[79] Ferenci P. Regional distribution of mutations of the ATP7B gene in patients with Wilson disease: impact on genetic testing. Hum Genet 2006;120:151–9.
[80] Vrabelova S, Letocha O, Borsky M, Kozak L. Mutation analysis of the ATP7B gene and genotype/phenotype correlation in 227 patients with Wilson disease. Mol Genet Metab 2005;86:277–85.
[81] Gojova L, Jansova E, Kulm M, Pouchla S, Kozak L. Genotyping microarray as a novel approach for the detection of ATP7B gene mutations in patients with Wilson disease. Clin Genet 2008;73:441–52.
[82] Nemeth D, et al. Clinical use of next-generation sequencing in the diagnosis of Wilson's disease. Gastroenterol Res Pract 2016;4548039:2016.
[83] Koppikar S, Dhawan A. Evaluation of the scoring system for the diagnosis of Wilson's disease in children. Liver Int 2005;25:680–1.
[84] Sintusek P, Dhawan A. Utility of serum zinc to predict severity of Wilson disease. In: Paper presentation at the American Association for the Study of Liver Disease (AASLD) liver meeting, Boston, the United States, 11–15 November; 2016.
[85] Ferenci P. Review article: diagnosis and current therapy of Wilson's disease. Aliment Pharmacol Ther 2004;19:157–65.
[86] Berman DH, Leventhal RI, Gavaler JS, Cadoff EM, Van Thiel DH. Clinical differentiation of fulminant Wilsonian hepatitis from other causes of hepatic failure. Gastroenterology 1991;100:1129–34.
[87] Milkiewicz P, Saksena S, Hubscher SG, Elias E. Wilson's disease with superimposed autoimmune features: report of two cases and review. J Gastroenterol Hepatol 2000;15:570–4.
[88] Sintusek P, Dhawan A. Aberrance of serum zinc and free copper level in Wilson disease. J Pediatr Gastroenterol Nutr 2016;62.
[89] El Balkhi S, et al. Relative exchangeable copper: a new highly sensitive and highly specific biomarker for Wilson's disease diagnosis. Clin Chim Acta 2011;412:2254–60.
[90] Lyon TD, et al. Use of a stable copper isotope (^{65}Cu) in the differential diagnosis of Wilson's disease. Clin Sci (Lond) 1995;88:727–32.
[91] Matthews WB. The absorption and excretion of radiocopper in hepato-lenticular degeneration (Wilson's disease). J Neurol Neurosurg Psychiatry 1954;17:242–6.
[92] Sternlieb I, Scheinberg IH. The role of radiocopper in the diagnosis of Wilson's disease. Gastroenterology 1979;77:138–42.

Chapter 12

Liver Pathology of Wilson Disease

Hans Peter Dienes* and Peter Schirmacher[†]
*Medical University of Vienna, Institute of Clinical Pathology, Vienna, Austria, [†]Institute of Pathology, University Hospital Heidelberg, Heidelberg, Germany

PATHOLOGY OF WILSON DISEASE

The symptoms and signs of Wilson disease are the result of copper overload in various organs. Impairment of copper transport in hepatocytes is associated with a reduced biliary excretion of copper and hepatic copper accumulation [1]. In the liver, excess copper causes hepatocellular necrosis, portal and periportal inflammation, and fibrosis. Massive hepatocellular necrosis may result in fulminant hepatic failure. The release of large amounts of copper from necrotic hepatocytes can induce hemolysis. At later stages of the disease, copper is taken up by a variety of nonhepatic tissues including the brain and cornea [2]. Liver injury and hemolysis tend to occur earlier in life compared with urological, neurological, or psychiatric manifestations, but the diagnosis is often missed or delayed, as it may be mild. The disease is rarely symptomatic before 3 years of age and is usually manifested by signs of liver disease in adolescence [1]. Occasionally, the diagnosis is even made in patients who are over 50 years of age. The diagnosis should be considered in any child or adult with hepatocellular disease of undetermined cause including elevated serum aminotransferases of fatty liver. Wilson disease may present as a noninfectious recurrent hepatitis or mimic autoimmune hepatitis including its extrahepatic manifestations, such as arthropathy, elevated gamma globulin levels, and positive antinuclear or antismooth muscle antibodies. Rarely, hepatocellular carcinoma and intrahepatic cholangiocarcinoma may occur in Wilson disease [3].

Correlation of phenotype with specific mutations (genotype) is difficult in Wilson disease because the vast majority of affected individuals are compound heterozygous possessing one copy of two different mutations [4]. In general, those mutations that abrogate the production of an intact functional Wilson ATPase protein cause more severe and earlier hepatic presentation [5]. The diagnosis is not necessarily easy even when the disease is actively been considered [6, 7]. A serum ceruloplasmin below 20 mg/dL is present in up to 95% of patients, but is not diagnostic of the disease since ceruloplasmin is an acute-phase protein and may be elevated in inflammatory disorders. In Wilson disease, the 24 h copper excretion is usually above 100 μg and almost always exceeds 40 μg. With compatible pathological changes, a copper concentration of over 250 μg per gram liver weight is diagnostic; however, lower values may be found if the sample is too small (due to possible error in measurement of the sample weight) or if there is variable distribution of parenchymal copper [8]. Hepatic copper content can in principle be determined using tissue extracted from paraffin-embedded liver biopsies.

Histopathologic patterns of Wilson disease in the liver are variable and range from almost normal liver, steatosis with few apoptotic bodies, acute liver failure, or extensive hepatocellular necrosis to chronic hepatitis [9] and cirrhosis (Figs. 1–6).

FIG. 1 Explant liver from a patient showing end-stage Wilson disease with complete cirrhosis.

FIG. 2 Histopathology from a patient with chronic Wilson disease and an acute flare up with confluent and bridging necrosis (H&E; ×200).

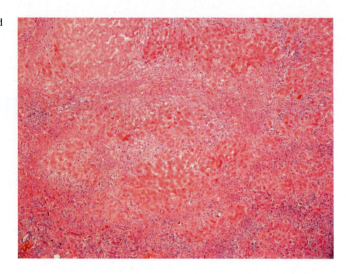

FIG. 3 Liver from a patient with chronic Wilson disease. Histopathology can mimic chronic hepatitis of another etiology, such as viral or autoimmune hepatitis with severe interface necroinflammation (H&E; ×400).

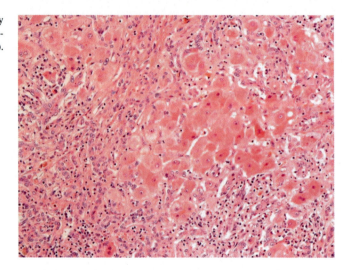

Oxidative stress appears to play an important role since copper is a redox-active prooxidant. In severe liver disease requiring liver transplantation, extensive functional abnormalities in hepatic mitochondria have been detected. In contrast to genetic iron storage disease, there is no direct relationship between the absolute hepatic copper concentration and the extent of liver damage [10].

Demonstration of copper deposits in hepatocytes is usually performed by rhodanine stain, which is specific but not very sensitive (Fig. 7). Timm's silver stain is more sensitive but rarely performed routinely [9]. In the precirrhotic stage of Wilson disease, cytochemically demonstrable copper is usually concentrated in periportal hepatocytes. Another helpful histochemical stain is the demonstration of copper-binding protein by orcein, but orcein becomes also positive under other cholestatic conditions. The negative staining for copper in the liver does not exclude Wilson disease, and the copper content of the tissue is determined best in liver tissue that is dried, weighed, and digested. Copper content can also be determined in liver tissue extracted from paraffin-embedded specimens.

Further cellular alterations in Wilson disease besides the deposits of copper are lipofuscin accumulations in periportal areas; some of the granules are large, irregular in shape, and even vacuolated. Mallory-Denk bodies may be numerous (Fig. 8), and hepatocellular ballooning is also frequently found in Wilson disease. There are many glycogenated nuclei and often a mixed macro- and microvesicular accumulation of fat droplets. Apoptotic bodies are seen to a variable degree. Cholestasis especially in zone 3 can occur.

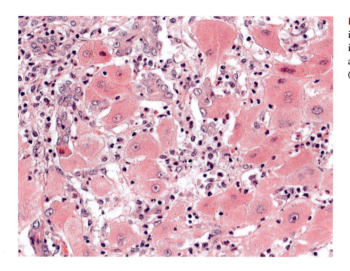

FIG. 4 Liver from a patient with active Wilson disease. A focal necrosis is present without zonal location showing lytic necrosis with abundant inflammatory infiltrates of mixed cell types: Lymphomonocytic cells are dominant, but there are also some polymorphonuclear granulocytes (H&E; ×400).

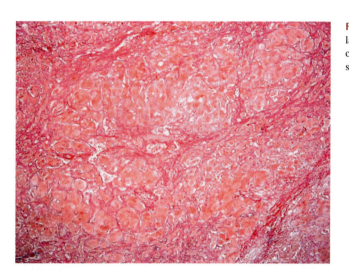

FIG. 5 Liver from a patient with long-standing Wilson disease: in the late stage, a cirrhotic transformation of the lobular architecture has occurred with portal-portal and portal-central fibrotic septa (Gomori's silver staining, ×300).

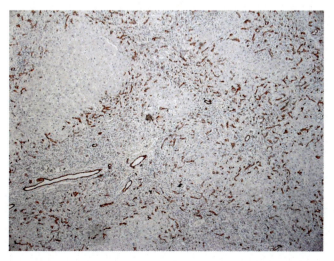

FIG. 6 Liver histology from a patient with the cirrhotic stage of Wilson disease: the immunostaining for cytokeratin 7 shows bile ducts and abundant proliferation of neoductules as a sign of regeneration (immunohistochemistry; ×100).

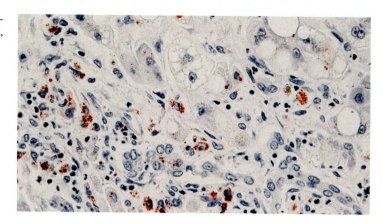

FIG. 7 Liver from a patient with chronic Wilson disease displaying a positive rhodanine stain for copper (rhodanine stain, ×400).

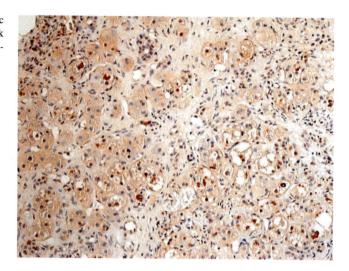

FIG. 8 Immunohistochemistry for ubiquitin in a patient with chronic Wilson disease. Staining for ubiquitin is helpful to detect Mallory-Denk bodies that are frequently found in Wilson disease (immunohistochemistry; ×300).

In the precirrhotic stage, liver tissue may present with mild polymorphism of hepatocytes and some apoptotic bodies. Steatosis in Wilson disease is rather frequent (Fig. 9) and can be seen as simple steatosis or steatohepatitis such that it can lead to the diagnosis of nonalcoholic fatty liver disease of other origin. The extent of steatosis does not correlate with the amount of copper but in some patients is linked to polymorphisms in the PNPLA3 "G" allele [11].

In the course of the disease, there is a progressive increase in periportal bile ductular reaction and fibrosis (Figs. 6 and 10). Variable numbers of inflammatory cells, mainly lymphocytes and plasma cells, are seen in the portal and septal areas. Changes may be indistinguishable from those of chronic hepatitis due to other etiologies, in particular autoimmune hepatitis.

Helpful differential clues in the diagnosis of Wilson disease in addition to portal inflammation are interface hepatitis and steatosis, including periportal glycogenated nuclei, moderate to marked copper storage, and the presence of Mallory-Denk bodies in periportal liver cells [9] (Fig. 8).

The development of cirrhosis in Wilson disease runs along the course of chronic hepatitis with increasing deposits of copper in the liver cells. Fibrosis and the formation of collagen fibers start in portal areas with portal to portal septa, finally separating pseudolobules. The cirrhosis of Wilson disease is usually macronodular but can be mixed or occasionally even micronodular in type (Fig. 5). Microscopically, it is characterized by varying-sized nodules separated by fibrous septa that may be broad or thin with minimal ductular reaction and variable inflammation. The number of Mallory-Denk bodies is frequently increased at the periphery of the cirrhotic nodules. Variable numbers of apoptotic bodies may be present. Clusters of large hepatocytes with a granular eosinophilic cytoplasm (oncocytic or oxyphilic cells) due to an increased number of mitochondria are often seen (Fig. 11). The distribution of copper is quite variable with some of the cirrhotic

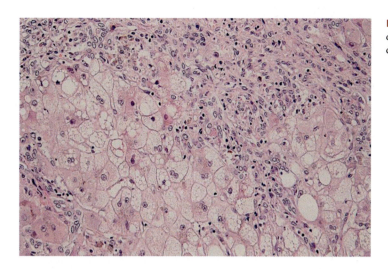

FIG. 9 Histopathology from a patient with chronic Wilson disease: steatosis is evident with micro- and macrovesicular fat droplets (H&E, ×300).

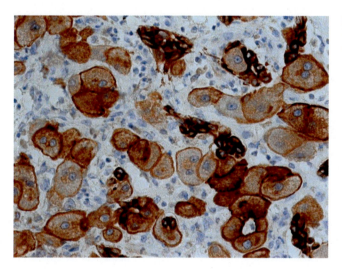

FIG. 10 Immunohistochemistry for cytokeratin 18 as a feature for ductular metaplasia in hepatocytes undergoing stress and damage (immunohistochemistry; ×450).

nodules containing a high amount and others containing little or none. Copper is rarely demonstrated in Kupffer cells or portal macrophages in the cirrhotic stage but is frequently encountered in cases of submassive or massive necrosis in patients who present with fulminant liver failure. Davies et al. also noted the presence of copper in hepatic parenchymal and mononuclear phagocytic cells in cases of Wilson disease, which although clinically presenting as fulminant hepatic failure, were cirrhotic [12].

Ultrastructural findings in the precirrhotic stage of Wilson disease have been described in detail by a number of investigators; of these, mitochondria, changes are the most distinctive and pathogenetically significant. They include heterogeneity of mitochondrial size and shape [13, 14]. In more detail, the inner membranes are separated by edema, the intercristal space is widened and slit-like, and sometimes, mitochondria show diamond-shaped or circular dilatations with crystalloid inclusions, granules, or dense deposits and may have lipid inclusions. Especially in zone 1, the copper is associated with lipofuscin, and lipolysosomes are numerous. Furthermore, the number of peroxisomes is increased. In the ER, there may be many glycogen bodies, but these may also appear in the cytoplasm and even in the nuclei. Here, they represent true inclusions in the glycogenated nuclei in zone 1. The microvilli of bile duct epithelia are elongated and more numerous. The Mallory-Denk bodies show misshapen and coagulated microfilaments. Singular ultrastructural features are not pathognomonic, but the ensemble is very typical and diagnostically highly informative.

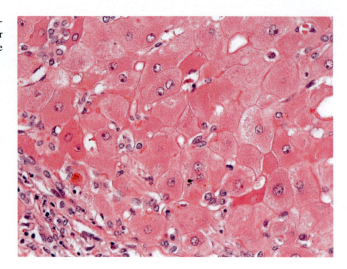

FIG. 11 Histology from a patient with chronic Wilson disease. Oncocytic transformation indicating mitochondrial damage and hepatocellular apoptosis is obvious in some hepatocytes as a sign of lethal cell damage (H&E; ×400).

After long-term effective therapy with copper chelators, the structural and functional alterations related to Wilson disease may disappear, and a normal histology and hepatocellular ultrastructure can be restored [15].

An important differential diagnosis to Wilson disease is Indian childhood cirrhosis. This disease, most prevalent in the Indian subcontinent, is caused by copper overload. Histopathology of this disease with respect to hepatic copper overload and other histological features is very similar to Wilson disease; however, it can be clearly distinguished due to its clinical and epidemiological background.

REFERENCES

[1] Roberts EA, Cox DW. Wilson disease. Baillieres Clin Gastroenterol 1998;12:237–56.
[2] Scheinberg IH, Sternlieb I. Wilson disease. Philadelphia: WB Saunders; 1984.
[3] Pfeiffenberger J, Mogler C, Gotthardt DN, Schulze-Bergkamen H, Litwin T, Reuner U, Hefter H, Huster D, Schemmer P, Członkowska A, Schirmacher P, Stremmel W, Cassiman D, Weiss KH. Hepatobiliary malignancies in Wilson disease. Liver Int 2015;35:1615–22.
[4] Thomas GR, Forbes JR, Roberts EA, et al. The Wilson disease gene: spectrum of mutations and their consequences. Nat Genet 1995;9:210–7.
[5] Panagiotataki E, Tzetis M, Manolaki M, et al. Genotype-phenotype correlations for a wide spectrum of mutations in the Wilson disease gene (ATP7B). Am J Med Genet A 2004;131A:168–73.
[6] Maier-Dobersberger T, Ferenci P, Polli C, et al. Detection of the His1069Gln mutation in Wilson disease by rapid polymerase chain reaction. Ann Intern Med 1997;127:21–6.
[7] Roberts EA, Schilsky ML. Diagnosis and treatment of Wilson disease: an update. Hepatology 2008;47:2089–111.
[8] Nuttall KL, Palaty J, Lockitch G. Reference limits for copper and iron in liver biopsies. Ann Clin Lab Sci 2003;33:443–50.
[9] Thompson RJ, Portmann BC, Roberts EA. Genetic and metabolic liver disease. In: Burt A, Portmann BC, Farell LD, editors. MacSween's pathology of the liver. Churchill-Livingstone; 2012. p. 157–260.
[10] Ferenci P, Steindl-Munda P, Vogel W, et al. Diagnostic value of quantitative hepatic copper determination in patients with Wilson's disease. Clin Gastroenterol Hepatol 2005;3:811–8.
[11] Stättermayer AF, Traussnigg S, Dienes HP, et al. Hepatic steatosis in Wilson's disease—role of copper and PNPLA3 mutations. J Hepatol 2015;63:156–63.
[12] Davies SE, Williams R, Portmann B. Hepatic morphology and histochemistry of Wilson's disease presenting as fulminant hepatic failure: a study of 11 cases. Histopathology 1989;15:385–94.
[13] Mandel H, Hartman C, Berkowitz D, et al. The hepatic mitochondrial depletion syndrome: ultrastructural changes in liver biopsies. Hepatology 2001;34:776–84.
[14] Sternlieb I. Fraternal concordance of types of abnormal hepatocellular mitochondria in Wilson's disease. Hepatology 1992;16:728–32.
[15] Sternlieb I, Feldmann G. Effects of anticopper therapy on hepatocellular mitochondria in patients with Wilson's disease: an ultrastructural and stereological study. Gastroenterology 1976;71:457–61.

Chapter 13

Neurological Wilson Disease

Tomasz Litwin*, Petr Dusek[†] and Anna Członkowska*
*2nd Department of Neurology, Institute Psychiatry and Neurology, Warsaw, Poland, [†]Department of Neurology and Center of Clinical Neuroscience, First Faculty of Medicine and General University Hospital, Charles University in Prague, Prague, Czech Republic

INTRODUCTION

The neurological symptoms of Wilson disease (WD) were first reported in 1912 in the dissertation of S.A. Kinnier Wilson, who described in detail the neuropsychiatric presentation of 12 WD patients with various combinations of movement disorders, drooling, dysarthria, and psychiatric symptoms [1]. However, probably the first ever neurological WD patient was described by Westphal in 1883 (described in detail in Chapter 1) [1]. Until then, the pseudosclerotic form of WD with dominant intentional tremor and cerebellar symptoms similar to multiple sclerosis was referred to as the "Westphal-Strumpell" variant. Neurological symptoms of WD can appear at different stages of the disease. As ATP7B, the protein that is defective in WD is predominantly expressed in the liver; the liver accumulates copper before other organs. The first symptoms of WD therefore are usually hepatic. Neurologic, psychiatric, and other symptoms arise when the liver storage capacity for copper is exhausted, and copper is released into the bloodstream and deposited in other organs. Consequently, neurological symptoms appear when not initial hepatic symptoms are unrecognized and there is a delayed diagnosis and in correctly diagnosed patients with poor compliance or treatment failure. Rarely, the neurological symptoms may be the initial leading to diagnosis [2–13]. Neurological manifestations of WD usually occur between the ages of 20 and 40 years; however, the range is very large. The youngest patients with neurological symptoms of WD were 6 years old and the oldest 72 years old [5, 13]. The most common neurological symptoms of WD are movement disorders (such as tremor, dystonia, and parkinsonism) along with ataxia, dysphagia, dysarthria, and drooling (Table 1).

It may be difficult to classify symptoms due to the wide spectrum of affected brain regions and symptom combinations. Additionally, there may be significant fluctuations in the neurological symptoms due to factors such as stress, concomitant medications, and general health conditions. In the past, several classification schemes of neurological WD, based on predominant symptoms, were suggested (Table 2). However, many patients do not fit into these classifications. Currently, apart from detailed descriptions of clinical symptoms, the clinical severity quantified by validated neurological scales for WD (Unified Wilson's Disease Rating Scale (UWDRS) or the Global Assessment Scale (GAS)) is used for assessing treatment outcome and recording neurological deterioration [20, 21].

The natural course of untreated WD (Wilson 1912) is progressive neurological disease leading to severe disability, immobilization, and death. By contrast, with correct treatment, 70%–80% of WD patients experience improvement or stabilization of symptoms during the first 3 years of treatment [5, 7, 11, 19, 22]. Persistent disabling symptoms should prompt special WD treatment analysis (assessments for compliance and adequacy of treatment) and discussions about additional neurological symptomatic treatments (see Chapter 20). In addition to clinical neurological examination, other supplementary tests, such as brain magnetic resonance imaging (MRI), ophthalmologic evaluation for Kayser-Fleischer (K-F) Rings, and neurophysiological exams, are valuable tools for WD diagnosis and treatment monitoring. Below, we present the clinical description and pathophysiology of the most frequent neurological symptoms of WD and the most important supplementary examinations.

PATHOPHYSIOLOGY OF NEUROLOGICAL WD SYMPTOMS

Central nervous system (CNS) damage leading to neurological and psychiatric symptoms (see Chapter 14) in WD is primarily a consequence of extrahepatic copper toxicity [23]. In WD, copper levels are profoundly increased in all brain regions and in the cerebrospinal fluid, reaching values that are up to 10-fold higher than in control subjects [23–28]. Macroscopic neuropathological abnormalities are mostly pronounced in the basal ganglia, thalamus, and upper brain stem, likely due to the high susceptibility of these brain regions to copper toxicity.

TABLE 1 Neurologic Features of Wilson Disease and Their Frequency

Symptom (Frequency in Neurological WD)	Details
Tremor (55%–88%)	Postural (provoked during special postures or movements, like the wing-beating and flapping tremor) Intention tremor (present during voluntary movement, when an object is nearly acquired) Action tremor (present during voluntary movements like writing and drinking) Rest tremor (parkinsonian) (tremor with low amplitude, observed in distal extremities)
Dystonia (38%–65%)	Focal (e.g., dystonic dysphonia; torticollis; blepharospasm; writer's cramp; tongue, hand, or leg dystonia; dystonic dropped jaw) Segmental (e.g., cranial and neck dystonia) Generalized (involving several parts of the body including axial muscles)
Parkinsonism (19%–62%)	Resting tremor, rigidity, bradykinesia, micrographia, postural imbalance, shuffling gait, freezing
Ataxia (30%)	Ataxic gait, action tremor, intention tremor, dysdiadochokinesia
Chorea (6%–16%)	Involuntary irregular movements of the face, head, trunk, hands, and legs that interrupt normal activity
Dysarthria (up to 97%)	Ataxic (cerebellar)—impaired coordination, explosive speech, accompanied by grimacing Dystonic (hyperkinetic)—for example, quiet enunciation and breathing difficulties (resulting from dystonia) Parkinsonian (hypokinetic)—for example, slowed monoloudness speech Spastic (pseudobulbar)—for example, hypernasality, slow rate of speech, monoloudness Mixed (unclassified)
Dysphagia (up to 50%)	Can involve each stage of swallowing: oral, preparation, transit, and swallowing
Drooling (46%)	In WD mainly caused by swallowing dysfunction or orofacial dystonia
Gait and posture disturbances (44%–75%)	Extrapyramidal (dystonic or parkinsonian) or ataxic

TABLE 2 The Neurological Forms of WD—Historically Proposed Classifications

The Author (Year of Proposition)	Classification
Hall (1921) [14]	(1) Classic form (progressive rigidity and tremor) with domination of rigidity and bad prognosis (2) Tremor form (Strumpell-Westphal, pseudosclerotic form) occurring in adults with a relatively good prognosis (3) Torsion spasms (dystonic form currently)
Konovalov (1960) [15]	(1) Arrhythmic-hyperkinetic form (hyperkinesias and dystonia) (2) Tremulous form (ataxia and postural tremor) (3) Tremulous-rigid form (rigidity and resting tremor—currently parkinsonian)
Denny-Brown (1962) [16]	(1) Juvenile form (onset before second decade of life, hyperkinetic/dystonic symptoms with putamen lesions) (2) Pseudosclerotic form (dominant intention and postural tremors with lesions in the cerebral cortex, thalamus, subthalamic regions, nucleus dentatus)
Marsden (1987) [17]	(1) Hyperkinetic-dystonic form (dystonia and choreoathetosis) (2) Ataxic form (postural and intentional tremor and ataxia) (3) Parkinsonian form (rigidity, rest tremor, and hypokinesia)
Oder et al. (1993) [18]	(1) Dyskinetic form (dystonia and choreoathetosis) (2) Pseudosclerotic (ataxia and postural tremor) (3) Pseudoparkinsonian form (rigidity, rest tremor, and abnormal cognition)
Czlonkowska (1996) [19]	(1) Dystonic form (2) Tremor form (3) Rigidity-tremor form (4) Rigidity form

Copper is brought to the CNS through the blood stream. Astrocytes, which are part of the blood-brain barrier, are the first brain cells affected by copper toxicity. Initially, astrocytes are able to buffer the toxic effects of copper, probably through the upregulation of metallothionein and glutathione [29, 30]. Later, due to chronic copper intoxication, astrocytes increase in numbers and undergo cellular swelling and morphological changes [30]. Upon neuropathological examination, several types of astrocytes with abnormal morphologies are found in the basal ganglia and less frequently in the cortex. These cells are referred to as Alzheimer type II glia (Fig. 1A) and Opalski cells [31]. After the astrocytes' storage capacities are met, copper toxicity affects other brain elements that are prone to oxidative damage, mainly oligodendrocytes and neurons. Neuronal dysfunction is pathologically manifested by axonal swelling and spheroid formation [27, 32]. Oligodendrocyte dysfunction and demyelination are typically present in the bundles passing through basal ganglia and in pontine fibers, while hemispheric white matter is rarely affected [33–35]. Demyelination of pontine fibers may lead to central pontine myelinolysis. Ultimately, nonselective damage to all parenchymal elements and tissue rarefaction is observed, and the putamen is typically the most affected structure. In the most severe cases, there is putaminal necrosis with iron-laden macrophages surrounding the necrotic cavity (Fig. 1B) [36]. Apart from copper toxicity, other factors may contribute to the pathophysiology of neurological symptoms in WD, namely, hepatic encephalopathy (HE) in patients with severe liver damage and portal hypertension with portosystemic shunting and hypocupremia from overtreatment leading to polyneuropathy and myelopathy, typically present only in patients on long-term anticopper treatment.

NEUROLOGIC SYMPTOMS OF WD

Tremor

The most common neurological symptom of WD is tremor. It is likely associated with lesions in the dentato-rubro-olivary triangle and/or its output pathways in the thalamus [37–39]. Different patients and even the same patient during the disease

FIG. 1 (A) Hematoxylin and eosin (200×); astrocytes with inconspicuous cytoplasm and large vesicular nuclei (Alzheimer's type II astrocytes, *arrows*) and a large reactive astrocyte *(arrowhead)*. (B) Ferritin immune staining (400×); putaminal necrosis with large numbers of iron-containing macrophages *(arrows)*.

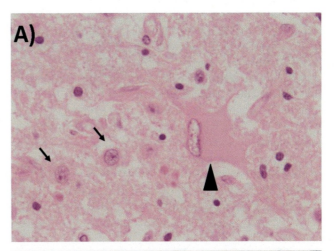
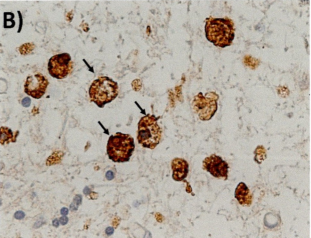

course may experience several types of tremor including resting, postural, action, and/or intention [3, 5, 9, 37–39]. Clinically, WD tremor could be an isolated symptom resembling essential (typically symmetrical, action tremor affecting the hands, head, and voice), dystonic (occurring in parts of the body affected by dystonia), rubral (slow, unilateral, proximal, and resting with action component), or parkinsonian (resting, asymmetrical, distal, and with low amplitude) tremors, but most frequently, it occurs in combination with other neurological symptoms. Tremor was described as the first neurological presentation of WD in up to 55% of patients [9]; during the disease course, it affects almost 88% of patients with neurological disease [9]. Importantly, the type of tremor in WD could change during the disease progression, mostly in untreated patients. It usually initially affects distal upper extremities; later, tremors with higher amplitude spread to other body parts such as the head and legs. The most spectacular type of tremor described in WD is the so-called "wing-beating" tremor, which is an irregular coarse tremor with gradually increasing amplitude involving the proximal parts of the upper extremities that appears in a sustained posture-holding position after a delay of 30–60 s. Asterixis, also referred to as a flapping tremor, manifests as irregular flexion-extension movements at the metacarpophalangeal and wrist joints. It is phenomenologically similar to a tremor, but pathophysiologically, it is described as negative myoclonus, which is indicative of HE [40]. Of the neurological WD symptoms, tremor has the most favorable prognosis for recovery [41] during correct anticopper treatment. In the case of treatment-refractory and disabling tremor, there are several pharmacological and neurosurgical options for symptomatic treatment with proved efficacy (described in detail in Chapter 20) [42].

Dystonia

Dystonia is defined as an involuntary movement disorder with sustained or intermittent muscle contractions resulting in repetitive, twisting movements or abnormal fixed postures [9]. The reported prevalence of dystonia ranges between 11% and 65% in neurological WD patients. It is one of the most severe neurological symptoms of WD and is frequently resistant to anticopper treatment [9, 18, 38, 42–47]. Dystonia has been associated with putaminal lesions [18, 46, 47]. Its localization and severity may vary in different patients; it could be focal, multifocal, segmental, and generalized, or there are even cases with status dystonicus. The most common presentation of dystonia in WD is abnormal facial expression, which can present as (1) "risus sardonicus" or a fixed smile (due to dystonic spasm of the risorius muscle), (2) "vacuous smile" (due to dystonic dropped jaw), and (3) open mouth smile. The other focal dystonia in WD may affect the neck (torticollis), hand (e.g., writer's cramp and "starfish" hand) (Fig. 2A), legs (Fig. 2B), tongue (dysarthria and dysphagia), vocal cords (dysarthria or dysphonia), etc. Focal dystonia usually spreads to other body regions during disease progression and ultimately affect axial muscles leading to dystonic posture (Fig. 3). Since the first description by Wilson, generalized dystonia has occurred in advanced neurological WD cases leading to immobilization and contractures that are associated with high mortality [5, 12, 18, 41–49]. Interestingly, dystonia can arise not only as a native symptom of copper toxicity but also as a complication of symptomatic treatment, for example, with dopamine antagonists [49]. Higher sensitivity to dopamine receptor antagonists with higher risk of drug-induced dystonia in WD patients may be caused by a decreased number of dopamine D2 receptors in the lesioned striatum [44, 45]. Therefore, avoiding or using extreme caution with dopamine antagonists is recommended in WD patients [50–53]. In the cases of dystonia resistant to anticopper treatment, there are additional symptomatic neurological treatment options, that is, botulinum toxin (BTX) for focal dystonia and anticholinergic agents or neurosurgical procedures for generalized dystonia (described in Chapter 20) [42].

 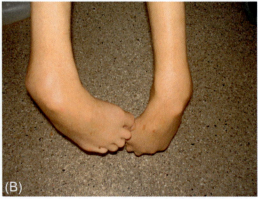

FIG. 2 (A) Focal hand dystonia and (B) leg dystonia.

FIG. 3 Axial dystonia in a WD patient.

Parkinsonism

Parkinsonism is reported in 19%–62% of WD patients [9, 18, 38, 42–48, 54]. It is caused by lesions in the substantia nigra and putamen that lead to concomitant impairment of the presynaptic and postsynaptic parts of the nigrostriatal pathway [34, 38, 44, 45]. The syndrome typically consists of symmetrical or asymmetrical slowing of movements, along with hypomimia, drooling, shuffling gate with postural imbalance, and cogwheel rigidity. In addition, the classical, asymmetrical resting tremor typically observed in Parkinson's disease can very rarely occur as an isolated symptom in WD [9, 42].

Ataxia

Cerebellar ataxia occurs in almost 30% of neurological WD patients, but as with parkinsonism or chorea, it usually manifests in combination with other neurological disorders, rarely as a solitary symptom [9, 38, 42, 54]. In fact, cerebellar ataxia may be "hidden behind" other overlapping neurological symptoms such as parkinsonism, dystonia, and gait apraxia. Careful examination focused on ataxia symptoms in WD can frequently reveal signs of damage to the cerebellum and/or its outflow pathways, such as (1) ataxic gait (wide stance and wide-based gait with impaired tandem walking), (2) impaired coordination of fine hand movements with overshooting and macrographia,(3) intentional tremor, and (4) dysdiadochokinesis. It is notable that other symptoms that may be present in hereditary ataxias like nystagmus are extremely rarely observed in WD (contrary to other oculomotor abnormalities such as impaired voluntary control of saccades with disturbed smooth eye movements likely related to lesions in the prefrontal cortex or frontonuclear pathways).

Chorea

Chorea and athetosis are rare in WD and are present in approximately 6%–16% of neurological patients [9, 18, 38, 42, 54]. Chorea is defined as involuntary, irregular, brief, "dance-like" movements of the face, head, trunk, and extremities that interrupt normal activity. Athetoid movements are slower and writhing and often accompany chorea in a movement disorder named choreoathetosis. Both chorea and athetosis are associated with lesions in the striatum and globus pallidus [18]. In WD, chorea occurs more often in young patients (20% in patients <16 years old) than in older patients (only 3% in patients >17 years old) and only very rarely can be observed as an isolated symptom of WD [9, 42].

Dysarthria

Dysarthria is one of the most frequent symptoms of WD, occurring in almost all (85%–97%) neurological patients [3, 9, 42]. It may result from lesions located in the basal ganglia (extrapyramidal features), cerebellar nuclei and associated tracts (cerebellar features), and corticobulbar tracts (pseudobulbar features). In most cases, dysarthria in WD is caused by lesions affecting more brain structures and their connections. Based on its predominant characteristics, it could be classified as (1) mixed, unspecific dysarthria arising from lesions in several structures and presenting with a combination of clinical features; (2) ataxic, caused by cerebellar dysfunction presenting with reduced coordination, overpronunciation, separated syllables, irregular articulatory breakdown, and explosive speech with grimacing; (3) dystonic, hyperkinetic dysarthria with quiet enunciation, breathing disturbances, and dystonic posturing during speech that may progress to anarthria in advanced cases; (4) parkinsonian, hypokinetic dysarthria manifested with slowed speech, monoloudness, and dysprosodia; and (5) pseudobulbar, spastic dysarthria with slow rate of speech, use of short phrases, harshness, hypernasality, and monopitch. Importantly, there is no specific dysarthria for WD [55]. The presented classification is based mainly on the dysarthric features related to the predominantly affected brain region. Similar speech disturbances may also be observed in other disorders (e.g., ataxic speech in multiple sclerosis or hypokinetic speech in Parkinson's disease) [42, 55].

Dysphagia

Dysphagia is defined as difficulties in any phase of swallowing and occurs in almost 50% of neurological WD and in 18% of all WD patients [3, 9, 42]. Due to coincident involvement of several brain structures in the WD pathology, any phase of swallowing can be affected, that is, oral, preparation, oral transit, or swallowing phase [56, 57]. Dysphagia significantly decreases quality of life and nutritional status. Furthermore, it may lead to difficulties with consuming anticopper drugs, silent aspirations, pulmonary infections, and cachexia, all of which worsen the prognosis of WD. Dysphagia should be actively screened for and adequately treated. There are general rules for dysphagia assessment and management (described in Chapter 20) [42]. The nutritional status of WD patients should be monitored, and appropriate treatment, for example, nutritional support, nasogastric tube, or percutaneous gastrostomy, should be initiated [57].

Drooling

The most prominent and characteristic symptoms of WD are drooling, dysarthria, and "wing-beating" tremor [9, 42]. Drooling is defined as involuntary flow of saliva from the mouth. Drooling affects 32%–46% of all WD patients and 68% of neurological WD patients [9, 42, 56, 58] and is a consequence of dysphagia and the inability to retain saliva within the mouth due to orofacial dystonia. Treatment options for drooling are described in Chapter 20 [42].

Gait and Posture Disturbances

Postural instability and gait disturbances are reported in 44%–75% of WD patients with neurological presentation. Gait problems are usually multifactorial and mainly involve extrapyramidal (e.g., dystonia, hypokinesia, freezing, and chorea) and cerebellar (e.g., ataxia and tremor) dysfunction [4, 9, 42, 58]. Falls due to postural instability are reported in 35% of neurological WD patients. Currently, for epidemiological purposes, dystonic, ataxic, and parkinsonian gaits can be distinguished based on the UWDRS clinical scale. The postural abnormalities are also secondary to extrapyramidal (e.g., trunk dystonia (Pisa sign)), parkinsonism (stooped semiflexed posture), or cerebellar symptoms. The UWDRS consists of separate points for posture assessment including trunk dystonia, ataxia of stance, and parkinsonism [42]. Posturography and posture analysis are currently also used in rehabilitation and the assessment of treatment outcomes [58, 59].

OTHER, SELECTED NEUROLOGICAL SYMPTOMS IN WD

In addition to the common symptoms already described, a variety of other rare neurological symptoms may occur in the course of WD. These include myoclonus, tics, muscle cramps, headache, taste dysfunctions, and olfactory dysfunctions [1–5, 42, 47, 54, 60, 61]. Due to space limitations, we have described only the more frequent neurological syndromes in WD that may be misdiagnosed or lead to therapeutic difficulties.

Epilepsy

Epilepsy is reported in 6.2%–8.3% of WD patients, a frequency that is 10-fold greater than for the general population [42, 62–64]. Seizures are mostly generalized, less frequently partial, and in 1% manifest as status epilepticus [64]. Seizures can appear at any stage of the disease but mostly at the beginning of anticopper treatment. Several risk factors for seizure development in WD were reported, including demyelinating lesions of frontal, parietal, or occipital white matter and cortical atrophy with gliosis [63–66]. Interestingly, white matter lesions may occur during the natural course of WD, as secondary complications of treatment and as a consequence of WD "overtreatment" resulting in an extremely negative copper balance [42, 66]. The prognosis of epilepsy treatment in WD is similar to idiopathic epilepsy. However, due to the liver damage associated with WD, the selection of antiepileptic drugs is limited, and hepatotoxic drugs (e.g., valproate) should be avoided [42].

Neuropathies

Although peripheral neuropathies are not "typical" in the clinical presentation of WD and their prevalence in WD is unknown, there are a few case reports documenting the occurrence of polyneuropathy in WD [42, 67–70]. In these reports, several different kinds of neuropathies are described: axonal, demyelination, autonomic (fainting spells and small fiber neuropathy), and sensory motor. Several possible etiologies may be involved, including copper toxicity, liver dysfunction, induction by penicillamine, or even copper deficiency due to overtreatment. In several case studies, electrophysiological and clinical examinations indicated improvement during anticopper treatment suggesting that either direct copper toxicity or hepatopathy was responsible for the neuropathy. Copper deficiency can cause axonal sensory or sensory-motor neuropathy associated with myelopathy and can mimic the subacute combined degeneration of the spinal cord. Copper deficiency in WD occurs mostly during long-term intake of high-dose zinc preparations but may occur with chelation therapy as well. This phenomenon is reversible after stopping the treatment and should always be considered in WD patients with sensory complaints who are treated chronically with anticopper agents [42].

Restless Leg Syndrome

Restless leg syndrome (RLS) is a disorder manifesting with unpleasant, uncomfortable sensations in the legs (itching, buzzing, pins, and needles), leading to the urge to move the legs and rarely arms; the symptoms are worse at rest, during relaxation, especially when lying or sitting, with worst symptoms in the evening [71]. It occurs more frequently in neurological WD, affecting around 30% patients than the general population where the prevalence is ~10% [71]. More importantly, clinically significant RLS is observed in 19% of WD patients and in only 2%–3% of the general population [71]. This observation is in concordance with the higher prevalence of RLS in other neurodegenerative disorders with disruption of dopaminergic transmission. RLS may significantly impair the quality of life. Currently, there are no data on RLS recovery after anticopper treatment in WD, but it can be successfully managed with dopaminergic agents [71].

Sleep Disturbances

Based on questionnaire studies, the prevalence of sleep disturbances in WD (regardless of the clinical form) is 40%–80% [72]. Disturbances include mainly increased nocturnal awakenings, decreased total sleep time, decreased sleep efficiency with daytime fatigue, tiredness, excessive daytime sleepiness, cataplexy-like episodes, and altered REM sleep functions. There are several case reports on WD patients with a reversible REM sleep behavioral disorder. During anticopper treatment, recovery of sleep disturbances and general improvement of sleep quality are usually observed. However, due to the significant impact on quality of life in affected patients, sleep disturbances in WD should be diagnosed and symptomatically treated based on their clinical significance [72].

NEUROLOGIC SYMPTOMS OF HEPATIC ENCEPHALOPATHY (HE) IN WD

Although we have presented the neurological signs and symptoms of WD, it is primarily a liver disease. In cases with clinically significant hepatic insufficiency or with portosystemic shunting, additional neurological symptoms due to HE can occur in WD [40]. This includes a wide spectrum of neurologic/psychiatric symptoms ranging from subtle cognitive dysfunctions, sleep-wake cycle disturbances, or altered consciousness (West-Haven criteria) [40]. A wide spectrum of neurological symptoms may also occur, including tremor (high frequency), difficulties with coordination, ataxia, asterixis (negative myoclonus), hypertonia, hyperreflexia, pyramidal signs (very rarely seen in WD) with focal neurological deficits, seizures, and finally coma. Additionally, some (minority) WD patients with chronic liver disease may present with extrapyramidal symptoms, so-called cirrhosis-related parkinsonism, progressive hypokinetic-rigid syndrome, dystonia, choreoathetosis, or even spastic paraparesis (due to liver failure that is independent from the etiology). The WD patients with advanced liver disease may have different neuropsychiatric symptoms and/or a complex etiology (WD and/or liver failure) [9, 40, 42].

Ophthalmologic Signs of WD

The most characteristic ocular signs of WD that arise from copper accumulation in the eyeball include the K-F ring (Fig. 4) and sunflower cataract (SC). The K-F ring was first described by Kayser in 1902 [73] and usually presents bilaterally as brown, green, or yellow rings, visible as peripheral pigment deposits in the corneal Descemet's membrane [74]. At diagnosis, it is present in almost 100% of neuropsychiatric WD patients, 50% of hepatic, and 20%–30% of presymptomatic individuals [5, 42]. However, neurological WD cases with clearly absent K-F rings are also reported. Initially described as pathognomonic for WD [5], the K-F ring was later found in other liver disorders (e.g., primary biliary cirrhosis, chronic cholestasis, and hyperbilirubinemia >20 mg/dL), in neoplastic disorders (multiple myeloma, lung cancer, and monoclonal gammopathies), and in estrogen uptake [75–77]. However, in these conditions, the distribution of pigment differs from the classic K-F ring. It is deposited in different corneal layers, and the ring may be composed of other metals or bilirubin deposits [75–77]. Therefore, the K-F ring should always be examined by an ophthalmologist with experience in WD using a slit lamp or other new techniques such as confocal microscopy or optical coherent tomography (OCT). The K-F ring initially forms in concordance with the vertical flow of aqueous fluid in the anterior chamber of the eye, from the superior part of the cornea, and later from the inferior part. Eventually, the ring closes from lateral parts. Its disappearance after several years of anticopper treatment follows a reverse order; it disappears first from lateral parts then inferior and superior parts of the cornea [3, 5, 78].

The SC was described for the first time in 1922 by Oloff and Siemerling [79]. It occurs relatively rarely, affecting only 2%–20% of WD patients [5, 79–81]. It is caused by copper deposits located under the lens capsule with a central disk with radiating petal-like fronds, similar to a sunflower.

The visual acuity is not affected by the SC, and it typically disappears with correct anticopper treatment [80, 81]. Current studies using new ophthalmologic techniques, including OCT and electrophysiological assessment of the retina, showed that not only the cornea and lens are involved in WD but also the neurodegenerative processes affect the retina and optic nerves (e.g., reduced thickness of the retinal nerve fiber layer). The clinical significance of these findings is very likely limited [82–84].

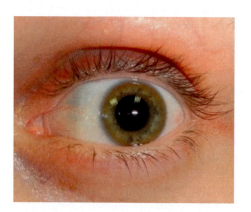

FIG. 4 Kayser-Fleischer Ring.

Neuroimaging in WD

In the early 1980s, brain computed tomography was used in the differential diagnosis of WD. However, its sensitivity and specificity for the WD diagnosis were limited as it showed mainly nonspecific changes such as brain atrophy and hypodensity in the basal ganglia in more advanced cases [85]. Currently, brain MRI plays a crucial role in the diagnosis and treatment monitoring and is considered in the Leipzig score algorithm [6]. In the course of the disease, pathological, symmetrical hyperintense lesions are observed in T2-weighted MR images in the putamina, globus pallidus, caudate nuclei, thalami, pons, and mesencephalon and further hypointensive usually in T1-weighted sequences (Fig. 5A and B) [86–99]. These neuroimaging abnormalities are present in almost 100% of neurological WD patients, in 40%–70% of hepatic cases, and even in 20% of presymptomatic cases [96].

Less frequently, demyelinating changes are observed in the hemispheric white matter and cerebellum. Brain atrophy may affect the entire brain, but the putamina, mesencephalon, cerebellum, and frontoparietal cortex are preferentially affected. In advanced neurological cases, T2 lesions also become visible in T1-weighted images as hypointense lesions. The most spectacular MRI changes are (1) "the face of giant panda" located in the mesencephalon (Fig. 6A) and (2) the "small panda" located in the pons (Fig. 6B). These occur infrequently, and their presence depends on the subjective evaluation of the radiologist [98].

Some WD patients with liver cirrhosis also present with symmetrical hyperintense lesions in the globus pallidus in T1-weighted images that probably correspond with manganese deposits [3–5, 40, 98, 99]. Recent studies involving new MRI techniques, such as susceptibility-weighted imaging, or high-resolution scans using ultrahigh field scanners suggest that paramagnetic iron species also accumulate in the basal ganglia in WD patients [36]. The clinical significance of cerebral iron accumulation in WD is currently unknown. During correct anticopper treatment or after liver transplantation, these MRI abnormalities improve or disappear [2–6]. This is not the case in advanced patients with irreversible neuropathological changes, for example, necrosis in the basal ganglia. Brain MRI could be helpful in treatment monitoring and predicting the outcome. Other neuroimaging techniques like brain MR spectroscopy (MRS, showing metabolic changes in the brain and neurons and glia status), single-photon emission computed tomography (SPECT), or positron emission tomography (PET) are mostly used in research investigating brain metabolism, perfusion, and density of neuronal receptors [99–106]. Another noninvasive neuroimaging technique currently explored in WD is transcranial sonography (TCS). A few studies have used TCS to document the hyperechogenicity of lenticular nuclei in neurological WD cases [107–111]. In the future, TCS could be helpful as a quick and easy screening test for the differential diagnosis of WD and other neurodegenerative disorders.

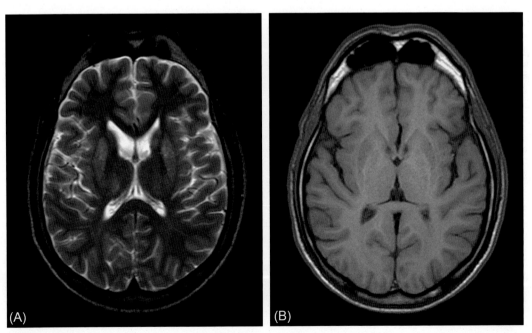

FIG. 5 Brain MRI of a WD patient showing (A) symmetrical hyperintense changes in T2-weighted sequences in the putamen, globus pallidus, and thalamus and (B) hypointensive in T1 sequences.

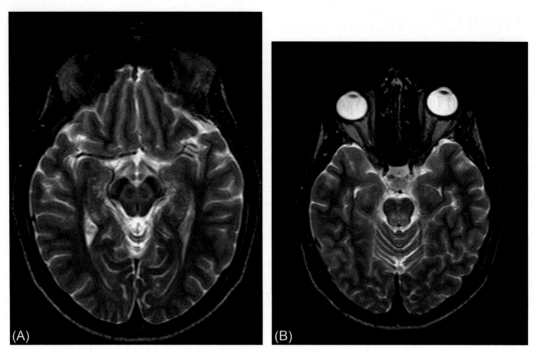

FIG. 6 Brain MRI of a WD patient: (A) "giant panda" sign and (B) "small panda" sign.

CONCLUSIONS

WD is a neurodegenerative disorder with a wide spectrum of neurological symptoms (mainly movement disorders). The disease is treatable with pharmacological agents and has a good prognosis if diagnosed and treated early and appropriately. Knowledge of these symptoms, their classifications, and scoring systems, as well as proper interpretation of supplementary examinations (brain MRI and ophthalmologic evaluation), enable an early and correct diagnosis and improve the WD treatment outcome. Familiarity with the neurological symptoms of WD may decrease the likelihood of diagnosis pitfalls, neurological deterioration on anticopper treatment, and WD "overtreatment" leading to copper deficiency.

REFERENCES

[1] Wilson SA. Progressive lenticular degeneration: a familial nervous disease associated with cirrhosis of the liver. Brain 1912;34:20–509.
[2] Bandmann O, Weiss KH, Kaler SG. Wilson's disease and other neurological copper disorders. Lancet 2015;14:103–13.
[3] Brewer GJ. Neurologically presenting Wilson's disease. CNS Drugs 2005;19:185–92.
[4] Dalvi A. Wilson's disease: neurological and psychiatric manifestations. Dis Mon 2014;60:460–4.
[5] Dusek P, Litwin T, Czlonkowska A. Wilson disease and other neurodegenerations with metal accumulations. Neurol Clin 2015;33:175–204.
[6] European Association for the Study of the Liver Disease. EASL Clinical Practice Guidelines: Wilson's disease. J Hepatol 2012;56:671–85.
[7] Pfeiffer R. Wilson's disease. Semin Neurol 2007;27:123–32.
[8] Roberts E, Schilsky M. Diagnosis and treatment of Wilson's disease an update. Hepatology 2008;47:2089–111.
[9] Lorincz MT. Neurologic Wilson's disease. Ann N Y Sci 2010;1184:173–87.
[10] Walshe JM, Yealland M. Wilson's disease: the problem of delayed diagnosis. J Neurol Neurosurg Psychiatry 1992;55:692–6.
[11] Walshe JM. Cause of death in Wilson disease. Mov Disord 2007;22:2216–20.
[12] Członkowska A, Tarnacka B, Litwin T, et al. Wilson's disease—cause of mortality in 164 patients during 1992-2003 observation period. J Neurol 2005;252:698–703.
[13] Ferenci P, Czlonkowska A, Merle U, et al. Late-onset Wilson's disease. Gastroenterology 2007;132:1294–8.
[14] Hall HC. La degenerescence hepato-lenticulare (Maladie de Wilson-pseudosclerose). Paris: Masson; 1921.
[15] Konovalov NV. Hepatocerebral dystrophy. Moscow: Russian Academy of Medical Sciences; 1960.
[16] Denny-Brown D. The basal ganglia. London: Oxford University Press; 1962.
[17] Marsden CD. Wilson's disease. QJM 1987;248:959–66.
[18] Oder W, Prayer L, Grimm G, et al. Wilsons' disease: evidence of subgroups derived from clinical findings and brain lesions. Neurology 1993;43:120–4.

[19] Członkowska A, Gajda J, Rodo M. Effects of long-term treatment in Wilson's disease with D-penicillamine and zinc sulphate. J Neurol 1996;243:269–73.
[20] Aggarwal A, Aggarwal N, Nagral A, et al. A novel global assessment scale for Wilson's diseases (GAS for WD). Mov Disord 2009;2494:509–18.
[21] Czlonkowska A, Tarnacka B, Moller JC, et al. Unified Wilson's diseases rating scale—proposal for the neurological scoring of Wilson's diseases patients. Neurol Neurochir Pol 2007;41:1–12.
[22] Aggarwal A, Bhatt M. The pragmatic treatment of Wilson's disease. Mov Disord Clin Pract 2014;1:14–23.
[23] Walshe JM, Potter G. The pattern of the whole body distribution of radioactive copper (67Cu, 64Cu) in Wilson's disease and various control groups. Q J Med 1977;46:445–62.
[24] Walshe JM, Gibbs KR. Brain copper in Wilson's disease. Lancet 1987;2:1030.
[25] Weisner B, Hartard C, Dieu C. CSF copper concentration: a new parameter for diagnosis and monitoring therapy of Wilson's disease with cerebral manifestation. J Neurol Sci 1987;79:229–37.
[26] Stuerenburg HJ. CSF copper concentrations, blood-brain barrier function, and coeruloplasmin synthesis during the treatment of Wilson's disease. J Neural Transm 2000;107:321–9.
[27] Mikol J, Vital C, Wassef M, et al. Extensive cortico-subcortical lesions in Wilson's disease: clinico-pathological study of two cases. Acta Neuropathol 2005;110:451–8.
[28] Litwin T, Gromadzka G, Szpak GM, et al. Brain metal accumulation in Wilson's disease. J Neurol Sci 2013;329:55–8.
[29] Scheiber IF, Dringen R. Copper-treatment increases the cellular GSH content and accelerates GSH export from cultured rat astrocytes. Neurosci Lett 2011;498:42–6.
[30] Pal A, Prasad R. Recent discoveries on the functions of astrocytes in the copper homeostasis of the brain: a brief update. Neurotox Res 2014;26:78–84.
[31] Bertrand E, Lewandowska E, Szpak GM, et al. Neuropathological analysis of pathological forms of astroglia in Wilson's disease. Folia Neuropathol 2001;39:73–9.
[32] Anzil AP, Herrlinger H, Blinzinger K, et al. Ultrastructure of brain and nerve biopsy tissue in Wilson disease. Arch Neurol 1974;31:94–100.
[33] Richter R. The pallial component in hepato-lenticular degeneration. J Neuropathol Exp Neurol 1948;7:1–18.
[34] Schulman S, Barbeau A. Wilson's disease: a case with almost total loss of white matter. J Neuropathol Exp Neurol 1963;22:105–19.
[35] Meenakshi-Sundaram S, Mahadevan A, Taly AB, et al. Wilson's disease: a clinico-neuropathological autopsy study. J Clin Neurosci 2008;15:409–17.
[36] Dusek P, Bahn E, Litwin T, et al. Brain iron accumulation in Wilson disease: a post-mortem 7 Tesla MRI—histopathological study. Neuropathol Appl Neurobiol 2016. https://doi.org/10.1111/nan.12341.
[37] Frucht S, Sun D, Schiff N, et al. Arm tremor secondary to Wilson's disease. Mov Disord 1998;13:351–3.
[38] Oder W, Prayer L, Grimm G, et al. Wilson's disease: evidence of subgroups derived from clinical findings and brain lesions. Neurology 1993;43:120–4.
[39] Sudmeyer M, Pollok B, Hefter H, et al. Synchronized brain network underlying postural tremor in Wilson's disease. Mov Disord 2006;21:1935–40.
[40] Ferenci P, Litwin T, Seniow J, et al. Encephalopathy in Wilson disease: copper toxicity or liver failure? J Clin Exp Hepatol 2015;(Suppl. 1):S88–95.
[41] Prashanth LK, Taly S, Sinha S, et al. Prognostic factors in patient presenting with severe neurological forms of Wilson's disease. Q J Med 2005;98:557–63.
[42] Litwin T, Dusek P, Czlonkowska A. Neurological manifestations in Wilson's disease—possible treatment options for symptoms. Expert Opin Orphan Drugs 2016;4:719–28.
[43] Starosta-Rubinstein S, Young AB, Kluin K, et al. Clinical assessment of 31 patients with Wilson's disease. Correlations with structural changes on magnetic resonance imaging. Arch Neurol 1987;44:365–70.
[44] Oder W, Brucke T, Kollegger H, et al. Dopamine D2 receptor binding is reduced in Wilson's disease: correlation of neurological deficits with striatal ^{123}I-iodobenzamide binding. J Neural Transm 1996;103:1093–103.
[45] Oertel WH, Tatsch K, Schwarz J, et al. Decrease of D2 receptors indicated by ^{123}I-iodobenzamide single-photon emission computed tomography relates to neurological deficit in treated Wilson's disease. Ann Neurol 1992;32:743–8.
[46] Svetel M, Kozic D, Stefanowa E, et al. Dystonia in Wilson's disease. Mov Disord 2001;16:719–23.
[47] Machado A, Chien HF, Deguti MM, et al. Neurological manifestations in Wilson's disease: report of 119 cases. Mov Disord 2006;21:2192–6.
[48] Burke JF, Dayalu P, Nan B, et al. Prognostic significance of neurologic examination findings in Wilson disease. Parkinsonism Relat Disord 2011;17:551–6.
[49] Litwin T, Chabik G, Czlonkowska A. Acute focal dystonia induced by tricyclic antidepressant in a patient with Wilson disease: a case report. Neurol Neurochir Pol 2013;47:502–6.
[50] Barthel H, Hermann W, Kluge R, et al. Concordant pre- and postsynaptic deficits of dopaminergic neurotransmission in neurologic Wilson disease. Am J Neuroradiol 2003;23:234–8.
[51] Litwin T, Gromadzka G, Samochowiec J, et al. Association of dopamine receptor gene polymorphisms with the clinical course of Wilson disease. JIMD Rep 2013;8:73–80.
[52] Litwin T, Dzieżyc K, Karliński M, et al. Early neurological worsening in patients with Wilson's diseases. J Neurol Sci 2015;355:162–7.
[53] Zimbrean PC, Schilsky ML. Psychiatric aspects of Wilson disease a review. Gen Hosp Psychiatry 2014;36:53–62.
[54] Oder W, Grimm G, Kollegger H, et al. Neurological and neuropsychiatric spectrum of Wilson's disease: prospective study of 45 cases. J Neurol 1991;238:281–7.
[55] Berry WR, Darley FL, Aronson AE. Dysarthria in Wilson's disease. J Speech Hear Res 1974;169–83.

[56] Trocello JM, Osmani K, Pernon M, et al. Hypersialorrhea in Wilson's disease. Dysphagia 2015;30:489–95.
[57] Da Silva-Junor FP, Carrasco AEAB, DaSilva Mendes AM, et al. Swallowing dysfunction in Wilson's disease: a scintigraphic study. Neurogstroenterol Motil 2008;20:285–90.
[58] Dziezyc K, Litwin T, Chabik G, Czlonkowska A. Frequencies of initial gait disturbances and falls in 100 Wilson's diseases patients. Gait Posture 2015;42:601–3.
[59] Ferrazzoli D, Fasano A, Maestri R, et al. Balance dysfunction in Parkinson's disease: the role of posturography in developing a rehabilitation program. Parkinsons Dis 2015;2015.
[60] Henkin R, Keiser HR, Jaeffe I, et al. Decreased taste sensitivity after D-penicillamine reversed by copper administration. Lancet 1967;290:1268–71.
[61] Mueller A, Reuner U, Landis B, et al. Extrapyramidal symptoms in Wilson's disease re associated with olfactory dysfunction. Mov Disord 2006;21:1311–6.
[62] Denning TR, Berrios GE, Walshe JM. Wilson's disease and epilepsy. Brain 1988;111:1139–55.
[63] Barbosa ER, Silveira-Moriyama L, Costa Machado A, et al. Wilson's disease with myoclonus and white matter lesions. Parkinsonism Relat Disord 2007;13:185–8.
[64] Pestana Knight EM, Gilman S, Selwa L. Status epilepticus in Wilson's disease. Epileptic Disord 2009;11:138–43.
[65] Aikath D, Gupta A, Chattppadhyay I, et al. Subcortical white matter abnormalities related to drug resistance in Wilson disease. Neurology 2006;67:878–80.
[66] Benbir G, Gunduz A, Ertan S, et al. Partial status epilepticus induced by hypocupremia in a patient with Wilson's disease. Seizure 2010;19:602–4.
[67] Brewer G, Fink J, Hedera P. Diagnosis and treatment of Wilson's disease. Semin Neurol 1999;3:261–9.
[68] Dziezyc K, Litwin T, Sobanska A, et al. Symptomatic copper deficiency in three Wilson's disease patient's treated with zinc sulphate. Neurol Neurochir Pol 2014;48:214–8.
[69] Hermann W, Vililmann T, Wagner A. Electrophysiological impairment profile of patients with Wilson's disease. Nervenarzt 2003;74:881–7.
[70] Kaveer N, Narayan SK. CNS demyelination due to hypocupremia in Wilson's disease from overzealous treatment. Neurol India 2006;54:110–1.
[71] Trindade MC, Bittencourt T, Lorenzi-Filho G, et al. Restless legs syndrome in Wilson's disease: frequency, characteristic, and mimics. Acta Neurol Scand 2016. https://doi.org/10.1111/ane.12585.
[72] Nevsimalova S, Buskova J, Bruha R, et al. Sleep disorders in Wilson's disease. Eur J Neurol 2011;18:184–90.
[73] Kayser B. Uber einen Fall von angeborner grunlicher Verfarbung der Kornea. Klin Monbl Augenheilkd 1902;40:22–5.
[74] Fleischer B. Zwei weiterer Falle von grunlicher Verfarbung der Kornea. Klin Monbl Augenheilkd 1903;41:489–91.
[75] Fleming CR, Dickson ER, Wahner HW, et al. Corneal pigmented rings in patients with primary biliary cirrhosis. Gastroenterology 1977;69:220–5.
[76] Tzelikis PF, Laibson PR, Ribeiro MP, et al. Ocular copper deposition associated with monoclonal gammopathy of undetermined significance; case report. Arq Bras Oftalmol 2005;68:539–41.
[77] Wiebers DO, Hollenhorst RW, Goldstein NP. The ophthalmologic manifestations of Wilson's disease. Mayo Clin Proc 1977;52:409–16.
[78] Loβner A, Loβner J, Bachmann H, et al. The Kayser-Fleischer ring during long-term treatment in Wilson's disease (hepatolenticular degeneration). Graefe's Arch Clin Exp Ophtalmol 1986;224:152–5.
[79] Siemerling E, Oloff H. Pseudoskleroses (Westphal-Strumpell) mit Cornealring (Kayser-Fleicher) und doppelseitiger Scheinkatarakt, die nur bei seitlicher Beleuchtnung sichtbar ist und die der nach Verletzung durch Kupfersplitter enstehenden Katarakt anlich ist. Klin Wochenschr 1922;1:1087–9.
[80] Cairns JE, Parry Wiliams H, Walshe JM. "Sunflower cataract" in Wilson's disease. Br Med J 1969;3:95–6.
[81] Langwinska-Wosko E, Litwin T, Dziezyc K, et al. The sunflower cataract in Wilson's disease: pathognomonic sign or rare finding? Acta Neurol Belg 2016;116:325–8.
[82] Albrecht P, Muller AK, Ringelstein M, et al. Retinal neurodegeneration in Wilson's disease revealed by spectral domain optical coherence tomography. PLoS One 2012;7.
[83] Grimm G, Madl C, Katzenschlager R, et al. Detailed evaluation of evoked potentials in Wilson's disease. Electroencephalogr Clin Neurophysiol 1992;82:119–24.
[84] Langwińska-Wośko E, Litwin T, Szulborski K, et al. Optical coherence tomography and electrophysiology of retinal and visual pathways in Wilson's disease. Metab Brain Dis 2016;31:405–15.
[85] Hall VW-v. Neuroimaging in Wilson disease. Metab Brain Dis 1997;12(1997).
[86] Alanen A, Komu M, Pentinen M, et al. Magnetic resonance imaging and proton MR spectroscopy in Wilson's disease. Br J Radiol 1999;72:749–56.
[87] Das M, Misra UK, Kalita J. A study of clinical, MRI and multimodality evoked potentials in neurologic Wilson disease. Eur J Neurol 2007;14:498–504.
[88] Dusek P, Roos P, Litwin T, et al. The neurotoxicity of iron, copper and manganese in Parkinson's and Wilson's diseases. J Trace Elem Med Biol 2015;31:193–203.
[89] Fritzsch D, Reiss-Zimmermann M, Trampel R, et al. Seven-tesla magnetic resonance imaging in Wilson disease using quantitative susceptibility mapping for measurement of copper accumulation. Invest Radiol 2014;49:299–306.
[90] Hermann W. Morphological and functional imaging in neurological and non-neurological Wilson's patients. Ann N Y Acad Sci 2014;1315:24–9.
[91] Hitoshi S, Iwata M, Yoshikawa K. Mid-brain pathology of Wilson's disease: MRI analysis of three cases. J Neurol Neurosurg Psych 1991;54:624–6.
[92] Ishida S, Doi Y, Yamane K, et al. Resolution of cranial MRI and SPECT abnormalities in a patient with Wilson's disease following oral zinc monotherapy. Intern Med 2012;51:1759–63.
[93] King AD, Walshe JM, Kendall BE, et al. Cranial MR imaging in Wilson's disease. Am J Roentgenol 1996;167:1579–84.

[94] Kozic D, Svetel M, Petrovic B, et al. MR imaging of the brain in patients with hepatic form of Wilson's disease. Eur J Neurol 2003;10:587–92.
[95] Magalhaes ACA, Caramelli P, Menezes JR, et al. Wilson's disease: MRI with clinical correlation. Neuroradiology 1994;36:97–100.
[96] Litwin T, Gromadzka G, Członkowska A, et al. The effect of gender on brain MRI pathology in Wilson's disease. Metab Brain Dis 2013;28:69–75.
[97] Prayer L, Wimberger J, Kramer J, et al. Cranial MRI in Wilson's disease. Neuroradiology 1990;32:211–4.
[98] Prashanth LK, Sinha S, Taly AB, et al. Do MRI distinguish Wilson's disease from other early onset extrapyramidal disorders? An analysis of 100 cases. Mov Disord 2010;25:672–8.
[99] Sinha S, Taly AB, Prashanth LK, et al. Sequential MRI changes in Wilson's disease with de-coppering therapy: a study of 50 patients. Br J Radiol 2007;80:744–9.
[100] Sinha S, Taly AB, Rvinshakar S, et al. Wilson's disease ^{31}P and ^{1}H MR spectroscopy and clinical correlation. Neuroradiology 2010;52:977–85.
[101] Hawkins RA, Mazziota JC, Phepls ME. Wilson's disease studied with FDG and positron emission tomography. Neurology 1987;37:1707–11.
[102] Schlaug G, Hefter H, Engelbrecht V, et al. Neurological impairment and recovery in Wilson's disease: evidence from PET and MRI. J Neurol Sci 1996;136:129–39.
[103] Sener RN. Diffusion MRI findings in Wilson's disease. Comput Med Imaging Graph 2003;27:17–21.
[104] Piga M, Murru A, Satta L, et al. Brain MRI and SPECT in the diagnosis of early neurological involvement in Wilson's disease. Eur J Nucl Med Mol Imaging 2008;35:716–24.
[105] Tarnacka B, Szeszkowski W, Gołebiowski M, et al. MR spectroscopy in monitoring the treatment of Wilson's disease patients. Mov Disord 2008;23:1560–6.
[106] Tarnacka B, Szeszkowski W, Gołebiowski M, et al. Metabolic changes in 37 newly diagnosed Wilson's disease patients assessed by magnetic resonance spectroscopy. Parkinsonism Relat Disord 2009;15:582–6.
[107] Walter U, Niehaus L, Probst T, et al. Brain parenchyma sonography discriminates Parkinson's disease and atypical parkinsonian syndromes. Neurology 2003;60:74–7.
[108] Walter U, Krolikowski K, Tarnacka B, et al. Sonographic detection of basal ganglia lesions in asymptomatic and symptomatic Wilson disease. Neurology 2005;64:1726–32.
[109] Walter U, Skowronska M, Litwin T, et al. Lenticular nucleus hyperechogenicity in Wilson's disease reflects local copper, but not iron accumulation. J Neural Transm 2014;121:1273–9.
[110] Skowronska M, Litwin T, Dziezyc K, et al. Does brain degeneration in Wilson diseases involve not only copper but also iron accumulation? Neurol Neurochir Pol 2013;47:542–6.
[111] Svetel M, Mijajlovic M, Tomic A, et al. Transcranial sonography in Wilson's disease. Parkinsonism Relat Disord 2012;18:234–8.

Chapter 14

Psychiatric Symptoms in WD

Paula C. Zimbrean
Department of Psychiatry and Surgery (Transplant), Yale University, New Haven, CT, United States

INTRODUCTION

Psychiatric symptoms have been included in description of Wilson disease (WD) since the condition was first reported. In his initial monograph, published 1912, Wilson mentioned that 8 out of the 12 cases reported had psychiatric symptoms. Traditionally, the psychiatric problems were reported in association with neurological findings, and this complex is typically referred to as the "neuropsychiatric" variant of WD. WD was typically cited in psychiatric textbooks as a classic example of subcortical dementia with irreversible cognitive decline as the expected outcome. In the past three to four decades, however, an abundance of clinical case reports and cohort studies described cognitive, psychiatric, and neurological findings as emerging and evolving separately in the longitudinal course of WD. Some authors have even postulated that the most common initial presentation of patients with WD is with neuropsychiatric symptoms rather than hepatic complaints [1]. Improved diagnostic tests allow easier confirmation of WD diagnosis when the presentation is atypical, such as nonspecific psychiatric findings.

This recent emphasis on psychiatric symptoms combined with the emergence of new and effective treatment for psychiatric disease in general has allowed successful treatment of psychiatric symptoms in WD, which in turn should lead to a better quality of life for patients. Prospective longitudinal studies have questioned the general view that cognitive impairment is irreversible in WD, especially with the availability of effective treatments for this disorder.

PREVALENCE OF PSYCHIATRIC SYMPTOMS IN WD

Early studies were not able to accurately capture the prevalence of psychiatric presentations in WD, since psychiatric symptoms were described cojointly with neurological findings. In addition, few of the earlier studies on WD included a formal mental health evaluation, and many of the initial descriptions of mental health concerns were based on patient's self-report, chart review, or nonpsychiatric clinicians. It was only over the past three decades that investigators started assessing these symptoms systematically, via detailed clinical examination or via structured assessment tools. This chapter summarizes information found in publications where psychiatric symptoms in WD patients were assessed either by trained mental health providers or by structured instruments. The structured assessment tools designed to measure psychiatric symptoms do come with limitations worth mentioning: results may be affected by motor or sensorial deficits, psychiatric scales have not been validated in this population, and most studies do not control for normal variations in mood and anxiety.

Also named the "great masquerader" [2], WD can present with a large variety of psychiatric and cognitive symptoms. In this section, we grouped the psychiatric presentations into the following categories: mood and anxiety disorders, psychotic disorder, sleep disturbances, cognitive problems, and others. Table 1 summarizes the prevalence of psychiatric symptoms in WD.

Mood and Anxiety Disorders

The prevalence of major depressive disorder, assessed via validated instruments, ranges between 4% [5] and 47.8% [18]. Once diagnosed, depression was found again at follow-up in 25% of cases [12]. Oftentimes, depression is found in the beginning of the WD illness and therefore cannot be attributed exclusively to the psychosocial consequences of a chronic medical problem. It is worth noting that even when assessed via the gold standard diagnosis tool for psychiatry, the Structured Clinical Interview for DSM (SCID), the rate of major depression varied widely, from 4% [5] to 35% [6].

TABLE 1 Prevalence of Psychiatric Presentations in WD Patients

Author, Year	Type of Study	N	Measure	Findings
Depression				
Rathbun 1996 [3]	CSec	34	MNB	26% had depression
Akil 1991 [4]	R	41	CE	27% had depression
Shanmugiah 2008 [5]	CSec	50	SCID, two psychiatrists	4% major depressive disorder 2% dysthymia
Svetel 2009 [6]	CSec	50	SCID NPI	36% had depression
Mercier-Jacquier 2011 [7]	R	19	CE	15.7% had depression
Carta 2012 [8]	CC	23	ANTAS MDQ	47.8% major depressive disorder
Mania/bipolar disorder				
Oder 1991 [9]	P	45	HDS ZDS ADS MMSE	26.6% mood symptoms 31.1% affective instability
Srinivas 2008 [10]	R	15 Out of cohort of 350	CE	2.5% had bipolar disorder
Shanmugiah 2008 [5]	CSec	50	SCID, two psychiatrists	18% had bipolar disorder
Carta 2012 [8]	CC	23	ANTAS MDQ	39% mania/hypomania
Anxiety				
Rathbun 1996 [3]	CSec	34	MNB	20% had anxiety
Svetel 2009	CSec	50	SCID NPI	62% had anxiety
Carta 2012 [8]	CC	23	ANTAS MDQ	8.7% had panic disorder
Psychosis				
Oder 1991 [9]	P	45	HDS ZDS ADS MMSE	6.7% had psychosis
Huang 1992 [11]	R	71	CE	11.3% had schizophrenia (schizophrenia was the initial diagnosis for 4.2%, prior to diagnosis of WD)
Rathbun 1996 [3]	CSec	34	MNB	2%
Srinivas 2008 [10]	R	15 Out of cohort of 350	CE	1.4% had schizophrenia 0.8% had schizoaffective disorder
Personality changes				
Huang 1992 [11]	R	71	CE	38.0%
Akil 1991 [4]	R	41	CE	45.9%

TABLE 1 Prevalence of Psychiatric Presentations in WD Patients—cont'd

Author, Year	Type of Study	N	Measure	Findings
Other				
Dening 1990 [12]	P	129	AMDP system	51% had psychopathologic features 20% had seen a psychiatrist prior to the diagnosis of WD 15% irritability 15% "incongruent behavior"
Akil 1991 [4]	R	41	CE	Catatonia, cognitive changes, anxiety, psychosis 64.8% had psych symptoms in the beginning of the illness
Oder 1991 [9]	P	45	HDS ZDS ADS MMSE	15.5% past suicide attempts 20% impaired social judgment 24.4% belligerence
Svetel 2009 [6]	CSec	50	SCID NPI	Irritability 26% 24% desinhibition; 24% apathy
Nevsimalova 2011 [13]	Csec	55	ESS RBD-SQ MSLT [14]	Lower sleep duration, decreased sleep efficiency, and increased wakefulness
Psychiatric symptoms in general				
Medalia 1989 [15]	CSec	24	MMPI	Patients with neurological impact scored higher on the schizophrenia and depression scale
Portala 2000 [16]	CSec	26	CPRS KSP	Autonomic disturbances, muscular tension, fatigability, decreased sexual interest, the lack of appropriate emotion, concentration difficulties, reduced sleep
Bono 2002 [17]	CR	21	CE	19% had initial symptoms psychiatric
Soltanzadeh 2007 [2]	R	44	CE	44% had psychiatric or sleep problems
Srinivas 2008 [10]	R	15 Out of cohort of 350	CE	4.2% had psychiatric findings 33.3% of psychiatric symptoms improved with decoppering treatment alone
Shanmugiah 2008 [5]	CSec	50	SCID, two psychiatrists	24% met diagnosis of a psychiatric disorder
Svetel 2009	CSec	50	SCID NPI	40% had >3 psychiatric diagnosis, 14% had two psychiatric diagnosis, 18% had one psychiatric diagnosis, 70% had one psychiatric symptom
Mercier-Jacquier 2011 [7]	R	19	CE	21% had psychiatric disorders

Abbreviations: *ADS*, adjective depression scale; *AMDP* system; *ANTAS*, advanced neuropsychaitric tools and assessment schedule; *APT*, automated psychological test; *BST*, Benton spatial test; *BVMT*, Benton visual memory test; *CC*, case control; *CE*, clinical Examination; *CPRS*, comprehensive psychopathologic rating scale; *CR*, case registry; *CSec*, cross sectional; *CSer*, case series; *ESS*, Epworth sleepiness scale; *GHQ*, General Health Questionnaire; *HDS*, Hamilton depression scale; *KSP*, Karolynskaya scale of personality; *MDQ*, mood disorders questionnaire; *MMPI*, Minnesota personality inventory; *MMSE*, Mini Mental Status Examination; *MNB*, Michigan Neuropsychological Battery; *MSLT*, multiple sleep latency test; *P*, prospective; *PAS*, Personality Assessment Schedule; *PSQI*, Pittsburgh Sleep Quality Index; *R*, retrospective; *RAVLT*, ray auditory-verbal learning test; *RBD-SQ*, rapid eye movement behavior disorder questionnaire; *RPM*, Raven progressive matrices; *SCID*, structured clinical diagnostic interview; *SDMT*, signal detection memory test; *SWT*, spot the word test; *WBISA*, Bellevue Intelligence Scale for Adults; *ZDS*, Zeissen depression scale.

Mood instability or conditions within the bipolar disorder spectrum are an interesting finding in some patients with WD. Within psychiatric conditions, disorders that have mood instability as the main finding vary in severity from bipolar I (severe manic episodes), to bipolar II (hypomanic, less severe "highs" and major depressive episodes), to cyclothymic disorder (mood variability that does not reach the severity of a hypomanic episode of major depression episode). Significant mood variability has been reported in up to 39% of patients with WD [8]. Up to 18% of patients with WD met criteria for bipolar disorder, when they were evaluated by two psychiatrists using SCID [5]. This number increased to 30% when a screening instrument (Mood Disorder Questionnaire (MDQ)) was used; 30% met criteria for bipolar disorder [18]. For comparison, in the general population, the lifelong prevalence of bipolar spectrum disorders, including subclinical presentations, is reported around 6.4% [19]. Even when mood instability does not meet all the formal criteria for hypomania or mania, the quality of life of WD patients can be negatively affected [20].

Significant anxiety can be found in 20%–62% of patients with WD [3, 6]; 8.7% of patients met criteria for panic disorder [18]. There is no information about the nature of the anxiety, if it is due to medications, adjusting to illness and disability or to an underlying anxiety disorder.

Psychotic Disorders

Psychotic symptoms in WD present significant diagnostic and treatment challenges. Firstly, if they present before hepatic or neurological symptoms, they may be misdiagnosed as primarily psychiatric illness, with subsequent delays in medical care and in the diagnosis of WD. A "classic" scenario is a patient that presents with psychotic symptoms and is prescribed antipsychotic medications and develops dystonic reactions, interpreted as a side effect of the antipsychotic medications. Often only after these neurological findings become very severe or atypical in presentation, a neurological consultation is requested and an evaluation for WD is initiated. There have been reports of patients who spent years in psychiatric units only to be diagnosed with WD and have their psychiatric symptoms resolved when chelation therapy was instituted. Some studies reported a 20% prevalence of psychotic disorders in patients with WD [21]. In a Chinese group, 11% of patients with WD met criteria for schizophrenia [11], while in another sample, 4.2% of patients seen in WD clinic reported symptoms of psychosis in the beginning of their illness [11]. Again for comparison, lifetime prevalence rates of narrowly and broadly defined psychosis in a general population were 0.2% and 0.7%, respectively [22].

It is important to note that psychotic symptoms may also occur during the pharmacological treatment of WD in up to 40% of cases [23, 24] or after liver transplantation [25]. Among all WD medical therapies, D-penicillamine in particular has been linked most frequently with new onset psychosis [23].

Sleep Disturbances

Sleep disturbances are frequently reported WD and can range from decreased sleep to hypersomnolence or parasomnia. The overall sleep quality in WD is reported as decreased, with frequent nocturnal awakening; parasomnias as sleep paralysis or cataplexy have also been described [26]. WD patients also report worse daytime somnolence compared with controls [27, 28]. The hypersomnolence has been objectively confirmed by multiple sleep latency test in some patients [28]. Severe hypersomnia can occur late in course of the illness; a case reported severe hypersomnia developed 3 years after initiation of depression and after 2 years of cognitive impairment [29]. One study of 26 patients with Wilson disease explored the frequency and the content of the dreams in this group of patients. Four out of the 26 patients had rapid eye movement sleep behavior disorder (RBD). Patients with WD and controls had a lower word count during dream recall than controls, which the authors attributed to decreased verbal memory previously described in WD patients. In general, patients with RBD tend to have more nightmares and more violent dreams than patients without RBD, and this was noted in this cohort of WD patients as well [30].

Other Psychiatric Symptoms

In addition to well-characterized psychiatric syndromes discussed above, patients with WD may present at various points in their lives with behavioral changes that do not meet criteria for a specific psychiatric disease. These behaviors are often interpreted as an expression of the underlying personality or character and can cause significant interpersonal difficulties and impairments in overall social functioning. These symptoms include irritability (prevalence 15%–25% [6, 12]), impaired social judgment or disinhibition [6,9], apathy [6], "belligerence" [9], or "incongruent behavior" [12]. A systematic assessment of personality changes in WD showed these changes were present in 57% of patients with Wilson disease [4, 12].

Lack of adherence with medical treatment is often a behavioral chief complaint that may prompt referral of patients with WD to mental health services. The lack of adherence with treatment is an independent factor for patients to develop symptomatic WD [14].

Other psychiatric presentations reported in WD include obsessive compulsive disorder (OCD) [31, 32] and catatonia [33–35].

Cognitive Deficits

The classical psychiatric textbooks listed WD as an example of subcortical dementia, and it was considered that left untreated, the disease would lead to irreversible cognitive decline, to the point of total dependency. Multiple studies found however that cognitive deficits in WD can range from mild to severe, but if properly treated, they do not lead to major cognitive impairment [36–38]. The significant pitfalls in assessing the cognitive impairment in WD patients have been extensively discussed elsewhere [39]. Cognitive impairment tends to be more pronounced in patients with active WD, either neurological or hepatic [38, 40, 41]. Impairment in attention, specifically in ability to shift attention, has been described in patients with WD with or without neurological findings [41]. In patients without hepatic or neurological involvement, subclinical cognitive deficits are often present, such as impairment in executive function or diminished learning abilities [42, 43]. Very specific aspects of the executive function have been studied in WD. For instance, an assessment of decision-making in patients with WD shows that patients were more likely to prefer the advantageous choices than healthy controls [44]. Executive inhibition, as assessed by the go-no go test, was significantly impaired in WD patients, even when the behavioral data did not reflect it [45]. These subtle cognitive deficits are important in the treatment of this patient since it may influence their options for care and may impact the adherence to medications. Social cognition, assessed by Facial Emotion Recognition Test, Movie for the Assessment of Social Cognition, and Ambiguous Intentions Hostility Questionnaire, was found to differ in patients with WD and neurological symptoms compared with patients with WD without neurological findings. Specifically, patients with neurological WD had difficulties with anger recognition and were more likely to react aggressively to ambiguous social situations [46].

THE DIAGNOSIS OF PSYCHIATRIC PRESENTATIONS IN WD

As discussed above, psychiatric symptoms in WD can vary widely and can present at any time during the disease. The question that often arises is as follows: are the symptoms due to WD or to a comorbid psychiatric disorder? DSM 5 has maintained the approach of previous versions, requiring that in order to confirm a psychiatric diagnosis, psychiatric symptoms cannot be due to a medical condition. When psychiatric symptoms are judged to be related to one medical condition (e.g., hypothyroidism), the diagnosis is psychiatric disorder (e.g., depression or mood disorder) secondary to medical condition. In this setting, the symptoms cannot be ascribed to a psychiatric diagnosis (e.g., major depressive disorder).

The literature supports the concept that WD affects the brain and can cause psychiatric symptoms. The literature also suggests psychiatric symptoms could occur independently from neurological or hepatic disease in WD. In general, symptoms are more likely to be due to a comorbid psychiatric condition when

- cognitive distortions suggesting a diagnosis are more significant than the neurovegetative symptoms,
- psychiatric symptoms persist in the absence of other organic disease,
- specific clinical features typical for psychiatric disorders are present (e.g., commenting and derogatory auditory hallucinations are more likely to suggest Schizophrenia than visual hallucinations with nonbizarre content).

Why is it important to distinguish the presence of psychiatric disease from psychiatric symptoms of WD? For short term, this distinction my not be important, since the treatment is still symptomatic as we discuss below. However, the long-term therapeutic approach may differ for psychiatric symptoms secondary to WD compared with primary psychiatric disorders. First of all, there are no FDA-approved treatments for psychiatric symptoms secondary to medical conditions, and that includes WD. Psychiatric presentations may be treatment symptomatically, when severe; however, the use of psychotropics in these cases is off-label (not FDA approved). Second, as we will discuss below, about half of the psychiatric symptoms secondary to WD are resolve with treatment of WD alone, without psychotropic agents. Third, it is reasonable to believe that if psychiatric symptoms are secondary to WD, psychiatric medications could be stopped when remission was achieved. This is particularly important with medications like antipsychotics, which can cause significant side effects if used for many years (e.g., tardive dyskinesia). On the other hand, the presence of a primary or chronic psychiatric disease may allow access to particular resources such as outpatient mental health centers, outreach programs, or community social support groups.

The structure clinical interview diagnosis (SCID) remains the goal standard to ascribe any psychiatric diagnosis. It is long and expensive, since it has to be purchased from the American Psychiatric Association and clinicians need specific training before being able to administer it. In clinical practice, psychiatrists, psychologists, or mental health clinicians use psychiatric initial templates for diagnosis, more or less structured. These typically include a history of present illness; review of psychiatric symptoms; a complete psychiatric history including psychiatric hospitalizations, medication trials (medication, doses, duration, response, and side effects) and experience with psychotherapy; family psychiatric history; a developmental history or psychosocial history; a detailed mental status examination; and a cognitive screen. Information is often obtained from prior records and from significant others, as well as prior mental health providers or medical providers. Since patients with WD are typically seen primarily in nonpsychiatric settings, such as hepatology or neurology clinics, a number of screening instruments may be useful to detect if a patient is at risk in having psychiatric complaints that would need further evaluation and possible treatment. These screening instruments include the following:

- Global Assessment Scale for Wilson disease [47]—a two-tier instrument screening for a wide range of symptoms in Wilson disease—has one item assessing behavioral problems and one question focusing on cognition. Its validity to detect psychiatric symptoms has not been studied extensively.
- Mood disorder questionnaire (MDQ), a screening instrument for bipolar disorders [48], has been used in WD patients [8].
- Hamilton depression scale—a six-item validated instrument for screening for depression [49]—is used in older studies looking at depression in WD [9]. This is a convenient screening tool that is easy to use in the office.
- Mini-mental State Examination [50] is a validated tool to assess cognitive impairment in various populations. Remains of one of the most popular cognitive screening instruments are used in clinical practice and have been used in the WD population [51].
- Neuropsychiatric inventory (NPI) is a clinical instrument for evaluating psychopathology in dementia [52] that has been implemented to assess cognitive status in WD patients [6].

Another question that often occurs in clinical practice pertains to what typical psychiatric presentations should lead to investigation for the presence of Wilson disease. Some others have suggested that all psychiatric patients with new onset of symptoms should be screened for WD by measuring serum ceruloplasmin [53]. Screening a wide psychiatric population by measuring ceruloplasmin has been attempted [54]. Fifty years ago, Cox and colleagues measured 336 persons hospitalized for psychiatric disorders. Out of them, two patients were found to have ceruloplasmin levels below normal limits but did not have WD. The interpretation of abnormal found findings in serum ceruloplasmin is still not clear [55]. Diagnostic algorithms have been developed to identify patients with atypical psychiatric symptoms that warrant investigation for organic disease at initial presentation [56, 57]. Bonnot et al. (2014) suggest that underlying medical causes should be investigated when psychosis is associated with confusion, catatonia, visual hallucinations, intellectual decline, and fluctuating symptoms [57]. Other authors have listed confusion, neurological findings, vision abnormalities, dysmorphia, cardiac abnormalities, and hepatosplenomegaly as features that should lead to investigations of an organic cause for new onset psychiatrist diseases [58].

An important take-home message of this chapter is that psychiatric symptoms can occur in any stage of WD. Psychiatric symptoms at the time of initial presentation, in the absence of hepatic or neurological manifestations, are particularly challenging, since they have been linked to delays in diagnosis. The average time between psychiatric symptoms and diagnosis of Wilson disease was 2.42 years (SD 2.97) [59]. For comparison, the median time between initial symptoms and diagnosis of Wilson disease was 1.5 years for neurological Wilson disease and 0.5 years for hepatic disease [9].

Some studies have shown that up to 64.8% of patients had psychiatric symptoms in the beginning of illness (with or without hepatic or neurological findings) [4]. When followed longitudinally, up to 70.7% of patients experience neuropsychological symptoms at some point in the course of the disease. [60]. Multiple psychiatric diagnoses are common: in a group of 50 WD patients assessed by SCID, 40% met criteria for more than three psychiatric diagnoses [6]. A fifth of patients had seen a psychiatrist prior to the formal diagnosis of WD [61].

Another challenging presentation is psychiatric symptoms that occur after the initiation of the cooperating treatment, most commonly with penicillamine. Manic episodes and psychosis have been described as occurring in this context [24, 62]. Many of these resolve with time; however, it may be challenging for the patient to accept that "it is going to get worse before is getting better" and sometimes symptomatic treatment may be necessary as we will discuss later in this chapter.

Psychiatric presentations of WD and children almost universally include not only deterioration of academic performance [63] but also irritability or depression [64], all nonspecific complaints. Especially when significant life stresses are present, behavioral changes can be interpreted as an adjustment reaction and delay the diagnosis [65]. Testing of asymptomatic siblings of patients affected by Wilson disease may lead in the future to helpful information about mild psychiatric complaints in this population.

Laboratory Data and Neuroimaging

Copper and Ceruloplasmin Levels

These are useful tests for assisting in the diagnosis of WD and are discussed in detail in Chapter 11. To date, there have been no systematic studies assessing the correlation between specific psychiatric symptoms and levels of copper or ceruloplasmin. Attempts at using serum ceruloplasmin levels as a screening test for WD in all patients with chronic psychiatric disease did now show a public health value [66, 67].

Electroencephalogram (EEG) and Evoked Potential Testing

Abnormal EEG findings have been described in patients with WD without a clear correlation between these findings and psychiatric symptoms. [68]. Other small studies indicated that pronounced psychiatric symptoms and cognitive deficits were more likely to be associated with an EEG generator more anteriorly localized [69], but the clinical relevance of this finding has not been elucidated.

The most relevant regarding the use of evoked potentials was a study that evaluated 19 patients with Wilson disease and mood disturbances. In this group, all subjects had abnormalities on the acoustic early evoked event-related potentials (ERPs), 58% had abnormalities on the somatosensory evoked potentials (SSEPs), and 53% had abnormalities in the visual and brain stem evoked potentials (VEP and BSEP) [68]. This finding is helpful clinically when patient with Wilson disease complained of significant sensitivity to loud sounds. This complaint may be interpreted as "functional" or "psychiatric," either as increased startle response suggesting a PTSD diagnosis or a deficit in coping with stress.

Neuroimaging

Neuroimaging, especially brain MRI, can help clarify the diagnosis inpatients with new onset of psychiatric symptoms who have not developed yet hepatic or neurological WD. The most common finding on the brain MRI in WD is T2 hyperintensity signal lesions in the basal ganglia and in the thalamus, brain stem, and cerebellum [70]. Seventy-six percent of WD patients show hypoperfusion in the superior frontal, prefrontal, parietal, occipital, caudate, and putamen on brain perfusion SPECT prior to beginning treatment [71]. Specific for WD is the sign of the "face of giant panda and her cub" [72–74]. Cognitive impairment has been associated with pathology within selective basal ganglia lesions [36] and with changes in the putamen [75]. Other studies have shown that abnormalities of the corpus callosum were present in about 23.4% of patients with WD with neurological presentation. The presence of these lesions in the post callosum correlated with the disability level as assessed by the Unified Wilson's Disease Rating Scale [76]. More recently, abnormalities of cortical gray matter in the form of widespread cerebral cortical paramagnetic signals detected by susceptibility-weighted imaging have been described in WD in the absence of any subcortical hyperintense lesions [77], which underlines that WD is not an exclusive subcortical disease process.

Neuroimaging may be helpful in monitoring the progress of the disease during treatment [78, 79]. A patient who presented with catatonia received initially a diagnosis of schizophrenia, but when severe neurological decline developed, serial computer tomography images showed necrosis of both putamen 15 months after the onset of disease [79].

Comment on Genetic Studies

These studies have attempted to find specific correlation between genetic profile and psychological/psychiatric findings. These studies are still limited due to the high number of genetic mutations found in WD, small samples of studies, and limited spectrum of psychiatric symptoms assessed. A study evaluated 25 patients with WD using extensive personality testing (CPRS and KPS) and analyzed the findings against the gene mutations found in *AP7B*. Patients homozygous for Trp779Stop mutation had the highest scores on the *psychopathy* versus *conformity* scale and *socialization* scale and low scores on the *impulsiveness*, *avoidant*, and *detachment scales*. Patients homozygous for the Thr977Met scored high on the *psychopathy* versus *conformity* scale as well. Patients with the compound heterogeneous mutation His1069Gln/Arg1319Stop had the lowest score on the *psychopathy* versus *conformity* scale. An important finding of this study warranting further research is that significant discrepancies were found between the patient self-ratings and interview-based ratings of personality traits in across all genetic groups [80].

Significance of Psychiatric Symptoms in WD

Many authors support the hypothesis that the presence of psychiatric symptoms in WD indicate a more severe or advanced process. The presence of psychiatric symptoms may therefore potentially signal irreversible brain damage or secondary to metabolic disturbances produced by the patient's liver disease, such as hyperammonemia associated with hepatic encephalopathy [3]. Psychiatric symptoms in WD tend to associate with lower quality of life [81, 82]. Symptoms of bipolar disorder seem to have lowered the quality of life in WD more than other symptoms [20]. Symptoms of bipolar disorder are also associated with a higher frequency of brain damage than for WD patients with depression or without psychiatric findings [83]. Like in most chronic medical illnesses, the coexistence of psychiatric symptoms may lead to nonadherence with medical care, further worsening the overall prognosis of WD [84]. Liver transplantation does not typically improve all neuropsychiatric symptoms [25, 85, 86]. For a long time, it was considered that patients with neuropsychiatric findings tend to do worse after liver transplantation compared with WD patients transplanted with only hepatic disease; however, that finding is now being disputed [87].

TREATMENT OF PSYCHIATRIC SYMPTOMS IN WD

Studies have suggested that up to 33% of patients with psychiatric symptoms will improve with decoppering treatment alone [10, 88–90]. Therefore, for WD patients with mild to moderate psychiatric symptoms with minimal impact on quality of life, a wait and see approach to their course on medical therapy aimed solely at WD may be reasonable. The challenge occurs when the psychiatric symptoms are interfering with the patient's safety or with their ability to function in a significant way. In these cases, psychiatric symptoms should be treated symptomatically with specific pharmacological interventions or psychotherapy. This approached is supported by the growing evidence about neurotransmitter abnormalities (such as abnormalities of the serotonin transporter) in patients with WD and depression [91, 92]. Although no psychotropic medications are FDA approved for WD, multiple case reports describing successful treatment with psychotropic agents support the use of psychopharmacology in WD.

Successful use of serotonin reuptake inhibitors [31, 93], tricyclic antidepressants [94, 95], methylphenidate [96], divalproex [97, 98], risperidone [97], haloperidol [99], and clozapine [100] has been described. Benzodiazepines are often prescribed for anxiety, insomnia, or agitation [97, 101, 102] as well as for muscle relaxation. Their long-term use must be weighed against the risk of worsening cognition in WD patients who are already at risk for cognitive impairment.

Lithium remains one of the preferred mood stabilizers in all patients with liver disease since it is not metabolized by the liver and is excreted exclusively through the kidneys. In addition, it does not directly impact the dopaminergic system, so it should not impact the neurological status of the patients. Lithium has been reported to be effective for treatment of mood instability or manic symptoms in WD in multiple studies [103–105]. The cognitive impact of lithium is still being debated [106]; therefore, we suggest repeated cognitive screening during the course of lithium therapy in patients with WD.

It is considered that patients with WD have increased sensitivity to neuroleptics due to direct connections between copper and dopamine metabolism [107, 108]. Patients with WD have been reported to develop neuroleptic malignant syndrome, a potentially fatal complication of antipsychotic exposure [94, 107, 109]. In this context, quetiapine has the advantage of reduced risk of extrapyramidal side effects [110]. In vitro, quetiapine has been reported to have protective effects against D-penicillamine-induced neurotoxicity (a possible mechanism of copper-related actions on the nervous system) [111, 112]; the clinical implications of these findings have not been explored.

Other biological interventions for psychiatric symptoms in WD are cyproterone in a patient with hypersexuality [113] and electroconvulsive therapy for depression with or without catatonic features [114, 115].

Psychotherapy can play a significant role in the management of WD with or without psychopharmacological treatment. Specific interventions have been reported to be successful in treatment of psychiatric presentations in WD cognitive behavioral therapy for OCD [31] or interpersonal therapy for depression [116]. In addition, psychotherapy can address issues related to coping with chronic medical illness, such as acceptance or denial of disease, treatment adherence, developing insight, increasing social support, and strengthening coping skills [117].

Liver transplantation does not typically improve the neuropsychiatric symptoms in all patients transplanted with WD who had these symptoms prior to their transplant [25, 85, 86]. A recent review of 107 patients with WD who receive liver transplantation showed that 18 had aggravation of neuropsychiatric symptoms after transplantation, and this included anxiety and depression [87]. Steroid medications, routinely used in the posttransplantation setting at significant doses, could contribute to psychiatric symptoms in these patients. Liver transplantation has been reported to lead to resolution of psychiatric symptoms in some cases [86, 118]. At this time, liver transplantation is not recommended for WD patients with psychiatric symptoms in the absence of liver disease as the driving indication for transplantation.

It is important to note that the resolution of psychiatric symptoms may take longer than the response expected in psychiatric diseases with similar symptoms. Multiple cases have reported that more than a year was necessary to obtain remission, but in many cases, the remission was complete [102, 119]. Practitioners should be encouraged to persist and aim for complete remission and maintain an attitude of caution when treating psychiatric symptoms in WD. The best treatment however remains prevention: current data suggest that if WD is treated early, neurological involvement in psychiatric disease does not develop [120].

CONCLUSION

Psychiatric symptoms are routinely present in the course in WD, and they can occur at any point in the disease history or during treatment. They can significantly alter quality of life and may often recur. There are however multiple reasons to approach these presentations confidently, since they no longer hold the negative prognostic significance they had in the past. This is due primarily to the availability of early testing for WD and availability of new and better tolerated treatments. A multidisciplinary approach is helpful in addressing these complaints promptly and safely.

REFERENCES

[1] Litwin T, Gromadzka G, Czlonkowska A. Apolipoprotein E gene (APOE) genotype in Wilson's disease: impact on clinical presentation. Parkinsonism Relat Disord 2012;18(4):367–9.

[2] Soltanzadeh A, Soltanzadeh P, Nafissi S, Ghorbani A, Sikaroodi H, Lotfi J. Wilson's disease: a great masquerader. Eur Neurol 2007;57(2):80–5.

[3] Rathbun JK. Neuropsychological aspects of Wilson's disease. Int J Neurosci 1996;85(3–4):221–9.

[4] Akil M, Schwartz JA, Dutchak D, Yuzbasiyan-Gurkan V, Brewer GJ. The psychiatric presentations of Wilson's disease. J Neuropsychiatry Clin Neurosci 1991;3(4):377–82.

[5] Shanmugiah A, Sinha S, Taly AB, Prashanth LK, Tomar M, Arunodaya GR, et al. Psychiatric manifestations in Wilson's disease: a cross-sectional analysis. J Neuropsychiatry Clin Neurosci 2008;20(1):81–5.

[6] Svetel M, Potrebic A, Pekmezovic T, Tomic A, Kresojevic N, Jesic R, et al. Neuropsychiatric aspects of treated Wilson's disease. Parkinsonism Relat Disord 2009;15(10):772–5.

[7] Mercier-Jacquier M, Bronowicki JP, Raabe JJ, Jacquier A, Kaminsky P. Wilson's disease in an adult. Rev Med Interne 2011;32(6):341–6.

[8] Carta MG, Sorbello O, Moro MF, Bhat KM, Demelia E, Serra A, et al. Bipolar disorders and Wilson's disease. BMC Psychiatry 2012;12:52.

[9] Oder W, Grimm G, Kollegger H, Ferenci P, Schneider B, Deecke L. Neurological and neuropsychiatric spectrum of Wilson's disease: a prospective study of 45 cases. J Neurol 1991;238(5):281–7.

[10] Srinivas K, Sinha S, Taly AB, Prashanth LK, Arunodaya GR, Janardhana Reddy YC, et al. Dominant psychiatric manifestations in Wilson's disease: a diagnostic and therapeutic challenge! J Neurol Sci 2008;266(1–2):104–8.

[11] Huang CC, Chu NS. Wilson's disease: clinical analysis of 71 cases and comparison with previous Chinese series. J Formos Med Assoc 1992;91(5):502–7.

[12] Dening TR, Berrios GE. Wilson's disease: a longitudinal study of psychiatric symptoms. Biol Psychiatry 1990;28(3):255–65.

[13] Nevsimalova S, Buskova J, Bruha R, Kemlink D, Sonka K, Vitek L, et al. Sleep disorders in Wilson's disease. Eur J Neurol 2011;18(1):184–90.

[14] Dziezyc K, Karlinski M, Litwin T, Czlonkowska A. Compliant treatment with anti-copper agents prevents clinically overt Wilson's disease in pre-symptomatic patients. Eur J Neurol 2014;21(2):332–7.

[15] Medalia A, Scheinberg IH. Psychopathology in patients with Wilson's disease. Am J Psychiatry 1989;146(5):662–4.

[16] Portala K, Westermark K, von Knorring L, Ekselius L. Psychopathology in treated Wilson's disease determined by means of CPRS expert and self-ratings. Acta Psychiatr Scand 2000;101(2):104–9.

[17] Bono W, Moutie O, Benomar A, Aidi S, el Alaoui-Faris M, Yahyaoui M, et al. Wilson's disease. clinical presentation, treatment and evolution in 21 cases. Rev Med Interne 2002;23(5):419–31.

[18] Carta MG, Farina GC, Sorbello O, Moro MF, Cadoni F, Serra A, et al. The risk of bipolar disorders and major depressive disorders in Wilson's disease: results of a case-control study. Int Clin Psychopharmacol 2012;28.

[19] Judd LL, Akiskal HS. The prevalence and disability of bipolar spectrum disorders in the US population: re-analysis of the ECA database taking into account subthreshold cases. J Affect Disord 2003;73(1–2):123–31.

[20] Carta MG, Mura G, Sorbello O, Farina G, Demelia L. Quality of life and psychiatric symptoms in Wilson's disease: the relevance of bipolar disorders. Clin Pract Epidemiol Ment Health 2012;8:102–9.

[21] Czlonkowska A, Rodo M. Late onset of Wilson's disease. Report of a family. Arch Neurol 1981;38(11):729–30.

[22] Kendler KS, Gallagher TJ, Abelson JM, Kessler RC. Lifetime prevalence, demographic risk factors, and diagnostic validity of nonaffective psychosis as assessed in a US community sample. The National Comorbidity Survey. Arch Gen Psychiatry 1996;53(11):1022–31.

[23] Aggarwal A, Bhatt M. Recovery from severe neurological Wilson's disease with copper chelation. Mov Disord 2012;27:S71.

[24] McDonald LV, Lake CR. Psychosis in an adolescent patient with Wilson's disease: effects of chelation therapy. Psychosom Med 1995;57(2):202–4.

[25] Al-Hilou H, Hebbar S, Gunson BK, Claridge LC, Syn WK, Holt AP. Long term outcomes of liver transplantation for Wilson's disease: a single centre experience. Gut 2012;61:A210.

[26] Portala K, Westermark K, Ekselius L, Broman JE. Sleep in patients with treated Wilson's disease. A questionnaire study. Nord J Psychiatry 2002;56(4):291–7.

[27] Netto AB, Sinha S, Taly AB, Panda S, Rao S. Sleep in Wilson's disease: questionnaire based study. Ann Indian Acad Neurol 2011;14(1):31–4.

[28] Amann VC, Maru NK, Jain V. Hypersomnolence in Wilson disease. J Clin Sleep Med 2015;11(11):1341–3.

[29] Kim SY, Chang Y, Jang IM, Eah KY. Wilson's disease presenting with the episodic hypersomnia, In: Neurodegenerative diseases conference: 10th international conference AD/PD Alzheimer's and Parkinson's diseases: advances, concepts and new challenges Barcelona Spain conference start; 2011. p. 8 [No pagination].

[30] Tribl GG, Trindade MC, Schredl M, Pires J, Reinhard I, Bittencourt T, et al. Dream recall frequencies and dream content in Wilson's disease with and without REM sleep behaviour disorder: a neurooneirologic study. Behav Neurol 2016;2016.

[31] Kumawat B, Sharma C, Tripathi G, Ralot T, Dixit S. Wilson's disease presenting as isolated obsessive-compulsive disorder. Indian J Med Sci 2007;61(11):607–10.

[32] Sahu JK, Singhi P, Malhotra S. Late occurrence of isolated obsessive-compulsive behavior in a boy with Wilson's disease on treatment. J Child Neurol 2013;28(2):277.

[33] Davis EJ, Borde M. Wilson's disease and catatonia. Br J Psychiatry 1993;162:256–9.

[34] Nayak RB, Shetageri VN, Bhogale GS, Patil NM, Chate SS, Chattopadhyay S. Catatonia: a rare presenting symptom of Wilson's disease. J Neuropsychiatry Clin Neurosci 2012;24(3):E34–5.

[35] Basu A, Thanapal S, Sood M, Khandelwal SK. Catatonia: an unusual manifestation of Wilson's disease. J Neuropsychiatry Clin Neurosci 2015;27(1):72–3.

[36] Seniów J, Bak T, Gajda J, Poniatowska R, Czlonkowska A. Cognitive functioning in neurologically symptomatic and asymptomatic forms of Wilson's disease. Mov Disord 2002;17(5):1077–83.
[37] Frota NAF, Caramelli P, Barbosa ER. Cognitive impairment in Wilson's disease. Demen Neuropsychol 2009;3(1):16–21.
[38] Wenisch E, De Tassigny A, Trocello JM, Beretti J, Girardot-Tinant N, Woimant F. Cognitive profile in Wilson's disease: a case series of 31 patients. Rev Neurol (Paris) 2013;169(12):944–9.
[39] Zimbrean P, Seniow J. Cognitive and psychiatric symptoms in Wilson disease. Handb Clin Neurol 2017;142:121–40.
[40] Han Y, Zhang F, Tian Y, Hu P, Li B, Wang K. Selective impairment of attentional networks of alerting in Wilson's disease. Plos One 2014;9(6).
[41] Iwański S, Seniów J, Leśniak M, Litwin T, Członkowska A. Diverse attention deficits in patients with neurologically symptomatic and asymptomatic Wilson's disease. Neuropsychology 2015;29(1):25–30.
[42] Tarter RE, Switala J, Carra J, Edwards N, Van Thiel DH. Neuropsychological impairment associated with hepatolenticular degeneration (Wilson's disease) in the absence of overt encephalopathy. Int J Neurosci 1987;37(1–2):67–71.
[43] Szutkowska-Hoser J, Seniów J, Członkowska A, Laudański K. Cognitive functioning and life activity in patients with hepatic form of Wilson's disease. Pol Psychol Bull 2005;36(4):234–8.
[44] Ma H, Lv X, Han Y, Zhang F, Ye R, Yu F, et al. Decision-making impairments in patients with Wilson's disease. J Clin Neuropsychol 2013;35(5):472–9.
[45] Stock AK, Reuner U, Gohil K, Beste C. Effects of copper toxicity on response inhibition processes: a study in Wilson's disease. Arch Toxicol 2016;90(7):1623–30.
[46] Peyroux E, Santaella N, Broussolle E, Rigard C, Favre E, Brunet AS, et al. Social cognition in Wilson's disease: a new phenotype? PLoS One 2017;12(4).
[47] Aggarwal A, Aggarwal N, Nagral A, Jankharia G, Bhatt M. A novel global assessment scale for Wilson's disease (GAS for WD). Mov Disord 2009;24(4):509–18.
[48] Miller CJ, Klugman J, Berv DA, Rosenquist KJ, Ghaemi SN. Sensitivity and specificity of the mood disorder questionnaire for detecting bipolar disorder. J Affect Disord 2004;81(2):167–71.
[49] O'Sullivan RL, Fava M, Agustin C, Baer L, Rosenbaum JF. Sensitivity of the six-item Hamilton depression rating scale. Acta Psychiatr Scand 1997;95(5):379–84.
[50] Folstein MF, Folstein SE, McHugh PR. "Mini-mental state". A practical method for grading the cognitive state of patients for the clinician. J Psychiatr Res 1975;12(3):189–98.
[51] Oder W, Prayer L, Grimm G, Spatt J, Ferenci P, Kollegger H, et al. Wilson's disease: evidence of subgroups derived from clinical findings and brain lesions. Neurology 1993;43(1):120–4.
[52] Kaufer DI, Cummings JL, Ketchel P, Smith V, MacMillan A, Shelley T, et al. Validation of the NPI-Q, a brief clinical form of the neuropsychiatric inventory. J Neuropsychiatry Clin Neurosci 2000;12(2):233–9.
[53] Bux CJ, Rosenbohm A, Connemann BJ. Wilson's disease in patients with mental disorders. Nervenheilkunde 2007;26(3):150–5.
[54] Cox DW. A screening test for Wilson's disease and its application to psychiatric patients. Can Med Assoc J 1967;96(2):83–6.
[55] Connemann BJ, Schonfeldt-Lecuona C, Maxon HJ, Kratzer W, Kassubek J. The role of ceruloplasmin in the differential diagnosis of neuropsychiatric disorders. Fortschr Neurol Psychiatr 2010;78(10):582–9.
[56] Bonnot O, Cohen D. Differential diagnosis in psychoses: the impact of NP-C and other organic disease. Eur Neuropsychopharmacol 2011;21:S627–8.
[57] Bonnot O, Klunemann HH, Sedel F, Tordjman S, Cohen D, Walterfang M. Diagnostic and treatment implications of psychosis secondary to treatable metabolic disorders in adults: a systematic review. Orphanet J Rare Dis 2014;9:65.
[58] Demily C, Sedel F. Psychiatric manifestations of treatable hereditary metabolic disorders in adults. Ann Gen Psychiatry 2014;13(1):1–9.
[59] Zimbrean PC, Schilsky ML. Psychiatric aspects of Wilson disease: a review. Gen Hosp Psychiatry 2014;36(1):53–62.
[60] Pan JJ, Chu CJ, Chang FY, Lee SD. The clinical experience of Chinese patients with Wilson's disease. Hepatogastroenterology 2005;52(61):166–9.
[61] Dening TR, Berrios GE. Wilson's disease. Psychiatric symptoms in 195 cases. Arch Gen Psychiatry 1989;46(12):1126–34.
[62] Kenar ANI, Menteseoglu H. Manic episode induced by discontinuance of D-penicillamine treatment in Wilson's disease. Klinik Psikofarmakoloji Bulteni 2014;24(4):381–3.
[63] Turnacioglu S, Gropman AL. Developmental and psychiatric presentations of inherited metabolic disorders. Pediatr Neurol 2013;48(3):179–87.
[64] Alam ST, Rahman MM, Islam KA, Ferdouse Z. Neurologic manifestations, diagnosis and management of Wilson's disease in children—an update. Mymensingh Med J 2014;23(1):195–203.
[65] Carr A, McDonnell DJ. Wilson's disease in an adolescent displaying an adjustment reaction to a series of life stressors: a case study. J Child Psychol Psychiatry Allied Discip 1986;27(5):697–700.
[66] Gedack M, Schonfeldt-Lecuona C, Kratzer W, Kassubek J, Connemann B. Ceruloplasmin serum concentrations in a randomized population sample of 2,445 subjects. In: European psychiatry conference: 19th European congress of psychiatry., 26:EPA; 2011 [No pagination].
[67] Demily C, Parant F, Cheillan D, Broussolle E, Pavec A, Guillaud O, et al. Screening of Wilson's disease in a psychiatric population: difficulties and pitfalls. A preliminary study. Ann Gen Psychiatry 2017;16:19.
[68] Arendt G, Hefter H, Stremmel W, Strohmeyer G. The diagnostic value of multi-modality evoked potentials in Wilson's disease. Electromyogr Clin Neurophysiol 1994;34(3):137–48.
[69] Dierks T, Kuhn W, Oberle S, Muller T, Maurer K. Generators of brain electrical activity in patients with Wilson's disease. Eur Arch Psychiatry Clin Neurosci 1999;249(1):15–20.
[70] da Costa Mdo D, Spitz M, Bacheschi LA, Leite CC, Lucato LT, Barbosa ER. Wilson's disease: two treatment modalities. Correlations to pretreatment and posttreatment brain MRI. Neuroradiology 2009;51(10):627–33.

[71] Piga M, Murru A, Satta L, Serra A, Sias A, Loi G, et al. Brain MRI and SPECT in the diagnosis of early neurological involvement in Wilson's disease. Eur J Nucl Med Mol Imaging 2008;35(4):716–24.

[72] Liebeskind DS, Wong S, Hamilton RH. Faces of the giant panda and her cub: MRI correlates of Wilson's disease. J Neurol Neurosurg Psychiatry 2003;74(5):682.

[73] Brito JC, Coutinho Mde A, Almeida HJ, Nobrega PV. Wilson's disease: clinical diagnosis and "faces of panda" signs in magnetic resonance imaging. Case report. Arq Neuropsiquiatr 2005;63(1):176–9.

[74] Hitoshi S, Iwata M, Yoshikawa K. Mid-brain pathology of Wilson's disease: MRI analysis of three cases. J Neurol Neurosurg Psychiatry 1991;54(7):624–6.

[75] Hegde S, Sinha S, Rao SL, Taly AB, Vasudev MK. Cognitive profile and structural findings in Wilson's disease: a neuropsychological and MRI-based study. Neurol India 2010;58(5):708–13.

[76] Trocello JM, Guichard JP, Leyendecker A, Pernon M, Chaine P, El Balkhi S, et al. Corpus callosum abnormalities in Wilson's disease. J Neurol Neurosurg Psychiatry 2011;82(10):1119–21.

[77] Lee JH, Yang TI, Cho M, Yoon KT, Baik SK, Han YH. Widespread cerebral cortical mineralization in Wilson's disease detected by susceptibility-weighted imaging. J Neurol Sci 2012;313(1–2):54–6.

[78] Chung EJ, Kim EG, Kim SJ, Ji KH, Seo JH. Wilson's disease with cognitive impairment and without extrapyramidal signs: improvement of neuropsychological performance and reduction of MRI abnormalities with trientine treatment. Neurocase 2016;22(1):40–4.

[79] Awada A, al Rajeh S, al Qorain A, al Ghassab G. Wilson's disease in a Saudi Arabian female patient. Rapid changes in cerebral X-ray computed tomography. Rev Neurol 1990;146(4):306–7.

[80] Portala K, Waldenstrom E, von Knorring L, Westermark K. Psychopathology and personality traits in patients with treated Wilson disease grouped according to gene mutations. Ups J Med Sci 2008;113(1):79–94.

[81] Petrovic I, Svetel M, Pekmezovic T, Tomic A, Kresojevic N, Potrebic A, et al. Quality of life in patients with Wilson's disease in Serbia. Mov Disord 2012;27:S76.

[82] Svetel M, Pekmezovic T, Tomic A, Kresojevic N, Potrebic A, Jesic R, et al. Quality of life in patients with treated and clinically stable Wilson's disease. Mov Disord 2011;26(8):1503–8.

[83] Carta MG, Saba L, Moro MF, Demelia E, Sorbello O, Pintus M, et al. Homogeneous magnetic resonance imaging of brain abnormalities in bipolar spectrum disorders comorbid with Wilson's disease. Gen Hosp Psychiatry 2015;37(2):134–8.

[84] Mrabet S, Ellouze F, Mrad MF. Wilson's disease and psychiatric disorders: Is there comorbidity or indicator of relapse? J Neurol Sci 2013;333.

[85] Boeka AG, Solomon AC, Lokken K, McGuire BM, Bynon JS. A biopsychosocial approach to liver transplant evaluation in two patients with Wilson's disease. Psychol Health Med 2011;16(3):268–75.

[86] Geissler I, Heinemann K, Rohm S, Hauss J, Lamesch P. Liver transplantation for hepatic and neurological Wilson's disease. Transplant Proc 2003;35(4):1445–6.

[87] Lankarani KB, Malek-Hosseini SA, Nikeghbalian S, Dehghani M, Pourhashemi M, Kazemi K, et al. Fourteen years of experience of liver transplantation for Wilson's disease; a report on 107 cases from Shiraz, Iran. PLoS One 2016;11(12).

[88] Modai I, Karp L, Liberman UA, Munitz H. Penicillamine therapy for schizophreniform psychosis in Wilson's disease. J Nerv Ment Dis 1985;173(11):698–701.

[89] Bachmann H, Lossner J, Kuhn HJ, Biesold D, Siegemund R, Kunath B, et al. Long-term care and management of Wilson's disease in the GDR. Eur Neurol 1989;29(6):301–5.

[90] Mitra S, Ray AK, Roy S. Control of mania with chelation-only in a case of Wilson's disease. J Neuropsychiatry Clin Neurosci 2014;26(1):E6.

[91] Eggers B, Hermann W, Barthel H, Sabri O, Wagner A, Hesse S. The degree of depression in Hamilton rating scale is correlated with the density of presynaptic serotonin transporters in 23 patients with Wilson's disease. J Neurol 2003;250(5):576–80.

[92] Hesse S, Barthel H, Hermann W, Murai T, Kluge R, Wagner A, et al. Regional serotonin transporter availability and depression are correlated in Wilson's disease. J Neural Transm 2003;110(8):923–33.

[93] Smit JP, De Graaf J, Tjabbes T, Kloek J. Depression as first symptom of Wilson's disease. A case study. Tijdschr Psychiatr 2004;46(4):255–8.

[94] Buckley P, Carmody E, Hutchinson M. Wilson's disease and the neuroleptic malignant syndrome. Ir J Psychotherapy Psychosom Med 1990;7(2):138–9.

[95] Benhamla T, Tirouche YD, Abaoub-Germain A, Theodore F. The onset of psychiatric disorders and Wilson's disease. Encephale 2007;33(6):924–32.

[96] Silva F, Nobre S, Campos AP, Vasconcelos M, Goncalves I. Behavioural and psychiatric disorders in paediatric Wilson's disease. BMJ Case Rep 2011;2011. https://doi.org/10.1136/bcr.05.2011.4249.

[97] Shah H, Vankar GK. Wilson's disease: a case report. Indian J Psychiatry 2003;45(4):253–4.

[98] Aravind VK, Krishnaram VD, Neethiarau V, Srinivasan KG. Wilson's disease—a rare psychiatric presentation. J Indian Med Assoc 2009;107(7):456–7.

[99] Chung YS, Ravi SD, Borge GF. Psychosis in Wilson's disease. Psychosomatics 1986;27(1):65–6.

[100] Krim E, Barroso B. Psychiatric disorders treated with clozapine in a patient with Wilson's disease, Presse Med 2001;30(15):738.

[101] Muller JL. Schizophrenia-like symptoms in the Westphal-Strumpell form of Wilson's disease. J Neuropsychiatry Clin Neurosci 1999;11(3):412.

[102] Zimbrean PC, Schilsky ML. The spectrum of psychiatric symptoms in Wilson's disease: treatment and prognostic considerations. Am J Psychiatry 2015;172(11):1068–72.

[103] Tatay A, Lloret M, Merino T, Harto MA. Mania as a first symptom of Wilson's disease: a case report. In: European psychiatry conference: 18th European congress of psychiatry Munich Germany conference start., 25; 2010 [No pagination].

[104] Loganathan S, Nayak R, Sinha S, Taly AB, Math S, Varghese M. Treating mania in Wilson's disease with lithium. J Neuropsychiatry Clin Neurosci 2008;20(4):487–9.
[105] Rybakowski J, Litwin T, Chlopocka- Wozniak M, Czlonkowska A. Lithium treatment of a bipolar patient with Wilson's disease: a case report. Pharmacopsychiatry 2013;46(3):120–1.
[106] Pfennig A, Alda M, Young T, MacQueen G, Rybakowski J, Suwalska A, et al. Prophylactic lithium treatment and cognitive performance in patients with a long history of bipolar illness: no simple answers in complex disease-treatment interplay. Int J Bipolar Disord 2014;2(1).
[107] Chroni E, Lekka NP, Tsibri E, Economou A, Paschalis C. Acute, progressive akinetic rigid syndrome induced by neuroleptics in a case of Wilson's disease. J Neuropsychiatry Clin Neurosci 2001;13(4):531–2.
[108] Tu J. The inadvisability of neuroleptic medication in Wilson's disease. Biol Psychiatry 1981;16(10):963–8.
[109] Kontaxakis V, Stefanis C, Markidis M, Tserpe V. Neuroleptic malignant syndrome in a patient with Wilson's disease. J Neurol Neurosurg Psychiatry 1988;51(7):1001–2.
[110] Kulaksizoglu IB, Polat A. Quetiapine for mania with Wilson's disease. Psychosomatics 2003;44(5):438–9.
[111] Wang HN, Liu GH, Zhang RG, Xue F, Wu D, Chen YC, et al. Quetiapine ameliorates schizophrenia-like behaviors and protects myelin integrity in cuprizone intoxicated mice: The involvement of notch signaling pathway. Int J Neuropsychopharmacol 2015.
[112] Zhang Y, Zhang H, Wang L, Jiang W, Xu H, Xiao L, et al. Quetiapine enhances oligodendrocyte regeneration and myelin repair after cuprizone-induced demyelination. Schizophr Res 2012;138(1):8–17.
[113] Volpe FM, Tavares A. Cyproterone for hypersexuality in a psychotic patient with Wilson's disease. Aust N Z J Psychiatry 2000;34(5):878–9.
[114] Avasthi A, Sahoo M, Modi M, Sahoo M, Biswas P. Psychiatric manifestations of Wilson's disease and treatment with electroconvulsive therapy. Indian J Psychiatry 2010;52:66–8.
[115] Negro PJ, Louza Neto MR. Results of ECT for a case of depression in Wilson's disease. J Neuropsychiatry Clin Neurosci 1995;7(3):384.
[116] Keller R, Torta R, Lagget M, Crasto S, Bergamasco B. Psychiatric symptoms as late onset of Wilson's disease: neuroradiological findings, clinical features and treatment. Ital J Neurol Sci 1999;20(1):49–54.
[117] Lauterbach EC. Wilson's disease. Psychiatr Ann 2002;32(2):114–20.
[118] Sorbello O, Riccio D, Sini M, Carta M, Demelia L. Resolved psychosis after liver transplantation in a patient with Wilson's disease. Clin Pract Epidemiol Ment Health 2011;7:182–4.
[119] Mehta S, Hindley P, Dhawan A, Samuel M, Hedderly T. Neuropsychiatry in paediatric Wilson's disease—need for surveillance and management. Eur J Paediatr Neurol 2009;13:S117.
[120] Dubbioso R, Ranucci G, Esposito M, Di Dato F, Topa A, Quarantelli M, et al. Subclinical neurological involvement does not develop if Wilson's disease is treated early. Parkinsonism Relat Disord 2016;24:15–9.

Part V

Treatment Decisions

Chapter 15

General Considerations and the Need for Liver Transplantation

Michelle Angela Camarata[*,†,‡], Karl Heinz Weiss[§] and Michael L. Schilsky[¶]

[*]Department of Surgery, Section of Transplant and Immunology, Yale School of Medicine, New Haven, CT, United States, [†]Department of Gastroenterology and Hepatology, Royal Surrey County Hospital, Guilford, United Kingdom, [‡]Department of Clinical and Experimental Medicine, University of Surrey, Guilford, United Kingdom, [§]Internal Medicine IV, University Hospital Heidelberg, Heidelberg, Germany, [¶]Division of Digestive Diseases and Transplant and Immunology, Department of Medicine and Surgery, Yale University School of Medicine, New Haven, CT, United States

INTRODUCTION

Wilson disease (WD) is a rare autosomal recessive disorder of copper metabolism. A combination of diet and medical management is often enough to improve symptoms and control the disease process. Treatment is lifelong. In about 5% of patients, the initial presentation of WD is with acute liver failure (ALF). In these patients, the only viable treatment option is liver transplantation (LT). LT restores the defective copper metabolic pathway for copper excretion from the liver and is thus curative [1]. For patients who have discontinued medical therapy and develop acute liver failure, in those with end-stage liver disease who have disease progression on medical therapy, LT is lifesaving. The use of LT in Wilson disease patients with purely neurological symptoms is controversial and associated with mixed outcomes. Due to the shortage of donor organs, bridging treatments are sometimes required to buy time when LT is not immediately available. Alternative transplant strategies including living-donor transplants and auxiliary transplants can be used to mitigate the strain on the donor pool and provide timely transplantation.

LIVER TRANSPLANT IN PATIENTS PRESENTING WITH ALF AS THEIR INITIAL PRESENTATION

The underlying etiology of liver disease in approximately 2%–5% of patients presenting with ALF is Wilson disease [2–4]. In 5% of patients with Wilson disease, ALF is their initial presentation [5]. The original description of acute liver failure encompasses features of acute liver injury of <26 weeks in duration presenting in patients without preexisting liver disease [6]. There are a subset of WD patients, approximately 5%, that present with the same syndrome without previously diagnosed WD that also have significant advanced fibrosis or cirrhosis that is often clinically silent. Similar to those with ALF of other etiologies, these patients are also at risk of cerebral edema and multiorgan failure [7].

Recognizing that ALF is secondary to WD can be difficult, particularly in patients without known preexisting liver disease, and a high index of suspicion is required. No single test is diagnostic, and a diagnosis is often established based on specific findings on physical examination, biochemical testing, and molecular analysis that support the diagnosis of WD. Moreover, in ALF, many parameters of copper metabolism, including serum and urinary copper and serum ceruloplasmin, are less reliable and specific [8, 9]. Biochemical findings often demonstrate a typical low alkaline phosphatase-to-bilirubin ratio and an elevated ratio of aspartate transaminase-to-alanine transaminase. Berman et al. were the first to describe that the ratio of alkaline phosphatase to bilirubin lower than 2 and aspartate aminotransferase (AST)-to-alanine aminotransferase (ALT) ratio >4 was highly sensitive and specific for fulminant WD [10]. Korman et al. [11] compared a cohort of patients with ALF to those with ALF due to WD and demonstrated a combined ratio of alkaline phosphatase to total serum bilirubin lower than 4, and AST to ALT >2.2 had a sensitivity and specificity of 100% in fulminant WD. These tests may be less useful in WD patients prior to full-blown ALF and in children [12–14]. WD patients are given top priority along with other patients with ALF on the transplant list in the United States and in Europe, even when they have acute-on-chronic liver failure. Therefore, establishing the diagnosis early is crucial for obtaining a timely transplant and for initiating bridging therapies to transplant (discussed below). The fatality rate of ALF due to WD without transplantation approaches 100%, and a timely transplant is critical to survival [3, 15].

THERAPIES USED TO BRIDGE LIVER TRANSPLANT

ALF in WD results in hepatic necrosis and hepatocellular apoptosis, resulting in the release of toxic copper from the liver cells during the acute injury. The rapid release of the copper into the circulation results in multiorgan complications, including acute kidney injury to the renal tubules from copper and copper-metallothionein complexes. Therefore, it was not surprising that in a single-center series reporting on LT in WD, these patients had higher creatinine levels and a higher need for renal replacement therapy perioperatively compared with patients with ALF secondary to other etiologies [16]. Other associated complications from ALF due to WD include circulatory failure that is compounded by the profound hemolytic anemia, infection, and hyperammonemia with intracranial hypertension with cerebral edema. These complications all contribute to the high mortality seen in ALF due to WD.

Reducing excess free circulating copper has been a focus of several supportive strategies used to bridge patients for LT and improve posttransplant outcomes. These strategies outlined in Table 1 include therapeutic plasma exchange transfusion [18], albumin dialysis [19], plasmapheresis [20], hemofiltration [17], fractioned plasma separation and absorption [27], liver dialysis with single-pass albumin dialysis (SPAD) [24], early institution of renal replacement therapy, and molecular adsorbents recirculating system (MARS) [26]. Each technique uses a different method to reduce the copper load and improve biochemical parameters with a resulting reduction in hemolysis and secondary organ damage.

TABLE 1 Bridging Treatments Prior to LT

Technique	Benefits	Disadvantages/Limitations	References
Postdilutional hemofiltration	Reduction in copper to normal in first 3 days Improved GCS	No risks reported Potential risk of volume overload	[17]
Plasma exchange	Normal copper after second exchange Improved hemolysis and coagulation	Risk of citrate toxicity reported	[18]
Albumin dialysis	Fall serum copper to normal D7 Easily available/inexpensive Improved hemolysis and encephalopathy	Thrombocytopenia and bleeding	[19]
Plasmapheresis +hemofiltration	Normalization of serum copper after first procedure Improved liver dysfunction, encephalopathy, hemolysis	None reported. Potential risk of citrate toxicity	[20]
Plasmapheresis +oral penicillamine	Rapid removal of copper, improved hemolysis and liver dysfunction including bilirubin	Acute respiratory symptoms, bilateral pulmonary infiltrates and thrombocytopenia, TRALI	[21]
TPE	Reduction in copper, improved coagulopathy	Citrate toxicity	[22, 23]
SPAD	Reduction in copper Improved multiorgan failure and encephalopathy	High cost and availability (device and cartridges) Large quantities of albumin used Increase in bile salts, no change in ammonia	[24]
MARS	Effective at removal of toxins Reduction in copper, ammonia, and bilirubin Improved hepatorenal and encephalopathy	High cost and availability	[25, 26]
FPSA	Improved biochemistry including bilirubin and ammonia Improved GCS	No risks reported	[27]

FPSA, fractionated plasma separation and absorption; *MARS*, molecular adsorbents recirculating system; *SPAD*, single-pass albumin dialysis; *TPE*, therapeutic plasma exchange.

Intensive plasma exchange used fresh frozen plasma replacement and removal of copper with the net copper removal being proportional to the serum level [18]. Jhang et al. [20] used hemofiltration alternating with plasmapheresis to remove copper using five daily therapeutic interventions. A collective multicenter apheresis registry was established to explore this further. In this report, prospective and retrospective data were collected from the year 2000. Ten WD patients who underwent 43 therapeutic plasma exchanges prior to LT were included; all had good outcomes after a 6-month period post-LT [22].

MARS has been shown to be another effective therapy for lowering the copper load and bridging patients with WD to LT [21, 25, 26]. In MARS, the WD patient undergrows hemofiltration with a recirculating circuit of albumin that binds copper and is removed through a specialized membrane into the albumin-rich dialysate. This dialysate is then recirculated through an activated charcoal chamber and an anion resin exchange chamber in which copper, bilirubin, and other toxins such as ammonia are adsorbed. The dialysate is regenerated, and binding sites on the albumin become available, so more toxins can be removed in another cycle (Fig. 1). MARS has demonstrated improved renal and liver function and improved encephalopathy and short-term survival [28]. While this technique is effective, cost and availability may limit its use.

These bridging techniques appear to be well tolerated by WD patients with ALF with few reported side effects and offer support to the patient awaiting LT, which is particularly useful when an immediate transplant opportunity is not available. Centers may have different devices available or experience with only one of these modalities, and given the rarity of these patients will often use what is available at their site. Recovery without liver transplantation is rare, and the use of these devices or interventions should be viewed with an eye toward optimizing patients for LT.

LIVER TRANSPLANT IN PATIENTS WITH ESTABLISHED WD

Acute liver failure can also be present in WD patients due to medication discontinuation in previously stable patients. Similar to those whose first presentation is ALF, these patients present with a deep jaundice, hemolytic anemia, severe coagulopathy, and thrombocytopenia. Further cohorts of WD patients that may require transplant are those with treatment nonresponse. Presentation in these patients is with severe hepatic insufficiency and decompensated cirrhosis with evidence of ascites, variceal bleeding, and/or hepatorenal syndrome. The modified Nazar Score is a prognostic score [29, 30] that can be used to predict whether a trial of medical therapy may reverse the hepatic deterioration and avoid the need for transplantation. The score has proved sensitive and specific in predicting mortality of WD with medical therapy without LT. It thus helps select appropriate candidates for transplant. Some reports that have questioned the negative predictive value of the score [31] report successful medical treatment of two out of three patients with scores above 10 who were predicted not to survive without transplant. It has been suggested that LT may only be appropriate after suitable medical therapy has been trialed [32]. A 3-month trial of treatment has been suggested in AASLD guidelines with the caveat that if clinical or biochemical deterioration occurs during this period, LT should be expedited [5]. Other factors such as continued variceal bleeding, resistant ascites, or intractable hepatic encephalopathy may all be independent drivers for LT independent of the overall modified Nazar Score.

FIG. 1 MARS therapy for bridging LT.

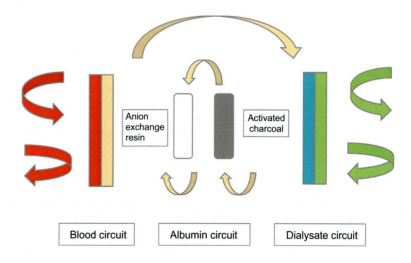

LIVER TRANSPLANTATION FOR NEUROPSYCHIATRIC MANIFESTATIONS OF WD

The defective excretion of copper into bile in WD can result in accumulation in the brain and resultant neuropsychiatric symptoms that can be debilitating. Neuropsychiatric disorders are more characteristic of late-onset WD when symptoms are present in about 74% of cases compared with 24% with a pure hepatic presentation [33, 34]. In early onset of WD before the age of 10, more hepatic disease (83%) compared with neuropsychiatric disease (17%) is present [33, 34]. This may possibly be explained by the types of mutation, missense mutations being associated with predominantly later and neurological presentations, while gene deletions are associated with earlier onset and hepatic presentations [33, 34]. In addition, about 11%–30% of patients with neurological manifestations of WD at the time of diagnosis develop paradoxical worsening of clinical symptoms, sometimes irreversible, with the initiation of D-penicillamine or trientine [35, 36]. The management of patients with significant neuropsychiatric burden, including the use of LT in patients with neuropsychiatric symptoms without the presence of liver failure, remains a topic of debate. The theory underlying the use of transplantation for neuropsychiatric symptoms is that transplantation restores normal copper metabolism and may thus help prevent further cerebral damage due to excess copper in WD. Preemptive LT would therefore potentially help reduce brain copper load and prevent further injury. On the other hand, LT for end-stage liver disease is associated with a high incidence of neuropsychiatric complications risking further deterioration in patients with preexisting disease [37].

Most of the data on LT for neurological WD are from case reports and case series (summarized in Table 2). Some of these report on LT for hepatic deterioration in patients who also had preexisting neuropsychiatric symptoms, while in others, neuropsychiatric symptoms were the main indication for LT. There are several reports that support improvement in neurological symptoms following LT [38–40, 42, 43, 45–48, 63]. There is, however, a lack of consistency in the literature as to whether LT is of clear benefit in patients with neurological WD. Increased morbidity and mortality following LT in WD patients with neurological features pretransplant have also been reported including worsening neurological symptoms and permanent disability [16, 32, 41, 44, 60–62, 69].

In a case series from France of 121 patients who underwent LT for WD, 7 patients were transplanted for neurological indications. Of those patients, three with severe axial parkinsonism died from infection without any neurological improvement. The other patients displayed major neurological improvements in symptoms that included dystonia, chorea, myoclonus, and ataxia [60]. In another report on LT in WD from Turkey, there were 9 of 42 patients with neurological symptoms that improved post-LT [61]. A more recent study from Iran demonstrated mixed results in a cohort of 60 patients with neuropsychiatric symptoms pretransplant. Improvement was seen in 40 patients, deterioration in 18, and 2 remained stable [62]. As with several studies, this study is limited due to the lack of objective scoring pre- and posttransplant to quantify the degree of neurological response.

The difficulty with drawing conclusions from current reports on neurological symptoms post-OLT is that several do not include objective measurements or scores pre- and post-OLT, and publication bias is difficult to eliminate. In addition, few studies include significant numbers of patients, and there is a lack of consistency between studies with some demonstrating poorer outcomes. Studies also have a lack of consistency in the time frame posttransplant that patients are reassessed for symptoms. Improvement in neurological symptoms may not be seen before 6 months after LT. In those with irreversible injury, they may persist beyond this time frame. There has also been a suggestion that overall survival post-LT may be significantly lower in those with a combination of neuropsychiatric and hepatic symptoms than in hepatic symptoms alone [32, 70]. For these reasons, transplantation of pure neuropsychiatric symptoms in WD remains a continued area of controversy, and appropriately, controlled studies comparing LT to medical therapy are warranted.

Medical therapy can also lead to neurological recovery, and it is still unclear what constitutes medical treatment nonresponse in patients with neuropsychiatric symptoms. The risks of transplant with unclear posttransplant outcomes and the shortage of donor organ supply mean that medical therapy is a more favorable option in this group. While living-donor liver transplantation (LDLT) would not impact on the donor organ pool, the operation is not without risks to the donor and recipient, and therefore, the same caution in candidate selection should apply. The use of LT in WD patients solely based on neurological symptoms still requires more data to provide confidence in its treatment effect.

OPTIONS FOR LIVER TRANSPLANTATION IN WILSON DISEASE

One of the first liver transplants performed by Thomas Starzl in the United States in the early 1960s was for a patient with WD with ALF [71]. Transplanting this patient with WD with a normal liver restored copper metabolism and cured the patient. Since then, >500 transplants have been performed in the United States for WD. WD was the primary indication for LT in 0.5% of adults and 1.5% of children [72]. Liver transplant for WD is lifesaving and curative, restoring biliary

TABLE 2 Neurological Improvement Posttransplant

Reference	Improved Outcome	Deterioration	Stable	Patients (n)	Evaluation Pre- and Post-LT
[38]	X			2	Physician assessed
[39]	X			1	Physician assessed
[7]	X (n=5)		X (n=1)	7	Physician assessed
[40]	X (n=6)			9	Physician assessed, performance status
[41]		X		1	Physician assessed, neuropsychological testing, and MRI
[42]	X			4	Physician assessment
[43]	X			7	Physician assessment and CT
[44]	X			1	Physician assessment
[45]	X			1	Neurological Hefter Score (1)
[46]	X			1	Physician assessment CT
[47]	X			1	Physician assessment, neuropsychological testing, MRI
[48]	X			1	Physician assessment, MRI
[49]	X			1	Physician assessment, MRI
[50]	X			3	Physician assessment
[51]	X			1	Physician assessment
[52]	X (n=3)			4	Physician assessment
[53]	X	X		2	Physician assessment
[54]	X			2	Physician assessment and MRI
[55]	X (n=7)	X (n=1)	X (n=3)	11	Medici score (2) for WD, neuropsychiatric score (3)
[56]	X			4	Physician assessment
[57]		X delayed		1	Physician assessment
[58]	X			4	Medici score (2) for WD, MRI
[59]	X			1	Physician assessment
[60]	X (n=4)	X (n=3)		7	Record review
[61]	X (n=5)	X (n=1)	(n=2)	8	Psychiatric score (3)
[61]	X (n=7)	X (n=1)		8	Neurological score (3)
[62]	X (n=40)	X (n=18)	X (n=2)	60	Physician assessment
[63]	X			4	UWDRS neurological score (4), modified Rankin (5), MRI

Physician assessed: Assessed through examination findings and symptoms without a scoring method to provide objective measurement pre- and post-OLT. References for scores: (1) Hefter et al. (1993) [64]; (2) Medici et al. (2005) [32]; (3) scores based on items from EuroWilson Registry and UWDRS [55, 65–67]; (4) Leinweber et al. (2008) [66] and Czlonkowska et al. (2007) [65]; (5) Uyttenboogaart et al. (2005) [68].

copper excretion and removal of toxic copper from extrahepatic sites. However, the scarcity of donor organs remains a global problem. With advances in liver transplant for pediatric patients requiring smaller-sized organs, split-liver or reduced-sized liver grafts have been used. This has led to advancements in surgical techniques using living-donor grafts.

Auxiliary Liver Grafts

Auxiliary liver grafts are partial grafts placed in a patient in addition to some or all of the patient's original liver. The auxiliary liver with functional ATP7B in hepatocytes is capable of clearing copper from the circulation and restoring normal excretion of copper in bile and protects copper-sensitive organs from damage. Withdrawal of immune suppression following recovery of the native liver results in the auxiliary graft undergoing rejection and elimination [73]. Auxiliary LT in patients with WD is a surgically demanding procedure fraught with potential posttransplant surgical complications including biliary strictures. There is also a potential increased risk for the development of HCC associated with leaving a cirrhotic organ in place. These problems limit its use.

Living Donor Liver Transplant

Due to the shortage of deceased-donor organs, living-donor transplantation has been explored as an alternative to orthotopic LT. It is particularly useful in some countries where cadaveric transplantation is not culturally or religiously accepted. It has proved beneficial in the pediatric patient population requiring reduced graft size. Surgical risks to the donor need to be considered as well as the risks of disease transmission. Assessments of the donor and recipient are required pretransplant. As most living donors are related, excluding WD in donors is essential. Careful clinical, biochemical, and molecular testing is required with occasional liver biopsy analysis of the donor liver. Heterozygosity for the WD gene mutation is associated with abnormal copper parameters in 28%–35% of subjects [74]. As most living donors are relatives, there has been interest in the safety of heterozygous donors. Data on LDLT for WD have demonstrated that the use of a donor who is a heterozygote carrier for WD is safe and provides effective function in the recipient with WD [75–77]. Although LDLT provides an interesting alternative to cadaveric transplantation, the procedure is not without risk, and careful consideration needs to be given to the need and timing of LT including indication and time on waiting list.

Outcomes Following LT

Excellent outcomes and long-term survival are seen in LT patients with WD. In terms of copper metabolism, an average period of 6 months results in normalization of parameters including serum ceruloplasmin, serum copper, and urinary copper excretion. Since the transplanted liver excretes the accumulated body copper over time, elevated 24 h urinary copper excretion can still be present during the early posttransplant period and up to 4 weeks after, reducing with time. Psychiatric symptoms did not appear significantly change, and as previously discussed, there is variability in the outcome of patients with neurological symptoms. In a study that included data collected by the US transplant registry for 170 pediatric patients, graft and patient 1-year survival were 90.1% and 89%, and for 400 adults with WD, these were 88% and 96%, respectively [72]. In a study in Europe, overall patient survival rates post-LT at 1 and 5 years after LT were 78% and 65% [55]. There was no significant difference in outcome according to age at transplant, gender, and blood group and whether LT was performed due to ALF or chronic liver failure. A major cause of postoperative mortality was sepsis. A study of 107 patients in Iran demonstrated similarly favorable outcomes at 1 year (86%) and 5 years (82%) with no difference in whether LT was performed due to CLF or ALF [62]. The favorable outcome in WD patients post-LT may be attributed to the relatively young age of patients at transplant, the low rate of comorbidities, and the lack of disease recurrence. Similarly, LDLT demonstrates positive posttransplant outcome results. In a large series, patient survival at 1 and 5 years after LDLT was 91.7 and 75%, respectively, and corresponding graft survival was 86.1% at 1 year and 75% at 5 years [78]. A second series provided data at 1-, 5-, and 10-year survival rates of 90.6%, 83.7%, and 80%, respectively [67]. A further series in Japan in LDLT in 59 children also demonstrated favorable patient survival outcomes 98.4% and 96.6% at 1 and 5 years [79].

Data on auxiliary LT are too limited to provide enough information on outcomes. Due to the difficult technical aspects of the procedure and the risks of portal hypertension and HCC in the remnant liver, the alternative transplant methods are favored [73, 80] (Table 3).

TABLE 3 Comparing Outcomes of LT

Type of Transplant	Region	Variable	Patients (n)	1-Year Patient Survival (%)	5-Year Patient Survival (%)
OLT/split graft/LDLT Treatment Decisions					
	The United States [72]	Adult (overall)	119	88.3	86
		Adult (FHF)	62	90.3	89.7
		Adult (CLF)	57	94.7	90.1
		Child (overall)	51	90.1	89
		Child (FHF)	41	90	87.5
		Child (CLF)	10	100	100
	Europe [55]	Adult and Child (Mean age 18.8±12 years)	19	78	65
		FLF	8	No difference to CLF	No difference to CLF
		CLF	11	No difference to ALF	No difference to ALF
	Iran [62]	Adult and Child Age range (5–59)	107	86	82
		FLF Median 16 (9–21.5)	21	No difference to CLF	No difference to CLF
		CLF Median 22 (15–26)	86	No difference to ALF	No difference to ALF
Living donor					
	China [78]	Overall	36	91.7	75
		Adult ≥14	26	92.3	80.8
		Child <14	10	90	60
		FLF	2	50	50
		CLF	34	91.2	76.5
	Japan [67]	Adult and Child Mean age 16±3.9 (range 6–40)	32	90.6	83.7
		FLF	21	100	94.4
		CLF	11	72.7	63.6
	Japan [79]	Children (11.4±2.8)	59	98.4	96.6

CLF, chronic liver failure; *FLF*, fulminant liver failure.

Immunosuppression

The degree of immunosuppression post-LT is a balance between the adverse effects of immunosuppression and the benefit of preventing rejection. WD patients with severe encephalopathy or neurological injury prior to LT may suffer complications from treatment more frequently than those without those signs or symptoms prior to LT. Tacrolimus has become the calcineurin inhibitor most frequently used in post-LT patients due to favorable outcomes in preventing acute cellular rejection; however, this medication unfortunately has a lower threshold for precipitating neurological complications, including seizures, and thus, its use in WD patients with preexisting neurological issues is not without risk. In WD patients, use of cyclosporine A is frequently considered a viable alternative in the early post-LT period. Prevention of post-LT neurotoxicity is also achieved by maintaining calcineurin inhibitor levels at the lower end of the therapeutic range and by replacing electrolytes such as magnesium. In those beyond 8 weeks after transplant, mTOR inhibitors such as sirolimus or everolimus may be considered. In patients with renal insufficiency or episodes of acute cellular rejection, antimetabolites such as azathioprine and mycophenolate mofetil as well as corticosteroids can be introduced. Treatment of rejection may involve the use of antibodies directed at cytokines involved in the rejection pathway and T-cell depleting antibodies [81].

There are some pediatric and adult patients that can wean and withdraw immunosuppression post-LT including some with WD. Protocols for achieving this and to determine which patients are likely to have favorable outcomes with weaning and withdrawal of immunosuppression are still being assessed [82].

CONCLUSION

A combination of copper chelation and diet stabilizes most patients with Wilson disease. In those with treatment failure or who present acutely with ALF, liver transplant continues to have a lifesaving role. Diagnosis remains a challenge, particularly in ALF where copper parameters may be less reliable. However, use of standard biochemical markers to determine ratios of alkaline phosphatase to bilirubin and AST to ALT and blood counts to look for hemolytic anemia may help with rapid diagnosis. Mortality is high without transplant, and organs may not be immediately available due to a global cadaveric organ shortage. In these circumstances, devices to bridge transplantation are useful in stabilizing patients, improving liver biochemistry, renal dysfunction, and hepatic encephalopathy. While hepatic decompensation is a valid indication for transplantation with successful outcomes, liver transplantation to manage the neuropsychiatric burden of the disease remains controversial and requires further studies. The neurological sequelae of WD may also provide challenges in the posttransplant period, and careful selection of immunosuppression is required to prevent further neurological complications. Techniques in living-donor transplantation are advancing, and its use is particularly prevalent in the pediatric cohort of WD patients requiring a reduced organ size. Adult-to-adult LDLT has been successfully performed for some with ALF due to WD, in particular in the Far East where there are few cadaveric organ donors. LDLT provides an alternative to cadaveric organ donation with the potential to mitigate the strain on the deceased-donor pool. Transplantation in WD is curative with excellent survival outcomes for both cadaveric and living-donor liver transplants.

REFERENCES

[1] Schilsky ML. Liver transplantation for Wilson's disease. Ann N Y Acad Sci 2014;1315:45–9.
[2] Lee WM, Squires RH, et al. Acute liver failure: summary of a workshop. Hepatology 2008;47(4):1401–15.
[3] Ostapowicz G, Fontana RJ, et al. Results of a prospective study of acute liver failure at 17 tertiary care centers in the United States. Ann Intern Med 2002;137:947–54.
[4] Reuben A, Tillman H, et al. Outcomes in adults with acute liver failure between 1998 and 2013: an observational cohort study. Ann Intern Med 2016;164(11):724–32.
[5] Roberts EA, Schilsky ML, American Association for Study of Liver Diseases (AASLD). Diagnosis and treatment of Wilson disease: an update. Hepatology 2008;47:2089–111.
[6] Trey C, Davidson CS. The management of fulminant hepatic failure. Prog Liver Dis 1970;3:282–98.
[7] Schilsky ML, Scheinberg IH, Sternlieb I. Liver transplantation for Wilson's disease: indications and outcome. Hepatology 1994;19:583–7.
[8] McCullough AJ, Fleming CR, et al. Diagnosis of Wilson's disease presenting as fulminant hepatic failure. Gastroenterology 1983;84:161–7.
[9] Schilsky ML, Sternlieb I. Overcoming obstacles to the diagnosis of Wilson's disease. Gastroenterology 1997;113:350–3.
[10] Berman DH, Leventhal RI, Gavaler JS, Cadoff EM, Van Thiel DH. Clinical differentiation of fulminant Wilsonian hepatitis from other causes of hepatic failure. Gastroenterology 1991;100:1129–34.
[11] Korman JD, Volenberg I, et al. Screening for Wilson disease in acute liver failure: a comparison of currently available diagnostic tests. Hepatology 2008;48:1167–74.
[12] Eisenbach C, Sieg O, Stremmel W, Encke J, Merle U. Diagnostic criteria for acute liver failure due to Wilson disease. World J Gastroenterol 2007;13:1711–4.
[13] Sallie R, Katsiyiannakis L, Baldwin D, Davies S, O'Grady J, Mowat A, Mieli-Vergani G, Williams R. Failure of simple biochemical indexes to reliably differentiate fulminant Wilson's disease from other causes of fulminant liver failure. Hepatology 1992;16:1206–11.
[14] Tissieres P, Chevret L, Debray D, Devictor D. Fulminant Wilson's disease in children: appraisal of a critical diagnosis. Paediatr Crit Care Med 2003;4:338–43.
[15] Sokol RJ, Francis PD, Gold SH, et al. Orthotopic liver transplantation for acute fulminant Wilson disease. J Pediatr 1985;107(4):549–52.
[16] Emre S, Atillasoy EO, et al. Orthotopic liver transplantation for Wilson's disease: a single center experience. Transplantation 2001;72:1232–6.

[17] Rakela J, Kurtz SB, et al. Fulminant Wilson's disease treated with postdilution hemofiltration and orthotopic liver transplantation. Gastroenterology 1986;90:2004–7.
[18] Kiss JE, Berman D, Van Thiel D. Effective removal of copper by plasma exchange in fulminant Wilson's disease. Transfusion 1998;38:327–31.
[19] Kreymann B, Seige M, Schweigart U, Kopp KF, Classen M. Albumin dialysis: effective removal of copper in a patient with fulminant Wilson disease and successful bridging to liver transplantation: a new possibility for the elimination of protein bound toxins. J Hepatol 1999;31:1085–90.
[20] Jhang JS, Schilsky ML, et al. Therapeutic plasmapheresis as a bridge to liver transplantation in fulminant Wilson disease. J Clin Apher 2007;22:10–4.
[21] Asfaha S, Almansori M, Qarni U, Gutfreund KS. Plasmapheresis for hemolytic crisis and impending acute liver failure in Wilson disease. J Clin Apher 2007;22:295–8.
[22] Pham HP, Schwartz J, et al. Report of ASFA apheresis registry study on Wilson's disease. J Clin Apher 2016;31:11–5.
[23] Reynolds HV, Talekar CR, Bellapart JB, Leggett BA, Boots RJ. Therapeutic plasma exchange as de-coppering technique in intensive care for an adult in a Wilson's crisis. Anaesth Intensive Care 2013;41(6):811–2.
[24] Collins KL, Roberts EA, Adeli K, Bohn D, Harvey EA. Single pass albumin dialysis (SPAD) in fulminant Wilsonian liver failure: a case report. Pediatr Nephrol 2008;23:1013–6.
[25] Chiu A, Tsoi NS, Fan ST. Use of the molecular adsorbents recirculating system as a treatment for decompensated Wilson disease. Liver Transpl 2008;14:1512–6.
[26] Sen S, Felldin M, Steiner C, Larsson B, Gillett GT, Olausson M, Williams R, Jalan R. Albumin dialysis and molecular adsorbents recirculating system (MARS) for acute Wilson's disease. Liver Transpl 2002;8:962–7.
[27] Grodzicki M, Kotulski M, et al. Results of treatment of acute liver failure patients with the use of the Prometheus FPSAA system. Transplant Proc 2009;41:3079–81.
[28] Mitzner SR, Stange J, et al. Improvement of hepatorenal syndrome with extracorporeal albumin dialysis MARS: results of a prospective, randomized, controlled clinical trial. Liver Transpl 2000;6:277–86.
[29] Dhawan A, Taylor RM. Wilson's disease in children: 37 year experience and revised King's score for liver transplantation. Liver Transpl 2005;37:1475–92.
[30] Nazer H, Ede RJ, et al. Wilson's disease: clinical presentation and use of prognostic index. Gut 1986;27:1377–81.
[31] Fischer RT, Soltys KA, et al. Prognostic scoring indices in Wilson disease: a case series and cautionary tale. J Pediatr Gastroenterol Nutr 2011;52:466–9.
[32] Medici V, Mirante VG, et al. Monotematica AISF 2000 OLT study group. Liver transplantation for Wilson's disease: the burden of neurological and psychiatric disorders. Liver Transpl 2005;11:1056–63.
[33] Walshe JM. Cause of death in Wilson disease. Mov Disord 2007;22:2216–20.
[34] Scheinberg IH, Sternlieb I. Wilson disease and idiopathic copper toxicosis. Am J Clin Nutr 1996;63:842S–5S.
[35] Czlonkowska A, Litwin T, Karliński M, Dziezyc K, Chabik G, Czerska M. D-Penicillamine versus zinc sulfate as first-line therapy for Wilson's disease. Eur J Neurol 2014;21:599–606.
[36] Litwin T, Dzieżyc K, et al. Early neurological worsening in patients with Wilsons's disease. J Neurol Sci 2015;355:162–7.
[37] Stracciari A, Guarino M. Neuropsychiatric complications of liver transplantation. Metab Brain 2001;16:3–11.
[38] Polson RJ, Rolles K, Calne RY, Williams R, Marsden D. Reversal of severe neurological manifestations of Wilson's disease following orthotopic liver transplantation. Q J Med 1987;64(244):685–91.
[39] Mason AL, Marsh W, et al. Intractable neurological Wilson's disease treated with orthotopic liver transplantation. Dig Dis Sci 1993;38:1746–50.
[40] Bellary S, Hassanein T, Van Thiel DH. Liver transplantation for Wilson's disease. J Hepatol 1995;23:373–81.
[41] Guarino M, Stracciari A, D'Alessandro R, Pazzaglia P. No neurological improvement after liver transplantation for Wilson's disease. Acta Neurol Scand 1995;92:405–508.
[42] Schumacher G, Platz KP, Mueller AR, Neuhaus R, Steinmüller T, Bechstein WO, Becker M, Luck W, Schuelke M, Neuhaus P. Liver transplantation: treatment of choice for hepatic and neurological manifestation of Wilson's disease. Clin Transplant 1997;11(3):217–24.
[43] Chen CL, Chen YS, Lui CC, Hsu SP. Neurological improvement of Wilson's disease after liver transplantation. Transplant Proc 1997;29(1–2):497–8.
[44] Kassam N, Witt N, Kneteman N, Bain VG. Liver transplantation for neuropsychiatric Wilson disease. Can J Gastroenterol 1998;12:65–8.
[45] Bax RT, Hässler A, et al. Cerebral manifestation of Wilson's disease successfully treated with liver transplantation. Neurology 1998;51:863–5.
[46] Lui CC, Chen CL, et al. Recovery of neurological deficits in a case of Wilson's disease after liver transplantation. Transplant Proc 1998;30:3324–5.
[47] Stracciari A, Tempestini A, Borghi A, Guarino M. Effect of liver transplantation on neurological manifestations in Wilson disease. Arch Neuro 2000;157:384–6.
[48] Wu JC, Huang CC, Jeng LB, Chu NS. Correlation of neurological manifestations and MR images in a patient with Wilson's disease after liver transplantation. Acta Neurol Scand 2000;102:135–9.
[49] Hermann W, Eggers B, Wagner A. The indication for liver transplant to improve neurological symptoms in a patient with Wilson's disease. J Neurol 2002;249:1733–4.
[50] Geissler L, Heinemann K, et al. Liver transplantation for hepatic and neurological Wilson's disease. Transplant Proc 2003;35(4):1445–6.
[51] Suzuki S, Sato Y, Ichida T, Hatakeyama K. Recovery of severe neurologic manifestations of Wilson's disease after living related liver transplantation: a case report. Transplant Proc 2003;35:385–6.
[52] Marin C, Robles R, et al. Liver transplantation in Wilson's disease: are its indications established? Transplant Proc 2007;39:2300–1.
[53] Senzolo M, Loreno M, Fagiuoli S, Zanus G, Canova D, Masier A, Russo FP, Sturniolo GC, Burra P. Different neurological outcome of liver transplantation for Wilson's disease in two homozygotic twins. Clin Neurol Neurosurg 2007;109:71–5.
[54] Duarte-Rojo S, Zepeda-Gómez S, et al. Liver transplantation for neurologic Wilson's disease: reflections on two cases within a Mexican cohort. Rev Gastroenterol Mex 2009;74:218–23.

[55] Weiss KH, Schäfer M, et al. Outcome and development of symptoms after orthotopic liver transplantation for Wilson disease. Clin Transplant 2013;27:914–22.
[56] Peedikayil MC, Al Ashgar HI, Al Mousa A, Al Sebayel M, Al Kahtani K, Alkhail FA. Liver transplantation in Wilson's disease: single center experience from Saudi Arabia. World J Hepatol 2013;5(3):127–32.
[57] Kim JS, Kim SY, Choi JY, Kim HT, Oh YS. Delayed appearance of wing-beating tremor after liver transplantation in a patient with Wilson disease. J Clin Neurosci 2014;21(8):1460–2.
[58] Mocchegiani F, Gemini S, et al. Liver transplantation in neurological Wilson's disease: is there indication? A case report. Transplant Proc 2014;46(7):2360–4.
[59] Sutariya VK, Tank AH, Modi PR. Orthotropic liver transplantation for intractable neurological manifestations of Wilson's disease. Saudi J Kidney Dis Transpl 2015;26(3):556–9.
[60] Guillaud O, Dumortier J, et al. Long term results of liver transplantation for Wilson's disease: experience of France. J Hepatol 2014;60:579–89.
[61] Yagci MA, Tardu A, et al. Influence of liver transplantation on neuropsychiatric manifestations of Wilson disease. Transplant Proc 2015;47:1469–73.
[62] Lankarani KB, Malek-Hosseini SA, et al. Fourteen years of experience of liver transplantation for Wilson's disease; a report on 107 cases from Shiraz, Iran. PLoS One 2016;11(12).
[63] Laurencin C, Brunet AS, et al. Liver transplantation in Wilson's disease with neurological impairment: evaluation in 4 patients. Eur Neurol 2017;77:5–15.
[64] Hefter H, Arendt G, et al. Motor impairment in Wilson's disease, I: slowness of voluntary limb movements. Acta Neurol Stand 1993;87:133–47.
[65] Czlonkowska A, Tarnacka B, et al. Unified Wilson's disease rating scale—a proposal for the neurological scoring of Wilson's disease patients. Neurol Neurochir Pol 2007;41(1):1–12.
[66] Leinweber B, Möller JC, et al. Evaluation of the unified Wilson's disease rating scale (UWDRS) in German patients with treated Wilson's disease. Mov Disord 2008;23:54.
[67] Yoshitoshi EY, Takada Y, et al. Long term outcomes for 32 cases of Wilson's disease after living donor liver transplantation. Transplantation 2009;87:261–7.
[68] Uyttenboogaart M, Stewart RE, Vroomen PC, De Keyser J, Luijckx GJ. Optimizing cutoff scores for the Barthel index and the modified Rankin scale for defining outcome in acute stroke trials. Stroke 2005;36:184–7.
[69] Brewer GJ, Askari F. Transplant livers in Wilson's disease for hepatic, not neurologic, indications. Liver Transpl 2000;6:662–4.
[70] Wang XH, Zhang F, et al. Eighteen living related liver transplants for Wilson's disease. A single center. Transplant Proc 2004;36:2243–5.
[71] Groth CG, Dubois RS, et al. Metabolic effects of hepatic replacement in Wilson's disease. Transplant Proc 1973;5:829–33.
[72] Arnon R, Annunziato R, et al. Liver transplantation for children with Wilson disease: comparison of outcomes between children and adults. Clin Transplant 2011;25:E52–60.
[73] Kasahara M, Takada Y, et al. Auxiliary partial orthotopic living donor liver transplantation: Kyoto University experience. Am J Transplant 2005;5:558–65.
[74] Okada T, Shiono Y, et al. Mutational analysis of ATP7B and genotype-phenotype correlation in Japanese with Wilson's disease. Hum Mutat 2000;15:454–62.
[75] Asonuma K, Inomata Y, et al. Living related liver transplantation from heterozygote genetic carriers to children with Wilson's disease. Pediatr Transplant 1999;3:201–5.
[76] Kobayashi S, Ochiai T, et al. Copper metabolism after living donor liver transplantation for hepatic failure of Wilson's disease from a gene mutated donor. Hepatogastroenterology 2001;48:1259–61.
[77] Wang XH, Cheng F, et al. Copper metabolism after living related liver transplantation for Wilson's disease. World J Gastroenterol 2003;9:2835.
[78] Cheng F, Li GQ, et al. Outcomes of living-related liver transplantation for Wilson's disease: a single-center experience in China. Transplantation 2009;87:751–7.
[79] Kasahara M, Sakamoto S, et al. Living donor liver transplantation for pediatric patients with metabolic disorders: the Japanese multicenter registry. Pediatr Transpl 2014;18:6–15.
[80] Park YK, Kim BW, Wang HJ, Kim MW. Auxiliary partial orthotopic living donor liver transplantation in a patient with Wilson's disease: a case report. Transplant Proc 2008;40:3808–9.
[81] Moini M, Schilsky ML, Tichy EM. Review on immunosuppression in liver transplantation. World J Hepatol 2015;7:1355–68.
[82] Feng S, Ekong UD, et al. Complete immunosuppression withdrawal and subsequent allograft function among pediatric recipients of parental living donor liver transplants. JAMA 2012;307:282–93.

Chapter 16

Chelation Therapy: D-Penicillamine

Peter Ferenci
Internal Medicine 3, Gastroenterology and Hepatology, Medical University of Vienna, Vienna, Austria

D-Penicillamine was the first oral drug introduced for therapy in Wilson disease by John Walshe. The first WD patient to use penicillamine regularly started treatment in 1955. In 1956, Walshe published this preliminary experience [1] and later the outcome of a small series of patients with neurological Wilson disease [2].

D-Penicillamine is a trifunctional organic compound, consisting of a thiol, an amine, and a carboxylic acid. It is very similar chemically to the α-amino acid cysteine but with geminal methyl groups α to the thiol (SH) group.

PHARMACOLOGY

D-Penicillamine is rapidly absorbed from the gastrointestinal tract with a double-peaked curve for intestinal absorption [3, 4]. If D-penicillamine is taken with a meal, its absorption is decreased overall by about 50%. Once absorbed, 80% of D-penicillamine circulates bound to plasma proteins. Greater than 80% of D-penicillamine is excreted via the kidneys. The excretion half-life of D-penicillamine is on the order of 1.7–7 h, but there is considerable interindividual variation.

Drug-Drug Interactions

Its enantiomer L-penicillamine inhibits the action of pyridoxine (vitamin B_6) [5].

Antacids and iron salts decrease penicillamine absorption and should be taken in separate administration times. Antimalarials, cytotoxic drugs, gold therapy, oxyphenbutazone, and phenylbutazone may cause serious hematologic and renal effects and should be avoided or if needed closely monitor administration together.

Mode of Action

The free sulfhydryl group of D-penicillamine acts as the chelator for many metals including copper. In vitro, it is more potent (C_{50} (to demetalate Cu_1Cox17) 14.70 μM; $K_d(M)$, 2.38×10^{-16}) that other chelators like EDTA, British anti-lewisite (BAL, C_{50} (μM), 9.38; $K_d(M)$, 1.52×10^{-18}), tetrathiomolybdate (C_{50} (μM), 2.1; $K_d(M)$, 2.32×10^{-20}), or triethylenetetramine (C_{50} (μM) 1080 $K_d(M)$, 1.74×10^{-16}) to chelate copper [6, 7].

D-Penicillamine also induces metallothionein, a cysteine-rich protein that acts as an endogenous chelator of metals, including copper [8]. D-Penicillamine when given to either healthy or diseased animals (LEC rats) prevented or reversed hepatitis, respectively. In diseased animals, the drug sequestered copper particularly from insoluble lysosomal particles [9]. In vitro, the mobilization of copper is likely to proceed through the solubilization of these particles. In contrast, D-penicillamine had only a minor effect on copper bound to metallothionein in the cytosol [9]. In Wilson disease, the role of D-penicillamine is to prevent the formation or to promote the solubilization of copper-rich particles that occur in lysosomes of hepatocytes and Kupffer cells in patients with Wilson disease. The drug particularly inhibited the disease-specific accumulation of copper in lysosomes of hepatocytes, tissue macrophages, and Kupffer cells. Once chelated with D-penicillamine, copper is excreted into urine.

Beside its action as metal chelator, D-penicillamine has many other effects like reducing numbers of T-lymphocytes, inhibiting macrophage function, decreasing IL-1, decreasing rheumatoid factor, and preventing collagen from cross-linking. Therefore, it has been used as second-line therapy in rheumatic diseases. Its immunologic effects are partly responsible for some unwanted side effects in Wilson disease patients.

USE OF D-PENICILLAMINE IN WILSON DISEASE

The initial dose of D-penicillamine in most patients is 1000–1500 mg/day in two to four divided doses. Dosing in children is initially 20 mg/kg/day rounded off to the nearest weight of the capsule (150–250 mg depending on the country) and given in two or three divided doses. The treatment is best taken 1 h before or 2 h after food, as absorption might only be 50% if it is taken with a meal. The initiation of treatment with lower initial doses, 250–500 mg/day, increasing over a few weeks, can increase tolerance to the drug. Adequacy of treatment can be monitored by measuring 24 h urinary copper excretion while on treatment. This is highest immediately after starting treatment and may exceed 16 µmol (1000 µg)/24 h. at that time. For long-term treatment, the most important sign of efficacy is a maintained clinical and laboratory improvement. In some patients, serum ceruloplasmin may markedly decrease after initiation of treatment, initially due to reduced hepatic inflammation and a decrease in the acute-phase state that can increase levels of ceruloplasmin [10, 11]. Urinary copper excretion should run in the vicinity of 3–8 µmol/24 h when measured on treatment. To document therapeutic efficiency, urinary copper excretion after 2 days of D-penicillamine cessation should be ≤1.6 µmol/24 h. Regular monitoring of full blood count and urinary protein (using dipsticks) is recommended because of possible adverse effects, which occur in 10%–20% of patients and can be severe enough to lead to the treatment being stopped. Furthermore, since penicillamine can affect pyridoxine metabolism, vitamin B6 should be substituted (50 mg weekly).

SIDE EFFECTS

Early side effects in the first 1–3 weeks may include sensitivity reactions with fever, rash, lymphadenopathy, neutropenia, thrombocytopenia, and proteinuria. These are immunologically mediated reactions.

Early neurological worsening during treatment initiation for Wilson disease is an unresolved problem. In patients with neurological Wilson disease, improvement of symptoms is slower and may be observed even after 3 years [12]. Worsening of neurological symptoms has been reported in 10%–50% of those treated with D-penicillamine during the initial phase of treatment [9]. In a recent series, neurological worsening occurred on all three treatments used for Wilson disease (D-penicillamine, trientine, and zinc) but mainly with D-penicillamine where 13.8% were adversely affected [13]. In 143 symptomatic patients diagnosed with WD early, neurological deterioration was observed in 11.1% [14]. There is no standard recommendation on how to start therapy. In my experience, tolerability of D-penicillamine may be enhanced by starting with lower doses (500–750 mg/day in adults), increased by 250 mg increments every 4–7 days to a maximum of 1000–1500 mg/day in 2–4 divided dosages. Based on common experience, administration of doses 1500 mg/day or higher at once may lead to rapid and often irreversible neurological deterioration in the initial treatment phase.

Late reactions include nephrotoxicity, usually heralded by proteinuria or the appearance of other cellular elements in the urine, for which discontinuation of D-penicillamine should be immediate. Other late reactions include a lupus-like syndrome marked by hematuria, proteinuria, and positive antinuclear antibody and with higher dosages of D-penicillamine Goodpasture's syndrome. Interestingly, these side effects are more commonly reported in patients treated with D-penicillamine for primary biliary cholangitis [15] or rheumatic diseases.

Very late side effects are rare and include nephrotoxicity, myasthenia gravis [16], polymyositis, the loss of taste, immunoglobulin A depression, and serous retinitis. Dermatologic toxicities reported include progeric changes in the skin and elastosis perforans serpiginosa [17], pemphigus or pemphigoid lesions, lichen planus, and aphthous stomatitis. Elastosis perforans serpiginosa is a rare degenerative skin disease characterized by a transepidermal elimination of elastic fiber aggregates [18]. Lysyl oxidase is a copper-dependent enzyme crucial to elastic fiber cross-linking, which is strongly affected by copper depletion. Clinically, the disease presents with multiple firm keratotic papules and nodules arranged in annular plaques over the neck, axillae, antecubital fossae, and forearms. The loss of elastic fibers may also lead to aortic dilation and aneurysm [19].

Overtreatment with D-penicillamine may lead to a reversible sideroblastic anemia and hemosiderosis.

EFFICACY

In general, in Wilson disease, there is not accepted marker of efficacy besides clinical improvement. There is a lack of high-quality evidence to estimate the relative treatment effects of all the available drugs in Wilson disease. Therefore, multicenter prospective randomized controlled comparative trials would be necessary.

Numerous studies attest to the effectiveness of D-penicillamine as treatment for Wilson disease [6, 20–22]. However, most studies define effectivity by different parameters. In patients with symptomatic liver disease, the recovery of synthetic liver function and improvement in clinical signs occurs typically during the first 2–6 months of treatment, but further recovery can occur during the first year of treatment and beyond. Due to the slow clinical response, in patients with acute Wilson disease, chelation therapy is of questionable value. In patients with neurological Wilson disease, improvement of symptoms is slower and may be observed even after 3 years [23]. In many patients, neurological deficits do not progress but do not improve either. As mentioned above, paradoxical neurological deterioration may be induced by chelation therapy.

To date, the chelator selected for therapy remains an individual decision because no head-to-head comparisons are available and costs and availability may play a role.

Some treaters combine D-penicillamine with zinc [6]. Based on limited evidence, there is no clear benefit in such combination [24]. In limited situations of unusually severe disease, reports on the use of these chelators in conjunction with zinc suggest a favorable outcome for combination therapy with DPA plus zinc or trientine plus zinc [25].

In summary, D-penicillamine is still an accepted and widely used drug for treatment of Wilson disease. Its use is limited by frequent side effects, which often necessitate a switch to other treatment modalities.

REFERENCES

[1] Walshe JM. Penicillamine, a new oral therapy for Wilson's disease. Am J Med 1956;21:487–95.
[2] Walshe JM. Treatment of Wilson's disease with penicillamine. Lancet 1960;(7117):188–92.
[3] Perrett D. The metabolism and pharmacology of D-penicillamine in man. J Rheumatol Suppl 1981;7:41–50.
[4] Kukovetz WR, Beubler E, Kreuzig F, Moritz AJ, Nimberger G, Werner-Breitnecker L. Bioavailability and pharmacokinetics of D-penicillamine. J Rheumatol 1983;10:90–4.
[5] Aronson JK. Meyler's side effects of analgesics and anti-inflammatory drugs. Amsterdam: Elsevier Science; 2010. p. 613.
[6] Walshe JM. Copper chelation in patients with Wilson's disease. A comparison of penicillamine and triethylene tetramine dihydrochloride. Q J Med 1973;42:441–52.
[7] Smirnova J, Kabin E, Järving I, Bragina O, Tõugu V, Plitz T, et al. Copper(I)-binding properties of de-coppering drugs for the treatment of Wilson disease. α-Lipoic acid as a potential anti-copper agent. Sci Rep 2018;8:1463.
[8] Scheinberg IH, Sternlieb I, Schilsky M, Stockert RJ. Penicillamine may detoxify copper in Wilson's disease. Lancet 1987;2:95.
[9] Klein D, Lichtmannegger J, Heinzmann U, Summer KH. Dissolution of copper-rich granules in hepatic lysosomes by D-penicillamine prevents the development of fulminant hepatitis in Long-Evans cinnamon rats. J Hepatol 2000;32:193–201.
[10] Steindl P, Ferenci P, Dienes HP, Grimm G, Pabinger I, Ch M, et al. Wilson's disease in patients presenting with liver disease: a diagnostic challenge. Gastroenterology 1997;113:212–8.
[11] Ferenci P. Wilson disease. In: Bacon B, DiBisceglie A, editors. Diagnosis and management. New York, Edinburgh, London, Philadelphia, San Francisco: Churchill Livingstone; 1999. p. 150–64 [Chapter 12].
[12] Brewer GJ, Terry CA, Aisen AM, Hill GM. Worsening of neurologic syndrome in patients with Wilson's disease with initial penicillamine therapy. Arch Neurol 1987;44:490–3.
[13] Merle U, Schaefer M, Ferenci P, Stremmel W. Clinical presentation, diagnosis and long-term outcome of Wilson disease—a cohort study. Gut 2007;56:115–20.
[14] Litwin T, Dzieżyc K, Karliński M, Chabik G, Czepiel W, Członkowska A. Early neurological worsening in patients with Wilson's disease. J Neurol Sci 2015;355:162–7.
[15] Gong Y, Frederiksen SL, Gluud C. D-Penicillamine for primary biliary cirrhosis. Cochrane Database Syst Rev 2004;4.
[16] Członkowska A. Myasthenia syndrome during penicillamine treatment. Brit Med J 1975;2:726.
[17] Becuwe C, Dalle S, Ronger-Savle S, Skowron F, Balme B, Kanitakis J, et al. Elastosis perforans serpiginosa associated with pseudo-pseudoxanthoma elasticum during treatment of Wilson's disease with penicillamine. Dermatology 2005;210:60–3.
[18] Pass F, Goldfischer S, Sternlieb I, Scheinberg IH. Elastosis perforans serpiginosa during penicillamine therapy for Wilson disease. Arch Dermatol 1973;108:713–5.
[19] Tilson MD, Davis G. Deficiencies of copper and a compound with ion-exchange characteristics of pyridinoline in skin from patients with abdominal **aortic** aneurysms. Surgery 1983;94:134–41.
[20] Czlonkowska A, Gajda J, Rodo M. Effects of long-term treatment in Wilson's disease with D-penicillamine and zinc sulphate. J Neurol 1996;243:269–73.
[21] Beinhardt S, Leiss W, Stättermayer AF, Graziadei I, Zoller H, Stauber S, et al. Long-term outcome of Wilson disease in Austria. Clin Gastroenterol Hepatol 2014;12:683–9.
[22] Dhawan A, Taylor RM, Cheeseman P, De Silva P, Katsiyiannakis L, Mieli-Vergani G. Wilson's disease in children: 37-year experience and revised King's score for liver transplantation. Liver Transpl 2005;11:441–8.
[23] Grimm G, Oder W, Prayer L, Ferenci P, Ch M. Prospective follow-up study in Wilson's disease. Lancet 1990;336:963–4.
[24] Brewer GJ, Yuzbasiyan-Gurkan V, Johnson V, Dick RD, Wang Y. Treatment of Wilson's disease with zinc: XI. Interaction with other anticopper agents. J Am Coll Nutr 1993;12:26–30.
[25] Askari FK, Greenson J, Dick RD, Johnson VD, Brewer GJ. Treatment of Wilson's disease with zinc. XVIII. Initial treatment of the hepatic decompensation presentation with trientine and zinc. J Lab Clin Med 2003;142:385–90.

Chapter 17

Trientine for Wilson Disease: Contemporary Issues

Eve A. Roberts[*,†,‡,§]

[*]Department of Paediatrics, University of Toronto, Toronto, Canada, [†]Department of Medicine, University of Toronto, Toronto, Canada, [‡]Department of Pharmacology & Toxicology, University of Toronto, Toronto, Canada, [§]History of Science and Technology Programme, University of King's College, Halifax, Canada

Wilson disease (WD), described in 1912, was the first chronic liver disease where pharmacological treatment was really successful. The discovery in the early 1950s that a metabolite of penicillin, namely, penicillamine, could serve as an oral chelator of copper completely changed the natural history of WD. Instead of progressive hepatic and neurological disease and early death, patients could live a near-normal life so long as they tolerated and actually took the drug. Improvement was dramatic. Switching from the racemic mixture to pure D-penicillamine solved some problems of tolerability; nevertheless, some patients developed life-threatening adverse effects. The need for alternative medical treatment became urgent. One promising candidate was the family of tetramine chelators. Trientine, also known by its formal chemical name as triethylenetetramine, often abbreviated to "trien" by inorganic chemists or as "TETA," was developed in the 1960s. A true orphan drug, trientine has proved highly effective as chronic treatment for WD. Its principal role is as an alternative chelator for patients intolerant of D-penicillamine. Additionally, it can serve as an effective first-line treatment. Importantly, it gained renewed scientific attention as a possible drug intervention to treat cardiomyopathy associated with diabetes mellitus. It has also been investigated for antioxidant properties and as an inhibitor of angiogenesis.

CHEMISTRY

A striking difference between trientine and D-penicillamine is structural. D-Penicillamine is the amino acid cysteine which has been substituted with two methyl groups to become β,β-dimethylcysteine. Trientine resembles the naturally occurring polyamines, spermidine, and spermine (see Fig. 1). Trientine was proposed as an oral chelator for copper by Dixon, who was investigating the utility of copper in structurally specific transamination reactions. John Walshe's specifications for an alternative drug to penicillamine were that it should be nontoxic and capable of forming a complex with copper that could be excreted via the kidneys. Dixon noted that triethylenetetramine (i.e., trientine) bound copper tightly and its resemblance to naturally occurring polyamines suggested it might not be toxic [1]; additionally, it lacked charged groups potentially interfering with renal excretion. The four nitrogen atoms set up a four-way coordination of the copper ion, producing a square-planar geometry, which in fact is very stable structure for capturing Cu^{2+}.

The immediate problem was to synthesize trientine suitable as a drug for clinical use, and the more distal problem was to produce it commercially. Issues of achieving a satisfactory and consistent yield of trientine suitable for producing as tablets were solved by inventing a new preparative method [2–4]. Trientine dihydrochloride was given a formal product license in the United Kingdom in 1985. Meanwhile in the United States, both D-penicillamine and trientine were being produced by Merck Pharmaceuticals. In 1985, trientine was granted formulary status in the United States under the new Orphan Drug Act of 1983.

The formulation in the United Kingdom is trientine dihydrochloride with a tablet size of 300 mg. In North America, the formulation is called trientine hydrochloride in the *United States Pharmacopeia*, but the chemical structure illustrated is a dihydrochloride [5]. In North America, the tablet size is 250 mg. This discrepancy can be confusing both in the medical literature and in actual prescribing. For the North American prescriber, the typical total daily dose for an adult is 1000 mg, typically divided into two equal doses, whereas for the European prescriber, it is 1200 mg, divided into two equal doses.

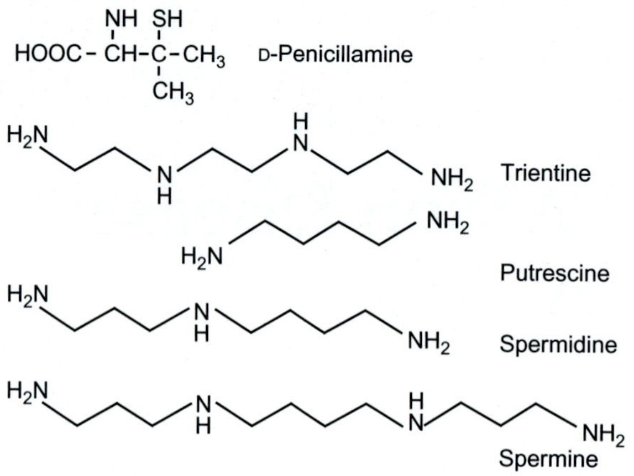

FIG. 1 Chemical structures of trientine, polyamines, and D-penicillamine. Similarities and contrasts are very obvious. Trientine and spermine both have four nitrogens. Trientine and D-penicillamine are completely different structurally.

Efficacy appears to be approximately the same. In 2017, a tetrahydrochloride formulation of trientine (Cuprior) was approved by the European Medical Agency, but so far, it is not yet marketed (EMA/267439/2017/EMEA/H/C/004005).

A different formulation whereby trientine is packaged in liposomes has been investigated for specific delivery to the CNS [6]. These studies are at a preliminary stage, limited to studies in animal models.

PHARMACOLOGY

As a true orphan drug developed quite differently from new orphan drugs for rare diseases nowadays, the pharmacology of trientine has been investigated only sporadically. Major pharmacological issues include biotransformation of trientine, its pharmacokinetics, and—for WD—comparison of its efficacy to that of D-penicillamine. Adequacy of analytic methods is critically important for assessing these studies that have been performed over several decades [7, 8].

Trientine appears to undergo extensive first-pass metabolism after intestinal absorption. Kodama and colleagues identified the main metabolite of trientine as acetyl-trientine in humans and showed that it had little chelating activity [9]. Further studies in healthy human volunteers confirmed this observation but detected an additional diacetylated metabolite [10]. The phase II enzymes mediating this acetylation are the acetyltransferases involved in polyamine catabolism [11, 12]. Studies in cell culture systems indicated that trientine alters polyamine metabolism, including interference with spermidine uptake and down-regulation of ornithine decarboxylase [13].

Limited data regarding the pharmacokinetics of trientine in humans are available [14]. Its intestinal uptake is mediated by polyamine carrier proteins [15, 16]. Polyamines thus compete for uptake. Absorption is comparatively slow, on the order of 1–3 h. (Consequently, trientine is taken *without* food.) The apparent volume of distribution is consistent with trientine being widely distributed [17]. A recent study in healthy nonsmoker volunteers examined acute dosing and then short-term steady-state chronic dosing with a dose of 600 mg, somewhat lower than that commonly prescribed for WD [18]. Samples were analyzed for trientine and its mono- and diacetylated metabolites via high-performance liquid chromatography (HPLC). Acetylator status for the N-acetyltransferase 2 (NAT2) polymorphism did not affect results. The terminal half-life (excretion $t_{1/2}$) was approximately 3 h. Peak concentrations were on the order of 0.8–1.2 mg/L (0.8–1.2 µg/mL). In another study, after steady-state dosing at 600 mg twice daily, the excretion $t_{1/2}$ was 10 h [17]. Excretion is mainly renal, with metabolites predominating.

The main comparison between trientine and D-penicillamine has to do with efficacy to correct copper overload. In general, trientine has been characterized as a weaker chelator of copper than D-penicillamine. This assessment is based on the extent of cupruria induced by trientine versus D-penicillamine in patients with WD [19] and in normal Sprague-Dawley rats [20]. However, in vitro data suggest that the comparison is somewhat more complicated [21]. Trientine appears to be a stronger chelator than D-penicillamine at physiological pH, and its copper complexes are more stable at that pH [22]. Previous characterization of copper complexes showed that copper-trientine is a small molecular complex whereas copper-D-penicillamine complexes are polymeric [23]. Trientine and D-penicillamine appear to mobilize copper from different pools, tissue versus plasma compartment. Whether trientine mobilizes copper from the liver has been disputed. A carefully designed study in normal Sprague-Dawley rats showed that D-penicillamine gets into liver tissue and mobilizes copper whereas trientine competes better than D-penicillamine for copper bound to albumin in the plasma compartment [23].

The impression is that trientine does not get into liver parenchyma. This is not the same as saying that it does not promote the loss of copper from the liver. Depletion of hepatic copper by trientine in WD patients was shown [24]. In the rat model of WD, the Long-Evans cinnamon rat, trientine treatment produced a decrease in hepatic copper concentration in 6-week-old rats, but not in 13-week-old rats, although marked cupruria was detected at both ages [25]. A study of hepatic parenchymal copper concentrations in the healthy liver of patients who had a small hepatocellular carcinoma showed that even limited treatment with trientine decreased hepatic parenchymal copper levels [26]. In terms of his clinical experience, Walshe recommended trientine for the patient with severe hepatic copper overload because it generates greater urinary excretion of copper [19, 27].

Thus, with respect to how trientine removes copper from the body, key features include chelation of non-ceruloplasmin-bound copper in the plasma compartment and renal excretion of that complex. Most of the copper is excreted within hours of taking the dose [17]. Moreover, in healthy volunteers, cupruresis on a dose of 1200 mg per day was 3- to 16-fold higher than in the placebo control [17]. Additionally, trientine appears to decrease intestinal absorption of copper when coadministered with food [28]. Trientine also increases urinary excretion of iron and zinc modestly [9]. Other recent pharmacological investigations reveal more complicated metabolic roles. Specifically, trientine affects polyamine metabolism. It might alter intracellular handling of iron and zinc. It can inhibit telomerase [29]. Additionally, it can inhibit angiogenesis and modulate certain cell-signaling pathways and anti-inflammatory mechanisms. However, the basis for some of these actions is not peculiar to trientine itself; instead, it reflects the diverse actions of copper in cells and tissues, in particular the contribution of copper to oxidative stress and its complicated critical role in angiogenesis [30].

THERAPEUTICS: EFFICACY/SAFETY

Trientine was developed as alternative treatment for WD patients who proved intolerant of D-penicillamine [31, 32]. It is highly effective in this role as a rescue drug [33]. This remains its prime indication, important because the rate of adverse reaction to D-penicillamine severe enough to require its discontinuation is in the range of 30%–40% in adults and children [34–36]. Trientine is appropriate "preemptive" first-line treatment for patients who appear to be at increased risk for adverse effects of D-penicillamine or those with thrombocytopenia due to congestive splenomegaly or the rare patient who has already had transient bone marrow failure due to a separate disease process. Many regard trientine as an excellent first-line treatment in WD patients generally because of its favorable safety and adequate efficacy; however, its suitability as first-line treatment has not been officially endorsed [37]. A systematic review of the literature relating to trientine therapeutics across all age groups discloses that the quality of the reported evidence falls short of current standards [38]. Current clinical research is focused on the utility of once-daily treatment with trientine in the clinically stable patient [39].

The general efficacy of trientine in adults and children with WD needs to be evaluated against two different treatment objectives. One is clinical stabilization after the development of severe adverse effects from D-penicillamine, principally renal impairment (proteinuria and nephrosis) and bone marrow dysfunction (leucopenia, thrombocytopenia, or aplastic anemia), as well as a lupus-like disorder and skin changes. The other objective is long-term disease control with trientine as the sole oral chelator administered. Data based on approximately 200–300 adults indicate that trientine works well in place of D-penicillamine [33, 40–46]. According to these reports, renal and bone marrow dysfunction became normal or else generally improved; the lupus-like syndrome was unpredictable; skin abnormalities did not always resolve. Patients tolerated trientine well and developed no new adverse effects. The experience is entirely similar in children including, notably, the first patient reported [31], but data are much more limited than in adults [35, 38, 43, 47, 48]. Follow-up in these children over 1–18 years indicated that they remained clinically stable so long as they were adherent to treatment.

Besides extended observations of patients switched to trientine from D-penicillamine, data attesting to long-term efficacy of trientine include some patients who received trientine as primary therapy [41, 45]. A report including 38 adult patients treated with trientine as first-line therapy indicated that 25 of 27 patients with hepatic WD improved and the other two remained unchanged [46]. A limited ($n=3$) but more conventional experience with trientine as primary treatment for children with WD found no untoward effects except some persisting abnormality of liver tests [38]. The experience with neurological WD was less straightforward; however, no new neurological symptoms developed in those patients treated with trientine, and none died [46]. Recent case reports indicated clinical and radiological improvement on trientine in two patients with neurological WD [49, 50].

Assessing long-term efficacy of trientine in some early reports is difficult because the drug was in relatively short supply. If the patient could be put back on D-penicillamine, resuming the standard drug had practical advantages. Moreover, there were pervasive worries about whether trientine was a potent enough chelator. These worries may have been misplaced. Just how much daily cupruresis is required over the long term of chronic therapy remains uncertain. A dependable relatively low-volume cupruresis may be adequate or possibly advantageous in the long term. A problem for more recent reports is the heterogeneity of clinical treatment strategies, rendering generalizable conclusions nearly impossible. The large American pediatric series employed trientine as primary treatment on a limited-duration basis followed by transition to trientine + zinc treatment and then transition to zinc monotherapy: thus, the data on trientine are not entirely evident [51]. Overall, the seemingly abundant clinical experience with trientine in WD is actually quite patchy. The general impression is that taking trientine during pregnancy entails no undue risks; again, experience is limited.

ADVERSE EFFECTS

Compared with D-penicillamine, trientine is remarkably free of adverse side effects. In patients with primary biliary cirrhosis, trientine caused acute gastritis in three patients and acute rhabdomyolysis in one [52], but neither has been prominent in patients with Wilson disease. Iron deficiency [41, 45] and sideroblastic anemia [53, 54] have been reported. This sideroblastic anemia is reversible and may be due to copper deficiency. Urinary excretion of zinc may be increased [45]. Dermatitis developed in one patient [55] and a fixed drug reaction in another (Roberts EA, unpublished observations). Colitis has been reported [45, 56]. Substituting zinc for trientine leads to improvement in colitis.

Almost no information about trientine overdose is available. Ingestion of 60 g of trientine was reported as causing dizziness and vomiting [57]. Impure preparations of trientine may be harmful [42].

With neurological WD, deterioration in clinical neurological status may occur after chelator treatment is commenced. The consensus is that trientine can be associated with the neurological worsening but the risk is less than with D-penicillamine. The first reports of trientine-associated neurological worsening are from the 1990s [44, 45], supported by more recent observations [46, 58].

EMERGING APPLICATIONS

Apart from its role in WD treatment, new therapeutic applications for trientine are being developed, mainly for disorders more prevalent than WD, namely, cancer and diabetes mellitus. Relating to cancer, one application has little to do with copper homeostasis: since polyamines are important for cell proliferation, an agent like trientine may interfere competitively with polyamine metabolism in neoplasia [13]. Potentiating effects of platinum-based antineoplastic drugs may be another antineoplastic effect [59]. A more general antineoplastic strategy involves the role of copper in angiogenesis. Although tetrathiomolybdate is preferred for inhibiting angiogenesis [60], trientine has been shown to inhibit murine hepatocellular carcinoma [61]. Copper chelation may also interrupt or enhance various pathways [62] mediating inflammatory response. The main focus relating to diabetes mellitus has been diabetic cardiomyopathy, which involves disruption of cardiac copper homeostasis. Treatment with trientine may improve it [11, 63, 64]. Oxidative stress associated with redox-active copper is also important. Tissue damage in diabetes may be related in part to oxidative changes in structural proteins, known as advanced glycation end products, to which copper can bind.

SOCIAL ISSUES: REAL OR EFFECTIVE NONAVAILABILITY OF TRIENTINE

Recently, trientine has been at the center of scrutiny relating to drug pricing of orphan drugs. This debate has often gotten conflated with discussion about how newly developed drugs for rare diseases ought to be priced. The confusion arises from ambiguous language calling rare diseases "orphan diseases" and the drugs aimed at treating those diseases "orphan drugs." Given the extensive interest in rare diseases currently, they are not orphans. The trientine debate has elicited a conceptual clarification about orphan drugs. There are "new" orphan drugs, an important area of expansion in the pharmaceutical industry worldwide, and there are "old" orphan drugs. (Both D-penicillamine and trientine are "old" orphan drugs, in regular beneficial, medically-sanctioned use for roughly 50–60 years.) Most new pharmacological treatments for rare diseases—"new" orphan drugs—fit the prevailing argument that drugs for rare diseases depend on emerging technologies and are costly to develop. "Old" orphan drugs are different. D-Penicillamine and trientine are archetypal examples; the critical research and development was done in the United Kingdom in the 1950s and 1960s. The practice of acquiring the license for an "old" orphan drug and then hiking its price many-fold, as opposed to along the lines of the rate of inflation, has become widespread in the past 7–10 years. National orphan drug policies are needed to address the status and dependable availability of such "old" orphan drugs. Contemporary orphan drug policy cannot focus exclusively on extremely expensive "new" orphan drugs and how to afford them. For "new" orphan drugs, rare-disease patients constitute potential markets. For "old" orphan drugs, the market is well-established. It consists of patients whose health and well-being depend on affordable availability of these standard treatments.

In North America, for decades, both D-penicillamine and trientine were manufactured by Merck Pharmaceuticals and marketed under the trade names Cuprimine and Syprine, respectively. In the United States, both Cuprimine and Syprine were licensed and added to the official formulary, and in fact, Syprine (trientine) was an early beneficiary of the 1983 Orphan Drug Act. This approval was based on Walshe's numerous papers plus the American reports on trientine as D-penicillamine rescue [33, 43]. In contrast, in Canada, Cuprimine went through formal Health Canada evaluation and received a Drug Identification Number (DIN), but Syprine never went through that process. The reason for the difference in registering these two drugs for WD is not known. This discrepancy in Canada subsequently exposed critical differences in drug availability and drug policy between the two countries. In the United States, both Cuprimine and Syprine can be prescribed without impediment, and both drugs are eligible for third-party coverage. In Canada, Cuprimine can be prescribed, and it is eligible for third-party cost coverage. Prescribing Syprine requires application to the Special Access Program of Health Canada; typically, its cost is covered exclusively by the patient. In Europe, in countries where trientine is available, the situation is similarly problematic, as the only EU-wide approved drug Cuprior (TETA 4HCl) is not on the market. Thus, reimbursement of trientine continues to be negotiated on an individual per-patient basis.

Regarding the North American market, Merck maintained relatively low prices for these chelators. However, effectively, Merck stopped manufacturing both oral chelators in 2004 when it acquired Aton Pharmaceuticals and transferred the licenses for both Cuprimine and Syprine to that company. At the time, there were vague worries that the cost of these drugs would increase, but little change occurred. In 2010, Aton was acquired by Valeant Pharmaceuticals. Valeant, an international pharmaceutical company based in the United States whose head offices were transferred to Canada in 2012, thus became the main manufacturer of Cuprimine and Syprine in North America. Almost immediately, the discrepancy between drug licensing in Canada and the United States was noteworthy. Initially, Valeant Canada ran a compassionate program in Canada whereby Syprine was provided through the Special Access Program free of charge. Apparently, it was cheaper to make Syprine available free of charge to Canadians than to seek a Canadian DIN for Syprine. This program likely had the effect that in clinical situations where it was difficult to decide which chelator was preferable, the decision tilted toward Syprine on economic grounds. Unexpectedly, in mid-November 2013, Valeant Canada announced informally and nonsystematically that as of 1 January 2014, the compassionate program was canceled. Canadian patients would have to purchase Syprine, and the price would be the American price, converted to Canadian dollars. Given the currency conversion at that time, this price was in the range of CAD$11,000 plus dispensing fees per 100 tablets. It represented a 13-fold increase in the previous price, approximately CAD$800 per 100 tablets, still quite a high price for a medication by Canadian standards. For the typical adult taking 500 mg BID, the annual cost would be roughly CAD $165,000. Importantly, the rate of inflation on North American currencies in the first decade or so of this century would have accounted for a price rise of approximately 40 cents on the dollar from the 2001 price.

In effect, Syprine was no longer available for treating WD in Canada. It is well known that not having trientine available imposes significant problems. The rationale for developing it in the first place was to have an alternative oral chelator. A single-center retrospective review of Wilson disease from Padua reported two deaths among their patients when D-penicillamine failed and trientine was not available [65]. Reported experience was similar from Malaysia where trientine is not available [66]. Although some patients previously treated with trientine can be switched to D-penicillamine or zinc, many cannot. Accumulating data indicate that zinc may be less effective long-term in hepatic WD than in neurological WD [67, 68]. In 2014, the strategy in Canada for dealing with this crisis involved finding an alternate source of trientine and initiating advocacy by hepatologists. Through the auspices of various advocates, the United Kingdom brand of trientine was added to the list of drugs available through the Special Access Program. In April 2014, Valeant Canada announced that the price for Syprine in Canada would revert to its 2013 Canadian price. Although this price is still high by Canadian standards, it restored trientine to the pharmacological armamentarium for WD. However, in other jurisdictions, the price of one or both oral chelators was hiked up. In the United States, both Cuprimine and Syprine were subjected to price inflation. In the United Kingdom, the price of Univar's trientine was greatly increased. (Between the increased price and the weak Canadian dollar, that drug became too expensive for Canadians.) Drug-pricing is highly influenced by what is permitted by law and what is apparently acceptable in terms of public or private payers for medications, for example, the dominance (or not) of the insurance industry.

This story is not limited to treatments for WD. Since 2014, the practice of acquiring "legacy" drugs (established effective drugs that are off-patent, generally defined as in use for ≥ 25 years or before 1962) and then increasing the price to what the market will bear has become obvious [69]. The practice has been around somewhat longer. Drug prices jacked up capriciously have been termed "extraordinary drug price increases" and defined—in retrospect, very conservatively—by Carpinelli and Schondelmeyer as any price increase $\geq 100\%$ at a single time point [70]. In the United States, drug brands limited to a single producer (i.e., a monopoly situation) and off-patent generic drugs were particularly prone to extraordinary price increases, as determined in the comprehensive 2008 study from the University of Minnesota. Importantly, orphan drugs were often subjected to extraordinary drug price increases, and some increases were in excess of 1000% [71]. Extraordinary price increases for the intravenous Indocin (indomethacin) and Acthar brand of corticotropin led to a Congressional hearing in 2008. Additional drugs affected in the United States include colchicine [72], albendazole [73], Daraprim brand of pyrimethamine, Isuprel brand of isoproterenol, Nitropress brand of nitroprusside, deflazacort, and the "EpiPen." In the United Kingdom, phenytoin has come under similar pressure. Importantly, although many of these drugs are for rare diseases, they do not involve contemporary, cost-intensive research. Indeed, the price is jacked up in the absence of recent research. An exception is phenylbutyrate for urea cycle disorders; recent research has led to the development of glycerol phenylbutyrate, which is more tolerable and effective than its legacy predecessor [74, 75], although priced substantially higher [76]. While attention from politicians may highlight the problem of extraordinary price increases for "old" orphan drugs, advocacy from professional organizations seems to be more effective in the short run, as found with Syprine in Canada [77] and colchicine in the United States. A common feature for all these drugs is that the manufacturer holds either a de facto or an actual monopoly in their market [78]. Forbidding that monopoly may be one mechanism to promote ordinary market forces to influence drug pricing and undermine capricious price inflation. Indeed, competition is well known to restrain drug pricing. It was apparent in the 1940s with penicillin [79]. Given that the economics and mechanics of how drugs are priced and purchased vary from country to country, different solutions may be needed in different jurisdictions. A general formulation of corporate social responsibility with respect to orphan drugs—both "old" and newly developed—remains elusive.

A key lesson from the Canadian experience with Syprine is that communication is all important. Communication about Syprine was handled badly. Specifically, no letter was sent to Canadian specialists treating WD nor to Canadian physicians generally. Many physicians learned of this change in Syprine availability indirectly: when their hospital pharmacies required replenished supply or their clinical team heard of these problems from patients. The timing of this change in drug availability and cost was also extremely awkward. Announcing in mid-November a change to take effect on January 1 of the following year gave physicians almost no time to organize therapeutic alternatives because of the Christmas holidays. The pharmaceutical company offered no transitional strategy or contingency plans for dealing with medical problems arising from the sudden dramatic change in drug availability. A financial assistance program from Valeant United States did not apply in Canada. Since most Canadian WD patients are outpatients, hospital pharmacy budgets do not pick up the cost of Syprine. Advanced comprehensive warning of such massive changes in drug pricing, as with sporadic drug unavailability, should be legally required.

A further lesson for WD patients and others with rare but eminently treatable diseases is that the loss of reliable or affordable supply of a life-saving drug creates tremendous emotional stress for these patients. In fact, these patients are highly vulnerable. Canadian WD patients stable on an effective regimen were faced with significant anxiety about whether they would be able to maintain their accustomed good health. Indeed, this anxiety now confronts all patients diagnosed with Wilson disease from the time of diagnosis onward. A medication for a rare disease should not get an inflated price just because the disease is rare [80]. Life-saving lifelong treatment must be affordable. This consideration extends to new drug treatments being developed for WD. Availability entails affordability. A drug treatment that is available but beyond the economic reach of patients is in fact not available. This is a problem for Western economies and also, *mutatis mutandis*, for emerging economies. It is ironic that trientine should be the focus of this discussion since one of its advantages, as Walshe pointed out in 1973, was that it was "cheap and easy" to manufacture and thus might be of especial utility in "underdeveloped countries" [19].

CONCLUSION

The history of trientine is one of drug development under duress. The life-threatening predicament of a single patient demanded accelerated innovative research to find a drug to substitute for D-penicillamine. In a somewhat less personal way, this has become a recurring scenario for rapid drug development in the latter half of the 20th century, for example, with AIDS and chronic hepatitis C, and for vaccine development. Given its structural biology, trientine has proved to be an effective treatment for WD. The alacrity of its adoption for treating WD reflects both the desperation of the circumstances and trientine's apparent safety/efficacy. With the prospect of broader applications, its pharmacology is being sorted out in greater detail.

One lesson from the trientine story is that WD requires availability of more than one medical treatment. The development of zinc salts and the prospect of further drug innovation [81] in the future are important. Trientine is arguably a more effective oral chelator than D-penicillamine because its marginally lower efficiency as a chelator, compared with D-penicillamine, is balanced by its greater tolerability. Nonavailability of trientine—or D-penicillamine—due to indefensible, avaricious pricing practices by pharmaceutical companies constitutes a serious danger to the well-being of individuals with WD. The risks of having effective monopolies on the production of well-established lifesaving "old" orphan drugs need to be addressed by social policy, lest the effective monopoly cease production. Broadening its therapeutic applications may ensure availability of trientine for WD patients, but the implications for pricing are hard to predict. More specific and concerted action in terms of public policy is required for all "old" orphan drugs. Otherwise, valued drugs whose availability we take for granted may disappear.

REFERENCES

[1] Dixon HBF. The chemistry of trientine. In: Scheinberg IH, Walshe JM, editors. Orphan diseases and orphan drugs. Manchester, UK: Manchester University Press; 1986. p. 23–32.

[2] Dixon HB, Gibbs K, Walshe JM. Preparation of triethylenetetramine dihydrochloride for the treatment of Wilson's disease. Lancet 1972;i:853.

[3] Humphries R, Purchase R. The development of trientine dihydrochloride, 1977-85. In: Scheinberg IH, Walshe JM, editors. Orphan diseases and orphan drugs. Manchester, UK: Manchester University Press; 1986. p. 53–5.

[4] Purchase R. The treatment of Wilson's disease, a rare genetic disorder of copper metabolism. Sci Prog 2013;96:19–32.

[5] Pharmacopeia US, http://www.pharmacopeia.cn/v29240/usp29nf24s0_m85228.html. Accessed 2 April 2017.

[6] Tremmel R, Uhl P, Helm F, Wupperfeld D, Sauter M, Mier W, et al. Delivery of copper-chelating trientine (TETA) to the central nervous system by surface modified liposomes. Int J Pharm 2016;512:87–95.

[7] Lu J, Chan YK, Poppitt SD, Cooper GJ. Determination of triethylenetetramine (TETA) and its metabolites in human plasma and urine by liquid chromatography-mass spectrometry (LC-MS). J Chromatogr B: Analyt Technol Biomed Life Sci 2007;859:62–8.

[8] Othman A, Lu J, Sunderland T, Cooper GJ. Development and validation of a rapid HPLC method for the simultaneous determination of triethylenetetramine and its two main metabolites in human serum. J Chromatogr B: Analyt Technol Biomed Life Sci 2007;860:42–8.

[9] Kodama H, Murata Y, Iitsuka T, Abe T. Metabolism of administered triethylene tetramine dihydrochloride in humans. Life Sci 1997;61:899–907.

[10] Lu J, Chan YK, Gamble GD, Poppitt SD, Othman AA, Cooper GJ. Triethylenetetramine and metabolites: levels in relation to copper and zinc excretion in urine of healthy volunteers and type 2 diabetic patients. Drug Metab Dispos 2007;35:221–7.

[11] Cooper GJ. Therapeutic potential of copper chelation with triethylenetetramine in managing diabetes mellitus and Alzheimer's disease. Drugs 2011;71:1281–320.

[12] Hyvonen MT, Weisell J, Khomutov AR, Alhonen L, Vepsalainen J, Keinanen TA. Metabolism of triethylenetetramine and 1,12-diamino-3,6,9-triazadodecane by the spermidine/spermine-N(1)-acetyltransferase and thialysine acetyltransferase. Drug Metab Dispos 2013;41:30–2.

[13] Hyvonen MT, Ucal S, Pasanen M, Peraniemi S, Weisell J, Khomutov M, et al. Triethylenetetramine modulates polyamine and energy metabolism and inhibits cancer cell proliferation. Biochem J 2016;473:1433–41.

[14] Lu J. Triethylenetetramine pharmacology and its clinical applications. Mol Cancer Ther 2010;9:2458–67.

[15] Tanabe R, Kobayashi M, Sugawara M, Iseki K, Miyazaki K. Uptake mechanism of trientine by rat intestinal brush-border membrane vesicles. J Pharm Pharmacol 1996;48:517–21.

[16] Tanabe R. Disposition behavior and absorption mechanism of trientine, an orphan drug for Wilson's disease. Hokkaido Igaku Zasshi 1996;71:217–28 [in Japanese].

[17] Cho HY, Blum RA, Sunderland T, Cooper GJ, Jusko WJ. Pharmacokinetic and pharmacodynamic modeling of a copper-selective chelator (TETA) in healthy adults. J Clin Pharmacol 2009;49:916–28.

[18] Lu J, Poppitt SD, Othman AA, Sunderland T, Ruggiero K, Willett MS, et al. Pharmacokinetics, pharmacodynamics, and metabolism of triethylenetetramine in healthy human participants: an open-label trial. J Clin Pharmacol 2010;50:647–58.

[19] Walshe JM. Copper chelation in patients with Wilson's disease. A comparison of penicillamine and triethylene tetramine dihydrochloride. Q J Med 1973;42:441–52.

[20] Jones MM, Singh PK, Zimmerman LJ, Gomez M, Albina ML, Domingo JL. Effects of some chelating agents on urinary copper excretion by the rat. Chem Res Toxicol 1995;8:942–8.

[21] Botha CJ, Naude TW, Swan GE, Guthrie AJ. Evaluation of the efficacy of D-penicillamine and trientine as copper chelators using an in vitro technique involving ovine red blood cells. Onderstepoort J Vet Res 1992;59:191–5.

[22] Riha M, Karlickova J, Filipsky T, Macakova K, Hrdina R, Mladenka P. Novel method for rapid copper chelation assessment confirmed low affinity of D-penicillamine for copper in comparison with trientine and 8-hydroxyquinolines. J Inorg Biochem 2013;123:80–7.

[23] Sarkar B, Sass-Kortsak A, Clarke R, Laurie SH, Wei P. A comparative study of in vitro and in vivo interaction of D-penicillamine and triethylenetetramine with copper. Proc R Soc Med 1977;70(Suppl 3):13–8.

[24] Gibbs K, Walshe JM. Liver copper concentration in Wilson's disease: effect of treatment with 'anti-copper' agents. J Gastroenterol Hepatol 1990;5:420–4.

[25] Iseki K, Kobayashi M, Ohba A, Miyazaki K, Li Y, Togashi Y, et al. Comparison of disposition behavior and de-coppering effect of triethylenetetramine in animal model for Wilson's disease (Long-Evans Cinnamon rat) with normal Wistar rat. Biopharm Drug Dispos 1992;13:273–83.

[26] Fukuda H, Ebara M, Okabe S, Yoshikawa M, Sugiura N, Saisho H, et al. Metal contents of liver parenchyma after percutaneous ethanol injection or radiofrequency ablation in patients with hepatocellular carcinoma before and after trientine hydrochloride therapy. J Lab Clin Med 2004;143:333–9.

[27] Walshe JM. Hepatic Wilson's disease: initial treatment and long-term management. Curr Treat Options Gastroenterol 2005;8:467–72.

[28] Siegemund R, Lossner J, Gunther K, Kuhn HJ, Bachmann H. Mode of action of triethylenetetramine dihydrochloride on copper metabolism in Wilson's disease. Acta Neurol Scand 1991;83:364–6.

[29] Liu J, Guo L, Yin F, Zheng X, Chen G, Wang Y. Characterization and antitumor activity of triethylene tetramine, a novel telomerase inhibitor. Biomed Pharmacother 2008;62:480–5.
[30] Urso E, Maffia M. Behind the link between copper and angiogenesis: established mechanisms and an overview on the role of vascular copper transport systems. J Vasc Res 2015;52:172–96.
[31] Walshe JM. Management of penicillamine nephropathy in Wilson's disease: a new chelating agent. Lancet 1969;ii:1401–2.
[32] Walshe JM. Triethylene tetramine. Lancet 1970;ii:154.
[33] Scheinberg IH, Jaffe ME, Sternlieb I. The use of trientine in preventing the effects of interrupting penicillamine therapy in Wilson's disease. N Engl J Med 1987;317:209–13.
[34] Walshe JM. Wilson's disease presenting with features of hepatic dysfunction: a clinical analysis of eighty-seven patients. Q J Med 1989;70:253–63.
[35] Dhawan A, Taylor RM, Cheeseman P, De Silva P, Katsiyiannakis L, Mieli-Vergani G. Wilson's disease in children: 37-year experience and revised King's score for liver transplantation. Liver Transpl 2005;11:441–8.
[36] Merle U, Schaefer M, Ferenci P, Stremmel W. Clinical presentation, diagnosis and long-term outcome of Wilson's disease: a cohort study. Gut 2007;56:115–20.
[37] Wiggelinkhuizen M, Tilanus ME, Bollen CW, Houwen RH. Systematic review: clinical efficacy of chelator agents and zinc in the initial treatment of Wilson disease. Aliment Pharmacol Ther 2009;29:947–58.
[38] Taylor RM, Chen Y, Dhawan A, EuroWilson C. Triethylene tetramine dihydrochloride (trientine) in children with Wilson disease: experience at King's College Hospital and review of the literature. Eur J Pediatr 2009;168:1061–8.
[39] Ala A, Aliu E, Schilsky ML. Prospective pilot study of a single daily dosage of trientine for the treatment of Wilson disease. Dig Dis Sci 2015;60:1433–9.
[40] Harders H, Cohnen E. Preparation of and clinical experiences with trien for the treatment of Wilson's disease in absolute intolerance of D-penicillamine. Proc R Soc Med 1977;70(Suppl 3):10–2.
[41] Walshe JM. Treatment of Wilson's disease with trientine (triethylene tetramine) dihydrochloride. Lancet 1982;i:643–7.
[42] Walshe JM. The management of Wilson's disease with triethylene tetramine 2HC1 (Trien 2HC1). Prog Clin Biol Res 1979;34:271–80.
[43] Dubois RS, Rodgerson DO, Hambidge KM. Treatment of Wilson's disease with triethylene tetramine hydrochloride (trientine). J Pediatr Gastroenterol Nutr 1990;10:77–81.
[44] Saito H, Watanabe K, Sahara M, Mochizuki R, Edo K, Ohyama Y. Triethylene-tetramine (trien) therapy for Wilson's disease. Tohoku J Exp Med 1991;164:29–35.
[45] Dahlman T, Hartvig P, Lofholm M, Nordlinder H, Loof L, Westermark K. Long-term treatment of Wilson's disease with triethylene tetramine dihydrochloride (trientine). QJM 1995;88:609–16.
[46] Weiss KH, Thurik F, Gotthardt DN, Schafer M, Teufel U, Wiegand F, et al. Efficacy and safety of oral chelators in treatment of patients with Wilson disease. Clin Gastroenterol Hepatol 2013;11(1028–1035):e1021–2.
[47] Haslam RH, Sass-Kortsak A, Stout W, Berg M. Treatment of Wilson's disease with triethylene tetramine dihydrochloride. A case report. Dev Pharmacol Ther 1980;1:318–24.
[48] Morita J, Yoshino M, Watari H, Yoshida I, Motohiro T, Yamashita F, et al. Wilson's disease treatment by triethylene tetramine dihydrochloride (trientine, 2HCl): long-term observations. Dev Pharmacol Ther 1992;19:6–9.
[49] Park HK, Lee JH, Lee MC, Chung SJ. Teaching NeuroImages: MRI reversal in Wilson disease with trientine treatment. Neurology 2010;74.
[50] Chung EJ, Kim EG, Kim SJ, Ji KH, Seo JH. Wilson's disease with cognitive impairment and without extrapyramidal signs: improvement of neuropsychological performance and reduction of MRI abnormalities with trientine treatment. Neurocase 2016;22:40–4.
[51] Arnon R, Calderon JF, Schilsky M, Emre S, Shneider BL. Wilson disease in children: serum aminotransferases and urinary copper on triethylene tetramine dihydrochloride (trientine) treatment. J Pediatr Gastroenterol Nutr 2007;44:596–602.
[52] Epstein O, Sherlock S. Triethylene tetramine dihydrochloride toxicity in primary biliary cirrhosis. Gastroenterology 1980;78:1442–5.
[53] Condamine L, Hermine O, Alvin P, Levine M, Rey C, Courtecuisse V. Acquired sideroblastic anaemia during treatment of Wilson's disease with triethylene tetramine dihydrochloride. Br J Haematol 1993;83:166–8.
[54] Perry AR, Pagliuca A, Fitzsimons EJ, Mufti GJ, Williams R. Acquired sideroblastic anaemia induced by a copper-chelating agent. Int J Hematol 1996;64:69–72.
[55] Rudzki E. Dermatitis from triethylenetetramine in Poland. Contact Dermatitis 1980;6:235–6.
[56] Boca M, Baran P, Boca R, Fuess H, Kickelbick G, Linert W, et al. Selective imidazolidine ring opening during complex formation of iron(III), copper (II), and zinc(II) with a multidentate ligand obtained from 2-pyridinecarboxaldehyde N-oxide and triethylenetetramine. Inorg Chem 2000;39:3205–12.
[57] Hashim A, Parnell N. A case of trientine overdose. Toxicol Int 2015;22:158–9.
[58] Kim B, Chung SJ, Shin HW. Trientine-induced neurological deterioration in a patient with Wilson's disease. J Clin Neurosci 2013;20:606–8.
[59] Fu S, Hou MM, Wheler J, Hong D, Naing A, Tsimberidou A, et al. Exploratory study of carboplatin plus the copper-lowering agent trientine in patients with advanced malignancies. Invest New Drugs 2014;32:465–72.
[60] Goodman VL, Brewer GJ, Merajver SD. Copper deficiency as an anti-cancer strategy. Endocr Relat Cancer 2004;11:255–63.
[61] Yoshii J, Yoshiji H, Kuriyama S, Ikenaka Y, Noguchi R, Okuda H, et al. The copper-chelating agent, trientine, suppresses tumor development and angiogenesis in the murine hepatocellular carcinoma cells. Int J Cancer 2001;94:768–73.
[62] Yin JM, Sun LB, Zheng JS, Wang XX, Chen DX, Li N. Copper chelation by trientine dihydrochloride inhibits liver RFA-induced inflammatory responses in vivo. Inflamm Res 2016;65:1009–20.

[63] Baynes JW, Murray DB. The metal chelators, trientine and citrate, inhibit the development of cardiac pathology in the Zucker diabetic rat. Exp Diabetes Res 2009;2009:696378.
[64] Zhang S, Liu H, Amarsingh GV, Cheung CC, Hogl S, Narayanan U, et al. Diabetic cardiomyopathy is associated with defective myocellular copper regulation and both defects are rectified by divalent copper chelation. Cardiovasc Diabetol 2014;13:100.
[65] Medici V, Trevisan CP, D'Inca R, Barollo M, Zancan L, Fagiuoli S, et al. Diagnosis and management of Wilson's disease: results of a single center experience. J Clin Gastroenterol 2006;40:936–41.
[66] Ping CC, Hassan Y, Aziz NA, Ghazali R, Awaisu A. Discontinuation of penicillamine in the absence of alternative orphan drugs (trientine-zinc): a case of decompensated liver cirrhosis in Wilson's disease. J Clin Pharm Ther 2007;32:101–7.
[67] Linn FH, Houwen RH, van Hattum J, van der Kleij S, van Erpecum KJ. Long-term exclusive zinc monotherapy in symptomatic Wilson disease: experience in 17 patients. Hepatology 2009;50:1442–52.
[68] Weiss KH, Gotthardt DN, Klemm D, Merle U, Ferenci-Foerster D, Schaefer M, et al. Zinc monotherapy is not as effective as chelating agents in treatment of Wilson disease. Gastroenterology 2011;140:1189–98.
[69] Kesselheim AS, Avorn J, Sarpatwari A. The high cost of prescription drugs in the United States: origins and prospects for reform. JAMA 2016;316:858–71.
[70] Carpinelli MM, Schondelmeyer SW. Extraordinary price increases in the pharmaceutical market. In: *Joint Economic Committee, United States Congress, Special Hearing: Small Market Drugs, Big Price Tags: Are Drug Companies Exploiting People with Rare Diseases*; 2008.
[71] Hemphill TA. Extraordinary pricing of orphan drugs: is it a socially responsible strategy for the U.S. pharmaceutical industry? J Bus Ethics 2010;94:225–42.
[72] Kesselheim AS, Solomon DH. Incentives for drug development—the curious case of colchicine. N Engl J Med 2010;362:2045–7.
[73] Alpern JD, Stauffer WM, Kesselheim AS. High-cost generic drugs—implications for patients and policymakers. N Engl J Med 2014;371:1859–62.
[74] Smith W, Diaz GA, Lichter-Konecki U, Berry SA, Harding CO, McCandless SE, et al. Ammonia control in children ages 2 months through 5 years with urea cycle disorders: comparison of sodium phenylbutyrate and glycerol phenylbutyrate. J Pediatr 2013;162:1228–34.
[75] Berry SA, Lichter-Konecki U, Diaz GA, McCandless SE, Rhead W, Smith W, et al. Glycerol phenylbutyrate treatment in children with urea cycle disorders: pooled analysis of short and long-term ammonia control and outcomes. Mol Genet Metab 2014;112:17–24.
[76] Guha M. Urea cycle disorder drug approved. Nat Biotechnol 2013;31:274.
[77] Chandok N, Roberts EA. The trientine crisis in Canada: a call to advocacy. Can J Gastroenterol Hepatol 2014;28:184.
[78] Roberts EA, Herder M, Hollis A. Fair pricing of "old" orphan drugs: considerations for Canada's orphan drug policy. CMAJ 2015;187:422–5.
[79] Gaynes R. The discovery of penicillin—new insights after more than 75 years of clinical use. Emerg Infect Dis 2017;23:849–53.
[80] Herder M. What is the purpose of the orphan drug act? PLoS Med 2017;14.
[81] Roberts EA. Broadening the implications of gene discovery. Hepatology 2016;63:1765–7.

Chapter 18

Tetrathiomolybdate (TTM)

Christian Rupp and Karl Heinz Weiss
Internal Medicine IV, University Hospital Heidelberg, Heidelberg, Germany

INTRODUCTION

The mainstay of current treatment regimen for WD focuses on achieving a negative copper balance to remove copper from affected organs and to avoid further toxic copper accumulation [1, 2]. The most common agents for treatment are chelators like D-penicillamine or triethylenetetramine dihydrochloride (trientine) and zinc salts [3, 4]. The orally administered chelator therapies bind copper that is removed from the circulation via the kidney. For that reason, this treatment option is at least in part dependent on sufficient renal function. By contrast, zinc salts reduce the absorption of copper by the intestinal mucosa, leading to a negative copper balance. Zinc salts are slower-acting agents with the risk of disease progression until adequate copper balance is achieved. Beside ceruloplasmin-bound copper, there is a fraction of free copper referred to as nonceruloplasmin-bound copper (NCC) that is freely available for cellular uptake and subsequently responsible for toxic damage [5, 6]. Especially at the beginning of therapy, rapid mobilization of copper from peripheral stores may cause a rapid increase of NCC leading to further disease progression and especially to neurological worsening. Chelators and zinc salt can be combined but have to be administered separately under fasting condition, requiring multiple-daily dosing. Several surveys indicate difficulties, especially in long-term adherence in about half of the patients treated with current available combination therapy. For these reasons, there is an unmet need for a fast-acting agent with the possibility of an easy dosing regimen, preferably single-daily dosing. Tetrathiomolybdate (TTM) was developed to address some of the abovementioned issues.

DEVELOPMENT AND EARLY CLINICAL STUDIES

It was primarily recognized in animals that the consumption of molybdenum causes copper deficiency. This copper deficiency could be even further exacerbated by dietary sulfate supplements [7, 8]. The resultant copper deficiency was explained by the formation of thiomolybdates with a high affinity to copper that form insoluble copper-thiomolybdate complexes. The synthetic compound ammonium TTM was first developed to treat copper-overload states in animals [9]. Sheep are highly susceptible to copper toxicosis caused by copper-containing food. TTM was able to prevent this copper-induced damage in animals and was subsequently investigated as a potential therapy option for WD.

The first clinical experiences with ammonium TTM in WD were already published by Walshe in 1986. Brewer published several studies using TTM in neurological WD a few years later [10–13]. These studies involved a limited exposure to TTM by the patient for a period of only few weeks and demonstrated a rapid control of NCC by ammonium TTM. Furthermore, there occurred almost no neurological deterioration, even with rapid escalation of doses up to 400 mg/day. However, adverse events (bone marrow suppression and aminotransferase elevation) were reported in association with rapid dose escalation of TTM.

In a randomized, double-blind study, trientine was compared with ammonium TTM in 48 WD patients with neurological symptoms. Of the 25 patients treated with TTM only, one experienced neurological worsening compared with six of 23 patients in the trientine arm ($P < 0.05$). Neurological improvement occurred in the treated patients over a period of years in both treatment groups. In this study, reversible anemia and transaminase elevations were reported in patients treated with TTM.

Another study focused on the control of free copper and compared efficiency of ammonium TTM with trientine [14]. TTM showed superior control of free copper levels compared with trientine in three different trials, but in the last study with a dose-reduced regimen of 60–120 mg/day ammonium TTM, this effect was less prominent than in studies using higher

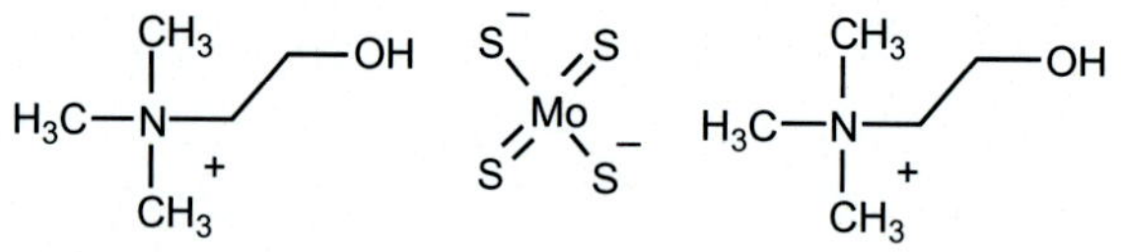

FIG. 1 Structural formula of bis-choline tetrathiomolybdate.

TTM dosing. Furthermore, ammonium TTM was found to be relatively unstable (highly susceptible to oxidation and inactivation) for use under routine clinical conditions.

These side effects, its limited pharmacological stability, and study data with only limited evidence meeting robust treatment end points hampered regulatory approval of ammonium TTM for the treatment of WD by the US Food and Drug Administration.

BIS-CHOLINE TTM (WTX-101)

Regarding the abovementioned limitations of earlier experiences with ammonium TTM in WD, a newly pharmacologically stabilized form of TTM, bis-choline TTM (WTX-101), is currently under clinical investigation (Fig. 1). Ammonium TTM and bis-choline TTM are pharmacologically equivalent. Bis-choline TTM has an improved stability than the ammonium TTM, with a superior safety profile. Furthermore, bis-choline TTM exhibits direct intracellular activity in hepatocytes where it forms copper-molybdenum multimetallic clusters and facilitates biliary copper excretion [15].

PHARMACOKINETICS

The pharmacokinetics of bis-choline TTM has been investigated under normal and copper-overload conditions in an animal study. Distribution and excretion of WTX-101 were observed for 168h in control and a WD rat model (LEC) rats [16, 17]. WTX-101 was administered intravenously at 0.75 and 1.5mg/kg, and excretion of molybdenum was assessed in different body fluids and tissues in the rats. Whereas almost full excretion of WTX-101 was achieved in control rats within 168h, WTX-101 was only partially excreted in the LEC WD rats. In control rats, WTX-101 was cleared from the body mainly due to a rapid renal excretion within the first 24h. In LEC rats, WTX-101 was only excreted partially within 168h, with only 29% excretion in the urine and 16% in bile fluid. These two phases occur at different time points. Renal clearance was mainly achieved within 48h, whereas fecal excretion took place at later time frame. In LEC rats, the proportion of renal to fecal molybdenum excretion was 6:4 compared with a ratio of 9:1 in control rats. The higher amount of excreted WTX-101 in control rats is consistent with higher biotransformation, in contrast to copper-overload conditions. Subsequently, in LEC rats less WTX-101 was converted into molybdate, resulting in reduced renal clearance and higher amount of fecal excretion, whereas in controls, only a limited amount of copper is available for chelating with WTX-101 and a larger fraction of WTX-101 is converted into molybdate, which is the terminal degradation product. Molybdate does not bind to copper and is cleared by renal excretion. Under copper-overload conditions, excess copper was removed from metallothionein in hepatocytes by WTX-101, and TTM-bound copper was formed in the circulation (Fig. 2). In the circulation, TTM-Cu binds to albumin, forming a highly stable copper-TTM-albumin tripartite complex. This complex is excreted via the biliary system into feces. In contrast to complexes formed with other chelators, copper trapped within the tripartite complex with WTX-101 cannot redistribute into the central nervous system, which may be the mechanism preventing neurological worsening that occurs with other chelating agents [18]. Additionally, significant differences with respect to tissue distribution between control and LEC rats were observed. Higher amounts of WTX-101 were found in the liver and kidney of LEC rats compared with controls. This is in line with the proposed mechanism of WTX-101 where copper is removed especially from hepatic metallothionein and stored in an insoluble and highly stable tripartite complex with subsequently less renal clearance and more biliary and fecal excretion. WTX-101 that does not bind to copper is released via the kidney as molybdate. This effect allows noninvasive drug monitoring of WTX-101. Preliminary results in WD patients indicate a saturation of WTX-101 at about 30mg/day with equal urinary molybdenum excretion with WTX-101 dosing of 30 or 60mg/day. Plasma copper content increased proportional to plasma molybdenum levels after administration of WTX-101 that is in accordance with the formation of the tripartite complex. Further pharmacokinetic analysis in WD patients is part of current clinical trials.

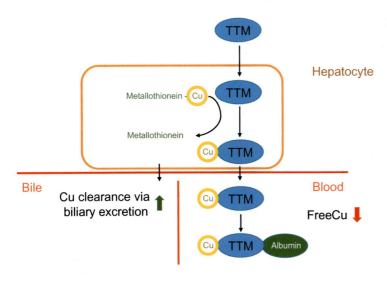

FIG. 2 Mode of action of tetrathiomolybdate.

MODE OF ACTION

WTX-101 has a high affinity for copper that is ~10,000-fold higher than that of penicillamine and trientine. Unlike penicillamine and trientine, WTX-101 does not appear to bind other divalent cations. The direct action of WTX-101 is thought to be due to the formation of a stable inert tripartite complex via four sulfur groups. The affinity of WTX-101 to copper is comparable with the major copper storage protein metallothionein. Thus, WTX-101 can remove excess copper even from intrahepatic metallothionein storage and dispose it into the bile. Furthermore, in the circulation, WTX-101 forms highly stable complexes with copper and albumin (TTM-Cu-albumin tripartite complex). Formation of this tripartite complex is the main mode of action of bis-choline TTM and differs from other chelators, which bind to copper only and do not form a protein complex. Accordingly, NCC in the blood circulation is caught in a TTM-Cu-albumin tripartite complex. This high-affinity and high-molecular-weight complex prevents copper-induced oxidative injury and additionally enhances biliary copper excretion. Usually, NCC is calculated by subtracting ceruloplasmin-bound copper from total serum copper. In the case of WTX-101, an additional step for NCC calculation has to be included, to subtract the copper bound by the TTM-Cu-albumin tripartite complex. Copper bound tightly to that complex does not belong to the free copper pool, available for cellular uptake. For that reason, NCC calculation must be adapted for this WTX-101 specific effect.

Another, less prominent effect of WTX-101 has been identified when the drug is administered together with food. Like other chelators, WTX-101 also binds copper contained in food and reduces its absorption. This mechanism was discovered in early clinical trials with ammonium TTM. In current studies, bis-choline TTM was administered under fasting conditions, to take advantage of its main mode of action and facilitate a more reliable dosing.

Beside decoppering effects, TTM seems to induce several other beneficial properties, like antifibrotic or antiinflammatory effects. Several studies indicate the inhibition of copper-dependent cytokines (e.g., TNF-α and TGF-β) by TTM, promoting antiinflammatory effects [9, 19, 20]. These properties of TTM have also been investigated in malignant or inflammatory disease models [21]. TTM was able to reduce significantly hepatic fibrosis in an animal study [22].

EFFICACY OF BIS-CHOLINE TTM

Safety and efficacy of bis-choline TTM monotherapy has been assessed in a "proof of concept" 24-week open-label single-arm phase II study (NCT02273596) [23]. The patients of that cohort had a significant neurological disease burden but limited hepatic impairment. Dosing of WTX-101 ranged between 15 and 60 mg/day adjusted to baseline NCC concentrations. Dosing was response-guided adapted individually, with most patients receiving 30 mg/day. The primary end point was the change in baseline NCC concentrations at 24 weeks. Secondary end points included change from baseline for neurological disability and status measured as Unified Wilson's Disease Rating Scale (UWDRS) part II and III and quality of life measured with the EuroQOL 5 dimensions visual analog scale (EQ VAS). Treatment of WTX-101 was associated with rapid improvement of NCC, with normalization within 12 weeks in the majority of patients. After 24 weeks, NCC levels were within normal range in 57% of patients; in further 14%, NCC levels were reduced by more than 25% from baseline. The average reduction of NCC after WTX-101 administration in this trial was more than 70%. Besides reduction of NCC, also significant improvement of the neurological status and grade of disease-related disability was observed. Both mean

UWDRS Part II scores and mean UWDRS Part III neurological examination scores improved significantly during the 24-week study period (6.6 at baseline to 4.1 at week 24 ($P<0.001$) and 22.8 at baseline to 16.6 at week 24 ($P<0.0001$), respectively). According to the significantly clinical improvements, mean EQ VAS scores increase ($+9.2\pm2.9$, $P=0.0024$). While no drug-related paradoxical neurological worsening occurred within the first 12 weeks after treatment initiation, liver function reflected by INR, albumin, and MELD score was stable throughout the study period. Taken together, WTX-101 displays a unique mechanism of action, resulting in rapid control of free copper, and detoxifies circulating copper in highly stable tripartite complexes. This makes the compound highly attractive, especially for WD patients with neurological symptoms who are at risk for drug-related paradoxical neurological deterioration.

SAFETY AND TOLERABILITY OF BIS-CHOLINE TTM

In the current trial, AEs were observed in 17 out of 28 patients. In seven patients, 11 serious adverse events (SAEs) were reported. The most frequent AE was asymptomatic reversible elevations in serum liver tests. It was reported in 11 patients treated with WTX-101 at doses of ≥ 30 mg/day and usually occurred after 4–10 weeks of treatment. In six patients, WTX-101 treatment was discontinued in three patients due to alanine aminotransferase (ALT) elevations. In one patient treated with 120 mg/day, a SAE of elevated ALT occurred. For that reason, the protocol was subsequently changed, and dosing was limited to a maximum of 60 mg/day. Liver test abnormalities were reversible in all patients and not associated with notable affection of other liver function parameters. In four patients with preexisting neurological/psychiatric disease manifestations, six psychiatric SAEs were reported. All SAEs were assessed as unlikely to be related to WTX-101 treatment. Within the 24-week study period, no paradoxical neurological worsening attributable to therapy with bis-choline TTM became overt. Preliminary results from 22 patients who completed the 24-week study phase and opted to continue in an extension study showed maintenance of improved neurological function and stable liver function until 72 weeks of treatment. Between 24 and 72 weeks, mean NCC remained stable. Patients had sustained improvements in mean UWDRS disability and neurological status scores at week 72. Mean INR, albumin, ALT levels, and MELD score improved or remained unchanged during the extension phase. Besides its beneficial pharmacological features, the option of a once-daily dosing regimen could further improve adherence to therapy and improve subsequential outcomes. First clinical experiences with WTX-101 argue for this compound as a new opportunity for a highly reliable initial therapy and long-term maintenance of WD patients.

CONCLUSION

In contrast to chelators that nonspecifically bind free copper in the blood circulation, bis-choline TTM displays a direct hepatic decoppering effect due to its high affinity for copper that removes copper from metallothionein in hepatocytes. Its main modes of action are the formation of a stable TTM-Cu-tripartite complex that makes free copper unavailable for cellular uptake and promotion of biliary copper excretion, resulting in a negative copper balance. First clinical experiences with bis-choline TTM demonstrate a rapid control of free copper with superior decoppering capabilities. The fast control of copper balance is associated with reduced disability, improved neurological status, stable liver function, and the opportunity for a simple once-daily dosing regimen. Taken together, currently available clinical data argue for a potential future role of bis-choline TTM as a new first-line therapy in WD. An ongoing phase III study of WTX-101 (NCT03403205) will add further evidence to the efficacy and safety of this promising therapy in WD.

REFERENCES

[1] Bandmann O, Weiss KH, Kaler SG. Wilson's disease and other neurological copper disorders. Lancet Neurol 2015;14:103–13.
[2] European Association for Study of Liver. EASL Clinical Practice Guidelines: Wilson's disease. J Hepatol 2012;56:671–85.
[3] Weiss KH, Thurik F, Gotthardt DN, Schafer M, Teufel U, Wiegand F, et al. Efficacy and safety of oral chelators in treatment of patients with Wilson disease. Clin Gastroenterol Hepatol 2013;11:1028–35. e1–2.
[4] Weiss KH, Gotthardt DN, Klemm D, Merle U, Ferenci-Foerster D, Schaefer M, et al. Zinc monotherapy is not as effective as chelating agents in treatment of Wilson disease. Gastroenterology 2011;140:1189–98. e1.
[5] Ferenci P, Steindl-Munda P, Vogel W, Jessner W, Gschwantler M, Stauber R, et al. Diagnostic value of quantitative hepatic copper determination in patients with Wilson's disease. Clin Gastroenterol Hepatol 2005;3:811–8.
[6] Squitti R, Siotto M, Cassetta E, El Idrissi IG, Colabufo NA. Measurements of serum non-ceruloplasmin copper by a direct fluorescent method specific to Cu(II). Clin Chem Lab Med 2017;55:1360–7.
[7] Suttle NF, Field AC. Effect of intake of copper, molybdenum and sulphate on copper metabolism in sheep. II. Copper status of the newborn lamb. J Comp Pathol 1968;78:363–70.

[8] Suttle NF, Field AC. Effect of intake of copper, molybdenum and sulphate on copper metabolism in sheep. I. Clinical condition and distribution of copper in blood of the pregnant ewe. J Comp Pathol 1968;78:351–62.

[9] Gooneratne SR, Howell JM, Gawthorne JM. Intravenous administration of thiomolybdate for the prevention and treatment of chronic copper poisoning in sheep. Br J Nutr 1981;46:457–67.

[10] Brewer GJ, Askari F, Lorincz MT, Carlson M, Schilsky M, Kluin KJ, et al. Treatment of Wilson disease with ammonium tetrathiomolybdate: IV. Comparison of tetrathiomolybdate and trientine in a double-blind study of treatment of the neurologic presentation of Wilson disease. Arch Neurol 2006;63:521–7.

[11] Brewer GJ, Hedera P, Kluin KJ, Carlson M, Askari F, Dick RB, et al. Treatment of Wilson disease with ammonium tetrathiomolybdate: III. Initial therapy in a total of 55 neurologically affected patients and follow-up with zinc therapy. Arch Neurol 2003;60:379–85.

[12] Brewer GJ, Johnson V, Dick RD, Kluin KJ, Fink JK, Brunberg JA. Treatment of Wilson disease with ammonium tetrathiomolybdate. II. Initial therapy in 33 neurologically affected patients and follow-up with zinc therapy. Arch Neurol 1996;53:1017–25.

[13] Brewer GJ, Dick RD, Johnson V, Wang Y, Yuzbasiyan-Gurkan V, Kluin K, et al. Treatment of Wilson's disease with ammonium tetrathiomolybdate. I. Initial therapy in 17 neurologically affected patients. Arch Neurol 1994;51:545–54.

[14] Brewer GJ, Askari F, Dick RB, Sitterly J, Fink JK, Carlson M, et al. Treatment of Wilson's disease with tetrathiomolybdate: V. Control of free copper by tetrathiomolybdate and a comparison with trientine. Transl Res 2009;154:70–7.

[15] Weiss KH, Czlonkowska A, Hedera P, Ferenci P. WTX101—an investigational drug for the treatment of Wilson disease. Expert Opin Investig Drugs 2018;27:561–7.

[16] Suzuki KT, Yamamoto K, Ogra Y, Kanno S, Aoki Y. Mechanisms for removal of copper from metallothionein by tetrathiomolybdate. J Inorg Biochem 1994;54:157–65.

[17] Ogra Y, Suzuki KT. Removal and efflux of copper from Cu-metallothionein as Cu/tetrathiomolybdate complex in LEC rats. Res Commun Mol Pathol Pharmacol 1995;88:196–204.

[18] Brewer GJ, Dick RD, Johnson VD, Fink JK, Kluin KJ, Daniels S. Treatment of Wilson's disease with zinc XVI: treatment during the pediatric years. J Lab Clin Med 2001;137:191–8.

[19] Brewer GJ. Tetrathiomolybdate anticopper therapy for Wilson's disease inhibits angiogenesis, fibrosis and inflammation. J Cell Mol Med 2003;7:11–20.

[20] Wei H, Frei B, Beckman JS, Zhang WJ. Copper chelation by tetrathiomolybdate inhibits lipopolysaccharide-induced inflammatory responses in vivo. Am J Physiol Heart Circ Physiol 2011;301:H712–20.

[21] Brewer GJ. Zinc and tetrathiomolybdate for the treatment of Wilson's disease and the potential efficacy of anticopper therapy in a wide variety of diseases. Metallomics 2009;1:199–206.

[22] Hou G, Dick R, Brewer GJ. Improvement in dissolution of liver fibrosis in an animal model by tetrathiomolybdate. Exp Biol Med (Maywood) 2009;234:662–5.

[23] Weiss KH, Askari FK, Czlonkowska A, Ferenci P, Bronstein JM, Bega D, et al. Bis-choline tetrathiomolybdate in patients with Wilson's disease: an open-label, multicentre, phase 2 study. Lancet Gastroenterol Hepatol 2017;2:869–76.

Chapter 19

Zinc Therapy of Wilson Disease

Roderick H.J. Houwen
Department of Pediatric Gastroenterology, University Medical Center Utrecht, Utrecht, The Netherlands

INTRODUCTION

Zinc was introduced in 1961 as an alternative for chelators in the treatment of Wilson disease by the Dutch neurologist Schouwink. He had learned from the literature that high dosages of zinc could be used to treat copper intoxication in sheep and therefore started using this medication in two patients [1]. When Hoogenraad, another Dutch neurologist, reexamined one of these patients after 14 years of continuous zinc therapy, it turned out that he was symptom-free and that the Kayser-Fleischer Rings that were present earlier had disappeared [2]. Given this observation and the lack of side effects, he started to treat more patients with zinc and communicated the results in several articles [3, 4]. The possible advantages of zinc therapy over chelators were soon picked up by several other groups, notably Brewer from the United States, Czlonkowska from Poland, and subsequently others. Consequently, a vast amount of data are now available, which allows us to give a good overview of zinc therapy in Wilson disease.

MECHANISM

Zinc induces a negative copper balance, whose effect was already described by Schouwink [1], and confirmed subsequently in larger studies [5, 6]. The negative copper balance under zinc therapy is a consequence of suppression of intestinal copper uptake, as can be visualized by ^{64}Cu [4]. It is directly related to increased metallothionein concentration of the enterocytes, as induced by zinc [7]. It is postulated that with high metallothionein levels, copper is sequestered in the enterocyte, making it unavailable for transfer into the portal circulation. Supposedly, copper is subsequently lost with the feces when the enterocytes at the end of their life cycle are sloughed into the lumen, and the copper cannot be reabsorbed as the copper-metallothionein complex is resistant to protease degradation [7].

The level of intestinal metallothionein induction is determined by body zinc status, which is reflected in urinary zinc level, whereby an excretion above 1800 μg/day correlates with an almost total block in copper uptake [7]. Zinc also induces metallothionein in the hepatocytes, and such may detoxify liver copper [8, 9].

DOSAGE

The standard prescription for zinc in adults is based on the dose found to be effective in the early papers from the Netherlands and the United States and amounts to 50 mg elemental zinc three times daily, taken separated from food by an hour [2–6]. Dosage in children aged 6–15 years is three times daily 25 mg elemental zinc, while the optimal dose for children below 5 years is recommended as 25 mg elemental zinc twice daily [10]. Children weighing >50 kg and adolescents should take the adult dose.

Several formulations of zinc are available, but zinc acetate is now mostly used, which is available in Europe as Wilzin and in the United States as Galzin.

EFFICACY

The use of zinc in Wilson disease has generated much controversy, initially mainly resulting in publications either fervently in favor or opposed to using this drug [11, 12], which opinions were generally only partially based on hard data, and sometimes not at all. The first attempt at systematically describing the efficacy of zinc was published in 2009 [13], methodically evaluating patients on zinc, as described in six reports, and comparing their outcome with patients on D-penicillamine, which was described in nine reports. This systematic review categorized patients into three main groups: those presenting with only mildly elevated transaminases or identified by family screening ("presymptomatic"), those with mainly hepatic symptoms, and those with mainly neurological symptoms. This initial review concluded that both zinc and D-penicillamine are equally effective when used in presymptomatic patients, that is, in both treatment groups, all patients remained asymptomatic. In neurological patients, zinc seemed to be more effective, that is, only 10% of patients deteriorated on zinc, as opposed to 19% on D-penicillamine. However, only 10 patients could be evaluated in the zinc group, as opposed to 72 in the D-penicillamine group. For patients with hepatic symptoms, D-penicillamine was clearly more effective than zinc, as 44% of the patients on zinc deteriorated, as opposed to 26% on D-penicillamine. This conclusion too was based on rather small numbers for the zinc group. In addition, there might have been a selection bias, as zinc might have been used in patients with less severe symptoms.

For the current chapter, we have updated this earlier analysis by once again systematically searching the MEDLINE, EMBASE, and Cochrane databases, starting from 1 January 2007 to 1 January 2017. This resulted in an additional eight papers describing the outcome on zinc as an initial therapy in 183 patients with Wilson disease [14–21]. The majority of these patients were presymptomatic; the others had neurological, hepatic, or mixed symptoms. Four of the above papers give sufficient detail to analyze the effect of zinc when used in presymptomatic patients (Table 1). In general, these four papers support our earlier conclusion that zinc is effective in presymptomatic patients, as transaminases generally improve substantially. Nevertheless, it may take over a year for the transaminases to fully normalize, while in some patients, this may never occur. Nonresponse, defined as a sustained increase of transaminases, or a failure of these enzymes to improve on therapy is generally caused by noncompliance, as evidenced by low urinary zinc [18, 19], although sometimes no clear cause could be identified. A similar phenomenon, that is, nonresponse to zinc without a relation to urinary zinc in some cases, was described by Weiss et al. in a mixed population of symptomatic and nonsymptomatic patients [20]. It seems sensible to switch such patients to a chelator. However, persistent abnormal transaminases in presymptomatic patients while on chelator is also present [18, 22] and is indeed more common on a chelator: 28% versus 12% on zinc (NS) [18], so zinc seems to be preferred in this group.

The early reports that showed a good outcome on zinc in symptomatic patients especially included neurological cases and only few with hepatic symptoms, which does not surprise, because they were generated from neurological clinics in Utrecht and Warsaw, respectively [3, 23]. A prospective paper from 1996, describing patients from the Warsaw clinic, with only 10% having hepatic symptoms, indeed showed improvement or normalization of symptoms in 23/29 (79%) patients on zinc as opposed to 13/18 (68%) patients on penicillamine (NS) [23]. A more recent paper from the same clinic showed deterioration or death in 5/35 (14%) of neurological patients on penicillamine, while this adverse outcome was recorded in 2/21 (9%) of neurological patients on zinc [21]. In addition, another prospective study from this clinic suggested that early neurological deterioration is more common on a chelator (12/42; 29%) than on zinc (4/28; 14%), although the difference was once again not significant [24]. Other recent papers also described a good effect of zinc in patients with neurological symptoms [14]. These data are in concordance with current guidelines that indicate that zinc is a viable option in patients with predominantly neurological symptoms [25, 26].

However, for hepatic patients, both the meta-analysis and the guidelines indicate that zinc is not a good option. Two more recent papers corroborate this opinion. Linn et al. describe deterioration of several patients with hepatic symptoms on zinc and a less efficient decoppering in this patient group as measured by 24h urinary copper excretion [14]. In a large cohort, consisting mainly of patients with hepatic symptoms, Weiss et al. describe that treatment failure, defined as

TABLE 1 Effect of Zinc in Presymptomatic Patients With Wilson Disease

Author (Reference)	Total (FU ≥ 1 year)	Age	AST and ALT Improved (Normalized)		
			At 1 year	At 2 years	At EOFU
Abuduxikuer [17]	N = 30 (12)	< 18 years	12/12 (7/12)	4/4 (3/4)	
Ranucci [18]	N = 15 (15)	< 16 years			13/15[a]
Santiago [19]	N = 9 (9)	< 20 years	8/9 (4/9)	5/8 (4/8)	
Mizuochi [16]	N = 4 (4)	< 7 years	4/4 (0/4)	3/3 (1/3)	

[a] End of follow-up at a median of 9.3 years (range 1.6–19.8 years).

insufficient response of transaminases, was more frequent on zinc than on a chelator [20]. Nevertheless, another paper showed equal outcomes for hepatic patients on zinc and penicillamine [21]. Patients in the zinc group had less severe symptoms however, which might have biased this outcome.

PREGNANCY

Treatment of Wilson disease should not be discontinued during pregnancy, as patients may deteriorate when treatment is temporarily stopped [25–27]. In addition, pregnant Wilson disease patients without therapy have a higher rate of spontaneous abortion [27]. However, both penicillamine and trientine are teratogenic in animals [28], and several patients, whose mothers had used penicillamine during pregnancy, have been reported with (transient) connective tissue disease [29, 30]. More serious anomalies have also been reported, although it is unclear whether these were due to direct toxicity of the drug or to copper and/or zinc deficiency [30]. Therefore, the chelator dose is often reduced during (part of) the pregnancy [25–27, 29]. It might however be better to use zinc, as this is not teratogenic in animals, and almost 30 pregnancies in patients with Wilson disease using zinc have now been reported without an anomaly that could possibly be related to zinc [28, 31]. Nevertheless, with this drug too, copper deficiency could ensue, resulting in birth defects. Therefore, when using zinc in pregnancy, urinary copper excretion should be carefully monitored, with dose adjustments when necessary [28].

SIDE EFFECTS

In the meta-analysis of all relevant papers up to 2008, almost 25% (50/205) of the patients on D-penicillamine had severe side effects, which necessitated withdrawal of penicillamine therapy in 26/205 (12.5%) [13]. On zinc therapy, however, side effects were significantly less frequent (28/224; 12.5%), while in only 2/224 (0.9%) patients, the therapy had to be discontinued. More recent papers show similar results. For example, Czlonkowska describe side effects of zinc in 2/72 (3%) of patients, while 11/71 (15%) of patients on penicillamine had side effects, which were also more severe, that is, bone marrow depression and proteinuria [21]. Adverse effects were also more common on a chelator than on zinc in the large cohort described by Weiss [20]. Gastrointestinal side effects are especially common on zinc sulfate and more so in children than in adults [10, 15, 23, 32]. In a group of 50 Polish patients, all below 18 years at diagnosis and on zinc sulfate, as much as 21 out of 50 (42%) had abdominal pain and/or nausea, while gastritis with ulceration or erosions were seen in the 7 who had an endoscopy [32]. These symptoms were treated by adding a PPI, an alternative dosing scheme for zinc sulfate, or a switch to penicillamine or zinc acetate, which was effective in alleviating gastrointestinal symptoms [32]. Indeed, gastrointestinal side effects are less frequently seen in patients on zinc acetate and if present are mild, so this formulation seems to be preferred [10, 15, 16].

OVERTREATMENT

Although Wilson disease is characterized by insufficient copper excretion and hence copper overload, intensive decoppering treatment can result in symptomatic copper deficiency. This has been described for patients on zinc monotherapy or on zinc in combination with a chelator. The resulting symptoms are reminiscent of patients with very mild mutations in ATP7A [33] and are characterized by leucopenia, especially neutropenia, and sometimes anemia and/or myeloneuropathy [34]. The latter may not always be reversible. To prevent this complication, it is essential to monitor hematologic parameters during the treatment of Wilson disease patients, as this may be the first sign of copper deficiency, and the 24 h copper excretion, which should not be below 30 µg/24 h [35]. If available and reliable, measurement of free serum copper might also be of value.

COMBINATION THERAPY

Combination therapy with zinc and a chelator, temporally separated, has been advocated as an option for patients with severe decompensated hepatic disease [36, 37]. While this may be correct for this specific situation, this regimen would be hard to maintain outside a clinical setting as compliance is already a problem with regular therapy, and adding a second drug, which also has be widely spaced in time to avoid interaction, seems problematic. A recent meta-analysis indeed suggests that either drug alone was superior to combination therapy [38].

WHERE DO WE GO FROM HERE

After its first use in the treatment of Wilson disease patients, now >50 years ago, zinc has earned its place in the treatment of presymptomatic patients and seems to be a good alternative for penicillamine in neurological patients too. Yet, some questions remain, such as the percentage of long-term normalization of transaminases in presymptomatic patients on zinc versus penicillamine. This can only be resolved when detailed patient series are published, with all data added, if necessary as a supplementary file. The article by Abuduxikuer [17] is a good example in this respect. Likewise, in addition to the articles by Czlonkowska [21, 23], more contemporary series on the outcome of different treatment modalities in neurological patients should be published.

No substantial dataset is available at present on treatment of presymptomatic patients who were diagnosed by the detection of two mutations in *ATP7B*. Many will argue that such patients should be treated as early as possible, although for many mutations, the pathogenicity is unclear [39] and unwarranted decoppering medication might result in copper deficiency [34]. An alternative might be to wait until urinary copper excretion rises, especially as Wilson disease will generally not become symptomatic before the age of 5 and never before the age of 3 years [40]. Also, *ATP7B* mutations seem to be more prevalent than thought earlier [39] and might never result in symptomatic copper overload or only at old age [41], for example, due to the effect of modifier genes and/or low copper intake. Therefore, case series devoted to patients diagnosed based on DNA sequencing of *ATP7B* are also urgently needed.

REFERENCES

[1] Schouwink G. De hepatocerebrale degeneratie. Met een onderzoek van de zinkstofwisseling. Amsterdam: Academisch Proefschrift; 1961.

[2] Hoogenraad TU, Koevoet R, de Ruyter Korver EGWM. Oral zinc sulphate as long-term treatment in Wilson's disease (hepatolenticular degeneration). Eur Neurol 1979;18:205–11.

[3] Hoogenraad TU, van Hattum J, van den Hamer CJA. Management of Wilson's disease with zinc sulphate. Experience in a series of 27 patients. J Neurol Sci 1987;77:137–46.

[4] Hoogenraad TU, van den Hamer CJA. 3 Years of continuous oral zinc therapy in 4 patients with Wilson's disease. Acta Neurol Scand 1983;67:356–64.

[5] Brewer GJ, Hill GM, Prasad AS, Cossack ZT, Rabbani P. Oral zinc therapy for Wilson's disease. Ann Intern Med 1983;99:3–20.

[6] Hill GM, Brewer GJ, Prasad AS, Hydrick CR, Hartmann DE. Treatment of Wilson's disease with zinc. I. Oral zinc therapy regimens. Hepatology 1987;7:522–8.

[7] Yusbasiyan-Gurkan V, Grider A, Nostrant T, Cousins RJ, Brewer GJ. Treatment of Wilson's disease with zinc. X. Intestinal metallothionein induction. J Lab Clin Med 1992;120:380–6.

[8] Chandok G, Schmitt N, Sauer V, Aggarwal A, Bhatt M, Schmidt HHJ. The effect of zinc and D-penicillamine in a stable human hepatome *ATP7B* knockout cell line. PLoS One 2014;9.

[9] Schilsky ML, Blank RR, Czaja MJ, Zern MA, Scheinberg IH, et al. Hepatocellular copper toxicity and its attenuation by zinc. J Clin Invest 1989;84:1562–8.

[10] Brewer GJ, Dick RD, Johnson VD, Fink JK, Kluin KJ, Daniels S. Treatment of Wilson's disease with zinc XVI: treatment during the pediatric years. J Lab Clin Med 2001;137:191–8.

[11] Lipsky MA, Gollan JL. Treatment of Wilson's disease: in D-penicillamine we trust—what about zinc? Hepatology 1987;7:593–5.

[12] Brewer GJ. Penicillamine should not be used as initial therapy in Wilson's disease. Mov Disord 1999;14:551–4.

[13] Wiggelinkhuizen M, Tilanus MEC, Bollen CW, Houwen RHJ. Systematic review: clinical efficacy of chelator agents and zinc in the initial treatment of Wilson disease. Aliment Pharmacol Ther 2009;29:946–58.

[14] Linn FHH, Houwen RHJ, Hattum van J, et al. Long-term exclusive zinc monotherapy in symptomatic Wilson disease; experience in 17 patients. Hepatology 2009;50:1442–52.

[15] Bruha R, Marecek Z, Pospisilova L, et al. Long term follow-up of Wilson disease: natural history, treatment, mutation analysis and phenotypic correlation. Liver Int 2011;31:83–91.

[16] Mizuochi T, Kimura A, Shimizu N, et al. Zinc monotherapy from time of diagnosis for young pediatric patients with presymptomatic Wilson disease. J Pediatr Gastroenterol Nutr 2011;53:365–7.

[17] Abuduxikuer K, Wang JS. Zinc monotherapy in pre-symptomatic Chinese children with Wilson disease: a single center, retrospective study. PLoS One 2014;9.

[18] Ranucci G, Di Dato F, Spagnuolo MI, et al. Zinc monotherapy is effective in Wilson's disease patients with mild liver disease diagnosed in childhood: a retrospective study. Orphanet J Rare Dis 2014;9:41.

[19] Santiago R, Gottrand F, Debray D, et al. Zinc therapy for Wilson disease in children in French pediatric centers. J Pediatr Gastroenterol Nutr 2015;61:613–8.

[20] Weiss KH, Gotthardt DN, Klemm D, et al. Zinc monotherapy is not as effective as chelating agents in treatment of Wilson disease. Gastroenterology 2011;140:1189–98.

[21] Czlonkowska A, Litwin T, Karlinski M, Dziezyc K, Chabik G, Czerska M. D-Penicillamine versus zinc sulfate as first-line therapy for Wilson's disease. Eur J Neurol 2014;21:599–606.
[22] Pena-Quintana L, Garcia-Luzardo MR, Garcia-Villarreal L, et al. Manifestations and evolution of Wilson disease in pediatric patients carrying ATP7B mutation L708P. J Pediatr Gastroenterol Nutr 2012;54:48–54.
[23] Czlonkowska A, Gajda J, Rodo M. Effects of long-term treatment in Wilson's disease with D-penicillamine and zinc sulphate. J Neurol 1996;243:269–73.
[24] Litwin T, Dziezyc K, Karlinski M, et al. Early neurological worsening in patients with Wilson's disease. J Neurol Sci 2015;355:162–7.
[25] European Association for Study of Liver. EASL Clinical Practice Guidelines: Wilson's disease. J Hepatol 2012;58:671–85.
[26] Roberts EA, Schilsky ML. American Association for Study of Liver Diseases (AASLD). Diagnosis and treatment of Wilson disease: an update. Hepatology 2008;47:2089–111.
[27] Pfeiffenberger J, Beinhardt S, Gotthardt DN, et al. Pregnancy in Wilson disease—management and outcome. Hepatology 2018;67:1261–9.
[28] Brewer GJ, Johnson VD, Dick RD, Hedera P, Fink JK, Kluin KJ. Treatment of Wilson's disease with zinc. XVII: treatment during pregnancy. Hepatology 2000;31:364–70.
[29] Sternlieb I. Wilsons's disease and pregnancy. Hepatology 2000;31:531–2.
[30] Pinter R, Hogge WA, McPherson E. Infant with severe penicillamine embryopathy born to a woman with Wilson disease. Am J Med Genet A 2014;128:294–8.
[31] Lee HJ, Seong WJ, Hong SY, Bae JY. Successful pregnancy outcome in a Korean patients with symptomatic Wilson's disease. Obset Gynecol Sci 2015;58:409–13.
[32] Wiernicka A, Janczyk W, Dadalski M, Avsar Y, Schmidt H, Socha P. Gastrointestinal side effects in children with Wilson's disease treated with zinc sulphate. World J Gastroenterol 2013;19:4356–62.
[33] Kennerson ML, Nicholson GA, Kaler SG, et al. Missense mutations in the copper transporter gene ATP7A cause X-linked distal hereditary motor neuropathy. Am J Hum Genet 2010;86:343–52.
[34] Dziezyc K, Litwin T, Sobanska A, Czlonkowska A. Symptomatic copper deficiency in three Wilson's disease patients treated with zinc sulphate. Neurol Neurochir Pol 2014;48:214–8.
[35] Socha P, Janczyk W, Dhawan A, et al. Wilson's disease in children: a position paper by the hepatology committee of the European Society for Paediatric Gastroenterology, Hepatology and Nutrition. J Ped Gastroenterol Nutr 2018;66:334–44.
[36] Askari FK, Greenson J, Dick RD, Johnson VD, Brewer GJ. Treatment of Wilson's disease with zinc. XVIII. Initial treatment of the hepatic decompensation presentation with trientine and zinc. J Lab Clin Med 2003;142:385–90.
[37] Santos Silva EE, Sarles J, Buts JP, Sokal EM. Successful medical treatment of severely decompensated Wilson disease. J Pediatr 1996;128:285–7.
[38] Chen JC, Chuang CH, Wang JD, Wang CW. Combination therapy using chelating agent and zinc for Wilson's disease. J Med Biol Eng 2015;35:697–708.
[39] Coffey A, Durkie M, Hague S, et al. A genetic study of Wilsons's disease in the United Kingdom. Brain 2013;136:1476–87.
[40] Wilson DC, Phillips J, Cox DW, Roberts EA. Severe hepatic Wilsons's disease in preschool aged children. J Pediatr 2000;137:719–22.
[41] Ala A, Borjigin J, Rochwarger A, Schilsky M. Wilson disease in septuagenarian siblings: raising the bar for diagnosis. Hepatology 2005;41:668–70.

Chapter 20

Symptomatic Treatment of Residual Neurological or Psychiatric Disease

Ana Vives-Rodriguez and Daphne Robakis
Department of Neurology, Yale School of Medicine, New Haven, CT, United States

The neurological manifestations of Wilson disease (WD) constitute a heterogeneous array of signs and symptoms that mainly originate from the basal ganglia but may also arise from the brainstem, cerebellar nuclei, and cortical and subcortical regions [1]. Due to its varied presentation, a high level of suspicion is required for a timely diagnosis of the disease [2].

Neurological symptoms are the dominant, initial manifestation of WD in 40%–70% of patients [3–7]. Patients that present initially with a neurological syndrome usually have a later onset (between 15 and 21 years of age) than patients with liver manifestations [3,8–10]. The psychiatric manifestations are thought to be underreported, but they have been described in 30%–64% of the patients and can occur prior to the diagnosis of WD [6,11].

Neurological improvement after decoppering therapy in WD is variable. The reported data are difficult to interpret, since most studies do not include a comprehensive neurological examination before and after chelating therapy that documents anatomical distribution and severity of symptoms. Nevertheless, some cohorts experience a 50%–75% improvement of the neurological symptoms after therapy [12]. Others, in contrast, have reported progressive deterioration or no therapeutic response in 50%–60% of patients with severe neurological disease [13,14]. Early diagnosis, prompt institution of treatment, and compliance with therapy appear to be the most important prognostic factors [13–15].

Initial worsening of neurological symptoms following the introduction of chelating therapy is a concern. This phenomenon has been described with both trientine and D-penicillamine but especially with D-penicillamine. The risk for neurological worsening after the initial treatment with D-penicillamine is about 10%–15% [8]. It usually occurs early after starting treatment with an average presentation at 2.3 months. Only 50% of those who deteriorate fully recover their previous baseline. Patients at higher risk of initial neurological deterioration appear to be those with neurological manifestations prior to the chelating therapy [15].

Decoppering therapy provides in most cases only a partial improvement of the neurological aspects of WD. Therapy for residual neurological manifestations is usually required. This chapter aims to review the therapeutic options available and the existing experience of their use in this specific population.

NEUROLOGICAL MANIFESTATIONS

The neurological manifestations of WD have been classically divided into three main types: a parkinsonian syndrome, a dystonic syndrome, and the so-called pseudosclerosis syndrome, which is characterized by ataxia, wing-beating tremor (present on abduction of the arms and flexion of the elbows), and dysarthria [2,8,16,17]. The above categorization is somewhat arbitrary, as individual patients often display symptoms that overlap categories, but it may be useful in guiding selection of symptomatic treatments. Besides these clinical syndromes, other common features documented in WD are postural and kinetic hand tremor and isolated dysarthria. Neurological signs that have been described with less frequency are pyramidal signs (mainly hyperreflexia and Babinski sign), chorea, athetosis, seizures, and myoclonus (Table 1).

PARKINSONISM

Parkinsonism is a clinical syndrome characterized by the combination of rigidity, resting tremor, bradykinesia, loss of postural reflexes, stooped posture, and freezing [16]. It is attributed to a dysfunction of the dopaminergic nigrostriatal projections that directly modulate movement. Dopaminergic presynaptic and postsynaptic deficits have been described in

TABLE 1 Neurological Manifestations of Wilson Disease

Features	Frequency Reported
Dysarthria	48%–91%
Tremor	39%–83%
Dystonia	11%–69%
Parkinsonism	38%–62%
Ataxia	28%–51%
Chorea	9%–16%
Pyramidal signs	16%
Athetosis	2%–14%
Myoclonus	3.3%
Seizures	2%–8%

Frequencies obtained from Refs. [2, 5, 6, 10, 12, 14, 15].

patients with neurological WD using SPECT imaging and have been attributed to a direct lesion of the substantia nigra and of the striatum itself [18].

Parkinsonism is present in about 38%–62% of WD cases. Masked face, bradykinesia, postural imbalance, and cogwheel rigidity are the most common parkinsonian features [8]. Symptoms and signs usually occur asymmetrically during the early stages of the disease. Resting tremor is rare, present in only 5% of patients in one series [19].

The medication that is most commonly used for parkinsonian features is levodopa. Levodopa is an amino acid precursor of dopamine that is actively transported into the brain and decarboxylated to dopamine. It replenishes the presynaptic dopaminergic deficiency, thereby enhancing dopaminergic nigrostriatal projections. It is prescribed as a combination of levodopa and carbidopa. Carbidopa is an inhibitor of aromatic amino acid decarboxylation that reduces the peripheral conversion of levodopa, thus improving tolerability. The high efficacy of levodopa in Parkinson disease has been strongly established since the 1960s. However, there are no reports that systematically assess the response of the parkinsonian features of WD to levodopa or dopamine agonists. Some observational reports have described the medication as ineffective or with minimal response [13,19]. The most likely reason for this poor response is that in WD, there are both a presynaptic and postsynaptic striatal neuron loss [18].

Nevertheless, in patients with significant parkinsonian signs, a trial of levodopa can be offered. A small dose of carbidopa/levodopa (25 mg/100 mg) three times a day can be initiated and slowly increased, at weekly intervals, to 1200 mg of levodopa per day to determine responsiveness [20]. The patient should be monitored for side effects, mainly nausea, orthostatic hypotension, hallucinations, and impulse control disorders.

DYSTONIA

Dystonia is a syndrome of sustained muscle contractions that causes twisting, repetitive movements, and abnormal postures [21]. It is usually classified by its distribution into focal (when it affects only a single area of the body), segmental (when it affects two adjacent body parts), multifocal (when it affects two nonadjacent body parts), or generalized (when it involves a segmental crural dystonia plus any other part of the body).

Dystonia has been described in 18%–65% of the patients with WD. It rarely occurs in isolation. Different patterns of dystonia have been described including generalized dystonia, multifocal, segmental, and bilateral foot dystonia [2,4,14,22].

Magnetic resonance imaging studies of patients with dystonic features have documented a significantly higher prevalence of putamen lesions as compared with patients without dystonia, suggesting a relationship between abnormalities in this brain region and dystonic postures in WD [4,23].

TREATMENT OPTIONS IN DYSTONIA

Medications that have been commonly used for the treatment of dystonia are botulinum toxin, anticholinergics, and baclofen.

Botulinum Toxin

Botulinum toxin is directly injected into the muscle to produce paralysis. It acts by cleaving the presynaptic soluble N-ethylmaleimide-sensitive factor attachment protein receptor (SNARE proteins) at the neuromuscular junction. These proteins are in charge of binding the acetylcholine vesicles to the cell membrane in order to release the acetylcholine to the neuromuscular synaptic cleft in a process called exocytosis. Only botulinum toxin types A and B have been approved for therapeutic purposes [24].

Botulinum toxin is the first line of treatment for focal dystonia because of its high efficacy and low rate of side effects. The therapeutic muscle paralysis usually starts within 2–3 days after the injection, with a peak effect at 2 weeks, and is completely reversible, requiring periodic injections every 3 months [16,24]. The target muscle selection and botulinum toxin dosage is highly individualized, usually varies over time, and depends on the distribution and severity of the dystonia.

In addition to its blocking effect at the neuromuscular junction, botulinum toxin directly injected to the salivary glands also impairs the cholinergic autonomic innervation [24]. Therefore, it is very helpful in controlling sialorrhea, a common symptom of patients with WD.

Oral Medications

Oral medications are recommended in patients with segmental and generalized dystonia where the total dose of botulinum toxin and the number of injections required would be too high. Nonetheless, the use of oral medication for dystonia is usually only partially effective, and it is highly limited by its systemic adverse effects. For this reason, a low starting dose and slow titration are commonly recommended [25]. There are no robust comparative studies between different oral antidystonic treatments, but anticholinergics are frequently used as first-line oral agents [26].

Anticholinergics

Trihexyphenidyl is the only anticholinergic with sufficient evidence supporting its use in dystonia [25,27]. Its mechanism of action is not completely clear. It is thought that a selective M1 muscarinic receptor antagonism offsets synaptic plasticity deficits in the striatum and provides dystonia relief [27]. Trihexyphenidyl should be started at a low dose of 1–2 mg daily and weekly titrated to a usual effective dose of 8–10 mg/day. Common side effects are dry mouth, constipation, blurred vision, cognitive impairment, and urinary retention. It is generally well tolerated in younger patients even at high doses, but its use in the elderly, especially in those with cognitive impairment or a history of retentive bladder, should be considered with caution.

Baclofen

Baclofen is a presynaptic gamma aminobutyric acid agonist that is effective for the treatment of dystonia due to its muscle relaxant effects. It should be started at a low dose of 10 mg daily and uptitrated slowly to an average dose of 60–120 mg daily [25,26]. Its main side effects are drowsiness, nausea, hypotonia, and dizziness. To minimize withdrawal symptoms, the dose should be tapered slowly when discontinuing treatment. Combined therapy with anticholinergics can be considered.

Surgical Treatment for Dystonia

Deep brain stimulation (DBS) at the internal segment of the globus pallidus has been established as an effective treatment for primary dystonia. In 2003, the FDA approved the procedure as a Humanitarian Device Exemption. Since then, substantial evidence has supported the long-term efficacy and safety of its use in idiopathic generalized dystonia [28–30].

The use of DBS in patients with structural lesions (symptomatic dystonia) is thought to be less effective. There is one case published in the literature about the use of bilateral globus pallidus internal segment DBS for the treatment of severe generalized dystonia in WD. The 20-week clinical follow-up demonstrated a 14% reduction in the Burke-Fahn-Marsden Dystonia Scale score and a 44% reduction of the Caregiver Burden Score [31]. More evidence is needed before this therapy can be recommended for this specific population.

TREMOR

Tremor in WD is a predominant neurological feature, present in 39%–83% of patients [5,8,12,19]. It is often characterized as a postural and kinetic tremor of the hands resembling the tremor in essential tremor. As the disease progresses, the tremor may adopt other features such as position dependence, an intention component, or the proximal, high amplitude tremor classically described as a wing-beating tremor. As discussed earlier, a unilateral resting tremor is infrequent in WD patients. Resting tremor, when present, usually is accompanied by concomitant postural and kinetic components [8,19].

There is a significant lack of evidence when analyzing the treatment options available for tremor in this group of patients. As noted above, the tremor presentation may vary, and it is usually accompanied by other neurological findings. For this reason, the treatment should aim to target the predominant and/or most disabling symptom. If the patient exhibits a postural and kinetic tremor with dystonic postures like spooning of the fingers with the arms outstretched or evidence of segmental dystonia in the arm or neck, anticholinergics or botulinum toxin injections should be the first line of treatment. If the tremor demonstrates a more postural and kinetic quality—similar to essential tremor—medications classically used for this condition might be considered.

Propranolol

Propranolol's antitremor mechanism is thought to involve an antagonistic effect on nonselective β-adrenergic receptors at the extrafusal muscle spindles. Some have posited an additional central nervous system mechanism where it limits the overall norepinephrine release [32].

The dose that has been typically used ranges from 40 to 240 mg daily [33]. Its most important side effects are bradycardia and hypotension. Hence, a slow titration and periodic monitoring of heart rate is recommended [32,34].

Primidone

Primidone acts by binding to the GABA-A receptor, increasing the duration of the channel opening, providing an increase in chlorine influx and secondary cell hyperpolarization. GABA-A receptors are more abundant in the prefrontal cortex, hippocampus, and cerebellum. GABA modulation in the cerebellum is thought to be the primary target in tremor control [32].

It is recommended that primidone be started at a low dose (25 mg daily) and uptitrated slowly at weekly intervals to avoid side effects. The dose may vary between 25 and 750 mg daily. The major side effects are drowsiness, ataxia, and nausea [33,34]. There have been no convincing reports of hepatotoxicity due to primidone in humans and no reports of its association with acute liver failure. Nevertheless, it is extensively metabolized by the liver; therefore, its use in the WD population should be considered with caution.

Other medications that have been reported as beneficial for the treatment of tremor are benzodiazepines, gabapentin, and topiramate (Table 2).

The usual benefit reported for primidone and propranolol in essential tremor is a mild to moderate decrease in the tremor amplitude in 30%–70% of the patients [34]. There are no reports of their efficacy in WD.

There has been one report of successful use of DBS therapy at the Vim nucleus of the thalamus for the treatment of refractory tremor in WD, but further experience is needed [35].

TABLE 2 Medications Used for Tremor

Drug	Usual Effective Dose (mg/d)
Propranolol	40–240
Primidone	50–750
Topiramate	100–300
Gabapentin	1200–1800
Clonazepam	0.5–4
Alprazolam	0.75–1.5

Effective dose obtained from Refs. [33, 34].

CEREBELLAR ATAXIA

In early descriptions of WD, a clinical syndrome of dysarthria, ataxia, titubation, and wing-beating tremor was described and given the name pseudosclerosis. Ataxia as a feature of WD has been documented in 28%–51% of the patients [6,8,19]. Dysarthria is a commonly encountered in this population, but its origin appears to be multifactorial with spastic, ataxic, dystonic, and hypokinetic features [10, 22].

At present, no pharmacological therapy has proved effective for the treatment of ataxia. Symptomatic treatments and physiotherapy-based interventions, such as balance and gait training, developing of postural control and use of compensatory orthotics and aids are usually offered to patients.

COGNITIVE IMPAIRMENT

Studies have documented the presence of cognitive deficits in patients with WD. Learning-related problems have been described in 8% of patients at diagnosis [36]. Significant differences from healthy controls were encountered in several domains such as attention, visuospatial perception, memory, and verbal and abstract reasoning [37]. A frontal or subcortical cognitive syndrome is the most described abnormality with greatest impairment in attention and visual-motor speed [8].

The use of acetylcholinesterase inhibitors such as donepezil, rivastigmine, and galantamine is still controversial for mild cognitive impairment and for other causes of subcortical and frontal cognitive impairment such as vascular disease. There is no literature at present that describes or evaluates the use of these medications in WD patients. For these reasons, their use is not recommended.

Even though there is no studies assessing the use of cholinesterase inhibitors in moderate to severe dementia secondary to WD, its use might be tried, especially in elderly patients where a mixed dementia syndrome could be possible.

PSYCHIATRIC MANIFESTATIONS

Psychiatric manifestations in WD are thought to be underreported, but they have been described in 30%–64% of patients at presentation [36]. In a retrospective study that included 195 patients, 51% developed psychiatric symptoms during the disease, and 20% developed psychiatric manifestations before the WD diagnosis [38]. Failure to recognize the disease may lead to delayed diagnosis, establishment of ineffective psychiatric therapy, or unnecessary treatment [39]. The average time described between the initial psychiatric symptoms and the WD diagnosis is 2.4 years, a considerable diagnostic delay as compared with the hepatic (0.5 years) and neurological predominant (1.5 years) WD [10,36].

The most common psychiatric symptoms described at presentation are psychosis (36%), depression (22%), and personality changes (8%). Exacerbation of psychiatric symptoms, such as depression, agitation, and anxiety, has been described after starting chelating therapy [36]. Later in the course of the disease, symptoms such as apathy, impulsivity, disinhibition, and mania can appear [11,36,40].

Anticopper therapy has been reported to improve cognitive and psychiatric symptoms in WD as well [39,41]. However, concomitant use of symptomatic treatment for psychiatric symptoms is often necessary. The psychotropic medications that have been used in WD include anxiolytics (such as benzodiazepines), antidepressants, lithium, antipsychotics, and even electroconvulsive therapy [36].

In several series, tricyclic antidepressants and serotonin reuptake inhibitors were the two most commonly used antidepressants in patients with WD [36], although no robust evidence exists to favor one group over another. Several case reports have described the successful use of lithium to treat hypomanic and manic states in patients with WD [42–45]. Lithium, as a mood modulator, has the advantage of not being metabolized by the liver, but more studies are needed to assess efficacy and safety of its use in this population.

Neuroleptic use in WD patients has been associated with early neurological deterioration [14]. Multiple cases of acute dystonia, severe parkinsonism, tardive dyskinesias, and even neuroleptic malignant syndrome have been reported after starting antipsychotics in WD [15,39,46,47]. Therefore, their use needs to be considered with caution. In addition, it is important to keep in mind that parkinsonism or dystonia could be the result of treatment with a dopamine-blocking agent. Before starting symptomatic treatment for these movement disorders, medications must be reviewed in order to rule out any drug-induced side effects. Likewise, if the use of antipsychotics is needed to control psychiatric symptoms, those with the least risk of inducing parkinsonism and dystonia are recommended. However, there are no robust studies supporting the safety or efficacy of second-generation antipsychotics for WD. Likewise, there are no data to guide the optimal duration of treatment with antipsychotics or antidepressants in WD after starting anticopper therapy. However, close follow-up of psychiatric symptoms is recommended, and the need of continuing psychiatric treatment needs to be reassessed periodically.

CONCLUSION

Despite the removal of excess copper by chelating agents, residual neurological and psychiatric symptoms often persist in WD patients. Moreover, some patients experience further worsening of symptoms, even after an appropriate therapy is initiated.

Treating residual neurological and psychiatric manifestations in WD is challenging. Evidence-based therapies are lacking, clinical improvement may be less than desired, and poor tolerability of many drugs due to central nervous system side effects may limit usefulness.

In a recent report, neurological symptomatic therapy of any kind had been offered to only one-third of patients with neurological residual symptoms [13]. This could be explained by the lack of evidence supporting these therapies in WD. Little has been published regarding specific treatments for the neurological and psychiatric manifestations of this condition. Consequently, neurologists and psychiatrists are often forced to extrapolate treatments described for other indications that have an uncertain effect on this population. If neuropsychiatric symptoms are mild and nondisruptive, treatment with anticopper therapy could be used in isolation with close follow-up of neurological improvement. However, if residual neuropsychiatric symptoms are significantly affecting the patient's quality of life and functionality, concomitant symptomatic therapy of these symptoms is recommended. Whenever possible, a low starting dose of the drug with slow titration should be used.

Overall, the neuropsychiatric therapeutic needs of WD patients must be properly addressed, and prospective studies that assess the symptomatic treatment of residual neuropsychiatric symptoms in WD are urgently required.

REFERENCES

[1] Algin O, Taskapilioglu O, Hakyemez B, Ocakoglu G, Yurtogullari S, Erer S, et al. Structural and neurochemical evaluation of the brain and pons in patients with Wilson's disease. Jpn J Radiol 2010;28(9):663–71.
[2] Walshe JM, Yealland M. Wilson's disease: the problem of delayed diagnosis. J Neurol Neurosurg Psychiatry 1992;55(8):692–6.
[3] Brewer GJ, Yuzbasiyan-Gurkan V. Wilson disease. Medicine (Baltimore) 1992;71(3):139–64.
[4] Starosta-Rubinstein S, Young AB, Kluin K, Hill G, Aisen AM, Gabrielsen T, et al. Clinical assessment of 31 patients with Wilson's disease. Correlations with structural changes on magnetic resonance imaging. Arch Neurol 1987;44(4):365–70.
[5] Dastur DK, Manghani DK, Wadia NH. Wilson's disease in India. I. Geographic, genetic, and clinical aspects in 16 families. Neurology 1968;18(1 Pt 1):21–31.
[6] Taly AB, Meenakshi-Sundaram S, Sinha S, Swamy HS, Arunodaya GR. Wilson disease: description of 282 patients evaluated over 3 decades. Medicine (Baltimore) 2007;86(2):112–21.
[7] Huang CC, Chu NS. Wilson's disease: clinical analysis of 71 cases and comparison with previous Chinese series. J Formos Med Assoc Taiwan Yi Zhi 1992;91(5):502–7.
[8] Lorincz MT. Neurologic Wilson's disease. Ann N Y Acad Sci 2010;1184:173–87.
[9] Merle U, Schaefer M, Ferenci P, Stremmel W. Clinical presentation, diagnosis and long-term outcome of Wilson's disease: a cohort study. Gut 2007;56(1):115–20.
[10] Oder W, Grimm G, Kollegger H, Ferenci P, Schneider B, Deecke L. Neurological and neuropsychiatric spectrum of Wilson's disease: a prospective study of 45 cases. J Neurol 1991;238(5):281–7.
[11] Dening TR, Berrios GE. Wilson's disease: a longitudinal study of psychiatric symptoms. Biol Psychiatry 1990;28(3):255–65.
[12] Stremmel W, Meyerrose KW, Niederau C, Hefter H, Kreuzpaintner G, Strohmeyer G. Wilson disease: clinical presentation, treatment, and survival. Ann Intern Med 1991;115(9):720–6.
[13] Hölscher S, Leinweber B, Hefter H, Reuner U, Günther P, Weiss KH, et al. Evaluation of the symptomatic treatment of residual neurological symptoms in Wilson disease. Eur Neurol 2010;64(2):83–7.
[14] Prashanth LK, Taly AB, Sinha S, Ravishankar S, Arunodaya GR, Vasudev MK, et al. Prognostic factors in patients presenting with severe neurological forms of Wilson's disease. QJM 2005;98(8):557–63.
[15] Litwin T, Dzieżyc K, Karliński M, Chabik G, Czepiel W, Członkowska A. Early neurological worsening in patients with Wilson's disease. J Neurol Sci 2015;355(1–2):162–7.
[16] Fahn S, Jankovic J. Hallett M. Elsevier Health Sciences: Principles and practice of movement disorders; 2011. 3330 p.
[17] Soltanzadeh A, Soltanzadeh P, Nafissi S, Ghorbani A, Sikaroodi H, Lotfi J. Wilson's disease: a great masquerader. Eur Neurol 2007;57(2):80–5.
[18] Barthel H, Hermann W, Kluge R, Hesse S, Collingridge DR, Wagner A, et al. Concordant pre- and postsynaptic deficits of dopaminergic neurotransmission in neurologic Wilson disease. AJNR Am J Neuroradiol 2003;24(2):234–8.

[19] Machado A, Chien HF, Deguti MM, Cançado E, Azevedo RS, Scaff M, et al. Neurological manifestations in Wilson's disease: report of 119 cases. Mov Disord Off J Mov Disord Soc 2006;21(12):2192–6.
[20] McFarland NR. Diagnostic approach to atypical parkinsonian syndromes. Contin Minneap Minn 2016;22(4 Movement Disorders):1117–42.
[21] Marsden CD, editor. Movement disorders: neurology 2. Butterworth-Heinemann; 2013. 394 p.
[22] Grimm G, Prayer L, Oder W, Ferenci P, Madl C, Knoflach P, et al. Comparison of functional and structural brain disturbances in Wilson's disease. Neurology 1991;41(2(Pt1)):272–6.
[23] Svetel M, Kozić D, Stefanova E, Semnic R, Dragasevic N, Kostic VS. Dystonia in Wilson's disease. Mov Disord Off J Mov Disord Soc 2001;16(4):719–23.
[24] Pellizzari R, Rossetto O, Schiavo G, Montecucco C. Tetanus and botulinum neurotoxins: mechanism of action and therapeutic uses. Philos Trans R Soc Lond B Biol Sci 1999;354(1381):259–68.
[25] Thenganatt MA, Jankovic J. Treatment of dystonia. Neurother J Am Soc Exp Neurother 2014;11(1):139–52.
[26] Dressler D, Altenmueller E, Bhidayasiri R, Bohlega S, Chana P, Chung TM, et al. Strategies for treatment of dystonia. J Neural Transm Vienna Austria 1996 2016;123(3):251–8.
[27] Maltese M, Martella G, Madeo G, Fagiolo I, Tassone A, Ponterio G, et al. Anticholinergic drugs rescue synaptic plasticity in DYT1 dystonia: role of M1 muscarinic receptors. Mov Disord Off J Mov Disord Soc 2014;29(13):1655–65.
[28] Cif L, Vasques X, Gonzalez V, Ravel P, Biolsi B, Collod-Beroud G, et al. Long-term follow-up of DYT1 dystonia patients treated by deep brain stimulation: an open-label study. Mov Disord Off J Mov Disord Soc 2010;25(3):289–99.
[29] Volkmann J, Wolters A, Kupsch A, Müller J, Kühn AA, Schneider G-H, et al. Pallidal deep brain stimulation in patients with primary generalised or segmental dystonia: 5-year follow-up of a randomised trial. Lancet Neurol 2012;11(12):1029–38.
[30] Vidailhet M, Vercueil L, Houeto J-L, Krystkowiak P, Lagrange C, Yelnik J, et al. Bilateral, pallidal, deep-brain stimulation in primary generalised dystonia: a prospective 3 year follow-up study. Lancet Neurol 2007;6(3):223–9.
[31] Sidiropoulos C, Hutchison W, Mestre T, Moro E, Prescott IA, Mizrachi AV, et al. Bilateral pallidal stimulation for Wilson's disease. Mov Disord Off J Mov Disord Soc 2013;28(9):1292–5.
[32] Ondo W. Essential tremor: what we can learn from current pharmacotherapy. Tremor Hyperkinetic Mov N Y 2016;6:356.
[33] Schneider SA, Deuschl G. The treatment of tremor. Neurother J Am Soc Exp Neurother 2014;11(1):128–38.
[34] Louis ED. Diagnosis and management of tremor. Contin Minneap Minn 2016;22(4 Movement Disorders):1143–58.
[35] Hedera P. Treatment of Wilson's disease motor complications with deep brain stimulation. Ann N Y Acad Sci 2014;1315:16–23.
[36] Zimbrean PC, Schilsky ML. Psychiatric aspects of Wilson disease: a review. Gen Hosp Psychiatry 2014;36(1):53–62.
[37] Seniów J, Bak T, Gajda J, Poniatowska R, Czlonkowska A. Cognitive functioning in neurologically symptomatic and asymptomatic forms of Wilson's disease. Mov Disord Off J Mov Disord Soc 2002;17(5):1077–83.
[38] Dening TR, Berrios GE. Wilson's disease. Psychiatric symptoms in 195 cases. Arch Gen Psychiatry 1989;46(12):1126–34.
[39] Srinivas K, Sinha S, Taly AB, Prashanth LK, Arunodaya GR, Janardhana Reddy YC, et al. Dominant psychiatric manifestations in Wilson's disease: a diagnostic and therapeutic challenge! J Neurol Sci 2008;266(1–2):104–8.
[40] Svetel M, Potrebić A, Pekmezović T, Tomić A, Kresojević N, Jesić R, et al. Neuropsychiatric aspects of treated Wilson's disease. Parkinsonism Relat Disord 2009;15(10):772–5.
[41] Modai I, Karp L, Liberman UA, Munitz H. Penicillamine therapy for schizophreniform psychosis in Wilson's disease. J Nerv Ment Dis 1985;173(11):698–701.
[42] Chand PK, Murthy P. Mania as a presenting symptom of Wilson's disease. Acta Neuropsychiatr 2006;18(1):47–9.
[43] Machado AC, Deguti MM, Caixeta L, Spitz M, Lucato LT, Barbosa ER. Mania as the first manifestation of Wilson's disease. Bipolar Disord 2008;10(3):447–50.
[44] Loganathan S, Nayak R, Sinha S, Taly AB, Math S, Varghese M. Treating mania in Wilson's disease with lithium. J Neuropsychiatry Clin Neurosci 2008;20(4):487–9.
[45] Rybakowski JK, Litwin T, Chlopocka-Wozniak M, Czlonkowska A. Lithium treatment of a bipolar patient with Wilson's disease: a case report. Pharmacopsychiatry 2013;46(3):120–1.
[46] Kontaxakis V, Stefanis C, Markidis M, Tserpe V. Neuroleptic malignant syndrome in a patient with Wilson's disease. J Neurol Neurosurg Psychiatry 1988;51(7):1001–2.
[47] Zimbrean PC, Schilsky ML. The spectrum of psychiatric symptoms in Wilson's disease: treatment and prognostic considerations. Am J Psychiatry 2015;172(11):1068–72.

Chapter 21

Other Treatment Regimens and Emerging Therapies

Christian Rupp and Karl Heinz Weiss
Internal Medicine IV, University Hospital Heidelberg, Heidelberg, Germany

RESPONSE GUIDED THERAPY

Correlation between a patient's clinical course and easy-to-determine biochemical response pattern are highly desirable for individualizing treatment and surveillance regimens. During the early phase of use of chelating agents in treating WD, it was noted that mobilized "free" copper essentially contributes to the copper toxicity. The increase of free copper levels that follows the successful mobilization of large stores of copper, derived mainly from the liver, may provoke neurological deterioration. Significantly increased levels of "free" copper were suggested as a predictor of neurological deterioration [1]. This fundamental study highlighted the importance for a rapid and substantial control of the free copper in neurologically affected WD patients. Adapted therapy according to this biochemical response pattern would enable regulation of therapy before neurological worsening or other adverse side effect occurs. One major limitation of this approach is the difficulty in calculating free or nonceruloplasmin-bound copper. This copper subfraction might be miscalculated under inflammatory conditions or by unpredictable cross-reactivity of degradation products in routinely used immunologic assays. This limitation might be circumvented by direct measurement of the noncovalently bound copper pool by ultrafiltration or other techniques, like exchangeable with chelator EDTA-bound copper (CuEXC) and its derived relative exchangeable copper (REC, ratio CuEXC/total copper %). Based on the exact determination of alterations in different copper pools, chelating therapy could be adjusted [2]. Though, the impact of peaking "free" copper levels, especially at the beginning of chelating therapy, for clinical deterioration is still unclear. When a high starting dose of chelating agent is used, there is more frequently neurological deterioration. For that reason, a slow uptitration of the dosage of chelating agents seems advantageous. Furthermore, estimation of nonceruloplasmin-bound copper or exchangeable copper may be useful markers for creating response-guided dosage regimens. So far, estimation of free copper demonstrated its benefit mainly in the diagnostic setting, but its value for therapy monitoring and response-guided therapy remains ambiguous. The rapid control of "free" copper related to the use of different chelating agents might become a major topic in the future. Until now, the most promising candidate for a biochemical-guided regimen is tetrathiomolybdate (see Chapter 18). Compared with other routinely used drugs, tetrathiomolybdate has a consistent biochemical and the clinical response pattern, especially in the neurologically affected WD [3, 4]. However, so far, no comprehensive clinical evidence for biochemical target "corridors" for nonceruloplasmin-bound copper and urinary copper excretion is available. Therefore, general recommendation for dosage adjustments cannot be made up to now.

OTHER DRUGS

Acid Sphingomyelinase

An essential study by Lang et al. revealed that copper-induced apoptosis in hepatocytes is caused by activation of acid sphingomyelinase (Asm) and subsequent release of ceramide [5]. This copper-triggered apoptosis could be circumvented by genetic modification and pharmacological inhibition with amitriptyline. The inhibition of acid sphingomyelinase showed protective effects against liver failure and death in an animal model of Wilson disease. These fundamental experiments suggest an alternative pathway of liver damage in Wilson disease, which is attributable by medical therapy. However, this approach has not been evaluated in Wilson disease patients to date.

Curcurmin

Several in vitro studies could show a partially enhanced protein expression of different *ATP7B* mutations by curcumin [6]. In patients with specific *ATP7B* mutations (that have residual copper excretion activity), curcumin might display a novel treatment strategy in Wilson disease by directly restoring mutant ATP7B protein expression. Furthermore, curcumin might have several other potential positive effects like antioxidant activity where it acts as an effective scavenger of reactive oxygen species and may have copper chelating activity. However, to date, no clinical data about curcumin therapy in Wilson disease patients are available.

Chinese Herbal Medicines

In China, various Chinese herbal medicines were evaluated as alternative treatment option of Wilson disease. A systematic review of about 600 WD patients indicates a potential benefit of concomitant therapy for Wilson disease [7]. Improvement of clinical symptoms, promotion of urinary copper excretion, and amelioration of liver function were demonstrated after administration of different Chinese herbal medicines. Compared with conventional medication, fewer adverse events were reported. However, well-designed, randomized, placebo-controlled clinical trials are not available for these medicines. For that reasons, general recommendations cannot be made regarding their use.

Plant Decapeptide OSIP108

Basic research studies with the plant-derived decapeptide OSIP108 showed significant reduction of copper-induced toxicity and apoptosis in various in vitro and in vivo models for Wilson disease [8]. Treatment with OSIP108 resulted in an increased viability of Cu-treated cell lines expressing mutant ATP7B. Furthermore, the treatment with OSIP108 prevented aberrancies in liver morphology of Cu-treated zebra fish larvae. This may indicate the potential of OSIP108 as a novel medical treatment option for WD.

Methanobactin

Recent research studies in rodents suggest that a posttranslationally modified peptide from the methanotrophic proteobacterium *Methylosinus trichosporium* OB3b—called Methanobactin—could be a promising anticopper agent for WD [9]. Methanobactin has a molecular weight of 1154 Da (metal free) and exhibits an exceptionally high copper-binding affinity. A further discriminator to currently used chelating agents, methanobactin, is available as an i.v. compound. However, clinical evaluation and development of this compound has not started yet.

THERAPEUTIC PLASMA EXCHANGE (TPE)

Several studies investigated the feasibility and effectiveness of therapeutic plasma exchange (TPE) to remove copper and provide a bridge to liver transplantation [10–12]. The ASFA apheresis registry on Wilson disease comprises a total of 10 patients (three males and seven females) with Wilson disease [13]. In total, these patients underwent 43 TPEs with a median number of TPE procedures per patient of 3.5. All 10 patients who underwent TPE had a positive outcome in terms of 6-month survival, although 9 of 10 patients had to undergo liver transplantation in the further course of disease. Furthermore, there are several case reports or series that emphasize TPE as valuable adjunctive therapy bridging to liver transplantation [14]. TPE was performed successfully in combination with chelating therapy or alone, whereas the effect of combination therapy seems to be superior to TPE alone. Especially in the setting of a Wilsonian crisis with severe hemolysis and impending acute liver failure, TPE provides the opportunity for rapid recovery. For that reasons, TPE should be considered as a therapeutic option to stabilize patients with Wilson disease by efficiently decreasing serum copper, reducing hemolysis, and helping to prevent renal tubular injury from copper and copper complexes until liver transplantation is possible.

MUTATION SPECIFIC THERAPY

Several *ATP7B* mutations were identified that cause misfolding of the transport protein leading to a reduced ATP7B expression with only impaired copper export capacity or even complete interruption of biliary copper excretion. This effect was at least partly resolved by pharmacological intervention with the chaperone 4-phenylbutyrate (4-PBA) or curcumin [6, 15].

Restoration of correct protein folding and subsequently regeneration of protein function by chaperones might represent a novel therapy regimen. This approach might especially be of value in WD with certain mutant ATP7B variants with at least residual copper export activity that can be pharmacologically augmented.

GENE TRANSFER

Another promising approach for the restoration of hepatic copper metabolism in Wilson disease is gene therapy [16]. By successful *ATP7B* gene transfer, hepatic or neurological damage could be prevented timely before the clinical manifestation of the disease. For cell or gene therapy in WD, the rationale for targeting the liver first and foremost is based on the fact that biliary copper excretion is primarily dependent on ATP7B expression in the liver. Even early gestational gene transfer might be considered to adjust impaired copper metabolism during fetal development. One of the foremost mandatory requirements for successful gene therapy is a stable and sustainable gene transfer, ensuring a stable long-term expression of the transferred gene. With this regard, HIV-derived lentiviral vectors (LV) show favorable capacity, whereas LV are able to integrate efficiently a stable gene replacement into the genome of nondividing cells. Animal studies gave proof of principle that transduction of ATP7B LV leads to hepatic expression of the transgene in hepatocytes resulting in lower liver copper levels and improvement of fibrosis compared with untreated animals [17]. In another murine model of Wilson disease, even prenatal gene transfer by injection of LV containing the human ATP7B resulted in decreased hepatic copper content. This study applied for the first time an in utero gene therapy in WD [18]. Another current approach of gene therapy used a small single-stranded adeno-associated virus (rAAV), to transfer ATP7B with hepatocyte-specific promoters into murine models for WD [19]. Gene transfer yielded a sustainable expression of ATP7B with a subsequently sufficient restoration of copper metabolism 6 month after only a single injection of rAAV. Another successful approach used adeno-associated vector serotype 8 (AAV8) encoding the human ATP7B cDNA placed under the control of the liver-specific alpha1-antitrypsin promoter (AAV8-AAT-ATP7B). Gonzalez-Aseguinolaza observed a dose-dependent therapeutic effect, manifested by the reduction of serum transaminases and restoration of physiological biliary copper excretion [19]. The application of this liver-directed gene therapy seems to be one of the most promising techniques in WD for the future.

HEPATOCYTE/TISSUE TRANSFER

In contrast to gene therapy, cell therapy is another approach that has been tested to achieve permanent cure of Wilson disease. Instead to restore the physiological ATP7B-dependent copper excretion into bile fluid by gene modification, healthy hepatocytes might be restored by cell therapy [20–22]. Essentially, the nonfunctional hepatocytes are exchanged by healthy liver cells, which are able to proliferate into functional hepatocytes and reconstitute the biliary canalicular network. So far, several studies about cell therapy in animal models revealed already huge differences of liver repopulation between Wilson disease and healthy controls. The outstanding advances in isolation, manipulation, and successful transplantation of healthy hepatocytes and other liver cell types in recent years point toward novel opportunities for cell therapy in these inherent diseases [23]. Despite the huge advances in stem cell biology, including a growing understanding about isolation, mechanisms of hepatic differentiation, and finally reconstitution of an entire functional liver by cell transplantation, it is still in an early experimental phase [24]. Besides transfer of healthy hepatocytes into the diseased liver itself, tissue engineering approaches also have evaluated transplantation of liver tissue into extrahepatic locations. However, excretion of excess copper depends necessarily on an intact hepatobiliary network that cannot be restored by transplantation of hepatocytes in locations other than the liver. Apparently, the combination of approved drugs with one or more of the abovementioned copper-mobilizing therapies might become a promising therapeutic approach in WD in the future.

REFERENCES

[1] Brewer GJ, et al. Treatment of Wilson's disease with tetrathiomolybdate: V. Control of free copper by tetrathiomolybdate and a comparison with trientine. Transl Res 2009;154(2):70–7.

[2] El Balkhi S, et al. Relative exchangeable copper: a new highly sensitive and highly specific biomarker for Wilson's disease diagnosis. Clin Chim Acta 2011;412(23–24):2254–60.

[3] Weiss KH, et al. Efficacy and safety of oral chelators in treatment of patients with Wilson disease. Clin Gastroenterol Hepatol 2013;11(8):1028–35. e1–2.

[4] Ala A, Aliu E, Schilsky ML. Prospective pilot study of a single daily dosage of trientine for the treatment of Wilson disease. Dig Dis Sci 2015; 60(5):1433–9.

[5] Lang PA, et al. Liver cell death and anemia in Wilson disease involve acid sphingomyelinase and ceramide. Nat Med 2007;13(2):164–70.

[6] van den Berghe PV, et al. Reduced expression of ATP7B affected by Wilson disease-causing mutations is rescued by pharmacological folding chaperones 4-phenylbutyrate and curcumin. Hepatology 2009;50(6):1783–95.
[7] Wang Y, et al. Clinical efficacy and safety of Chinese herbal medicine for Wilson's disease: a systematic review of 9 randomized controlled trials. Complement Ther Med 2012;20(3):143–54.
[8] Spincemaille P, et al. The plant decapeptide OSIP108 prevents copper-induced toxicity in various models for Wilson disease. Toxicol Appl Pharmacol 2014;280(2):345–51.
[9] Lichtmannegger J, et al. Methanobactin reverses acute liver failure in a rat model of Wilson disease. J Clin Invest 2016;126(7):2721–35.
[10] Morgan SM, Zantek ND. Therapeutic plasma exchange for fulminant hepatic failure secondary to Wilson's disease. J Clin Apher 2012;27(5):282–6.
[11] Collins KL, et al. Single pass albumin dialysis (SPAD) in fulminant Wilsonian liver failure: a case report. Pediatr Nephrol 2008;23(6):1013–6.
[12] Jhang JS, et al. Therapeutic plasmapheresis as a bridge to liver transplantation in fulminant Wilson disease. J Clin Apher 2007;22(1):10–4.
[13] Pham HP, et al. Report of the ASFA apheresis registry study on Wilson's disease. J Clin Apher 2016;31(1):11–5.
[14] Asfaha S, et al. Plasmapheresis for hemolytic crisis and impending acute liver failure in Wilson disease. J Clin Apher 2007;22(5):295–8.
[15] Zhang S, et al. Rescue of ATP7B function in hepatocyte-like cells from Wilson's disease induced pluripotent stem cells using gene therapy or the chaperone drug curcumin. Hum Mol Genet 2011;20(16):3176–87.
[16] Roy-Chowdhury J, Schilsky ML. Gene therapy of Wilson disease: a "golden" opportunity using rAAV on the 50th anniversary of the discovery of the virus. J Hepatol 2016;64(2):265–7.
[17] Merle U, et al. Lentiviral gene transfer ameliorates disease progression in Long-Evans Cinnamon rats: an animal model for Wilson disease. Scand J Gastroenterol 2006;41(8):974–82.
[18] Roybal JL, et al. Early gestational gene transfer with targeted ATP7B expression in the liver improves phenotype in a murine model of Wilson's disease. Gene Ther 2012;19(11):1085–94.
[19] Murillo O, Luqui DM, Gazquez C, Martinez-Espartosa D, Navarro-Blasco I, Monreal JI, Guembe L, Moreno-Cermeño A, Corrales FJ, Prieto J, Hernandez-Alcoceba R, Gonzalez-Aseguinolaza G. Long-term metabolic correction of Wilson's disease in a murine model by gene therapy. J Hepatol 2016;64(2):419–26.
[20] Gupta S. Cell therapy to remove excess copper in Wilson's disease. Ann N Y Acad Sci 2014;1315:70–80.
[21] Filippi C, Dhawan A. Current status of human hepatocyte transplantation and its potential for Wilson's disease. Ann N Y Acad Sci 2014;1315:50–5.
[22] Malhi H, et al. Development of cell therapy strategies to overcome copper toxicity in the LEC rat model of Wilson disease. Regen Med 2008;3(2):165–73.
[23] Park SM, et al. Hepatocyte transplantation in the long Evans Cinnamon rat model of Wilson's disease. Cell Transplant 2006;15(1):13–22.
[24] Sauer V, et al. Repeated transplantation of hepatocytes prevents fulminant hepatitis in a rat model of Wilson's disease. Liver Transpl 2012;18(2):248–59.

Part VI

Monitoring

Chapter 22

Monitoring of Medical Therapy and Copper End Points

France Woimant and Aurélia Poujois
French National Reference Centre for Wilson Disease, Neurology Department, Lariboisière Hospital, Assistance Publique-Hôpitaux de Paris, Paris, France

In Wilson disease, a regular clinical, biological, and radiological follow-up is essential to ensure good efficiency and tolerance of treatment and observance. Monitoring of Wilson disease patients is quite different according to the phase of the treatment. Indeed, the major difficulties are the risk of paradoxical worsening of neurological symptoms during the initial phase and the adherence to the treatment during the maintenance phase. The follow-up must also be adapted to the severity of the hepatic and neurological features. So, the regular evaluation is often carried out by a multidisciplinary team of specialists: pediatrician, hepatologist, neurologist, psychiatrist, rheumatologist, dermatologist, physiotherapist, speech therapist, psychologist, dietician, and social worker.

THE INITIAL PHASE

At the initial stage, close and regular clinical and biological monitoring is necessary to assess the efficacy of treatments. In most patients, liver function stabilizes and then improves during the first 2–6 months, and recovery is generally observed during the first year of treatment [1]. The improvement of neurological symptoms may take longer and continue over many years.

During this initial phase of treatment, the major risk is a paradoxical worsening of the neurological presentation, reported in up to 20% of patients with initial neurological signs [2]. This deterioration is reversible in only 66% of patients [3]. The reasons for this phenomenon are not entirely understood. Different studies report an increase of free copper in serum and cerebrospinal fluid in WD patients with neurological manifestations during D-penicillamine (D-pen) therapy [4, 5]. In toxic milk mice, during D-pen administration, free copper markedly increased in serum and striatum and then gradually decreased, which means that the copper is mobilized [6, 7]. So, one proposed explanation is that too rapid introduction of D-pen could lead to an important free copper mobilization from tissues with an increase of toxic free copper in the serum and cerebrospinal fluid. Free copper induces cellular distress leading to an accentuation of brain tissue damage [8]. But, this risk of neurological aggravation is not limited to D-pen but may occur with other treatments. In a retrospective analysis on 288 WD patients with a median follow-up time of 17.1 years, German colleagues reported a neurological deterioration in 9.1% of patients under D-pen, 8.8% under trientine, and 7.3% under zinc therapy [9]. So, another explanation is proposed; the deterioration could be in part due to the natural progression of very acute forms of the disease, as treatment is too slow to act [10].

During this initial phase of treatment, it is also important to look for early side effects of treatments, especially for patients on D-pen. Adverse effects are relatively common with D-pen and are the cause for discontinuation of this drug in approximately 20%–30% of patients [11]. Early sensitivity reactions with fever, rash and lymphadenopathy may occur a few days after introducing D-pen while proteinuria or bone marrow depression with neutropenia and/or thrombocytopenia may appear after a few weeks or months; disturbance of taste and fatigue are also common. With zinc salts, dyspepsia can be a problematic side effect; an increase of serum lipase and/or amylase may occur after a few weeks of treatment, but these are only biochemical changes not associated with actual pancreatitis. Frequency of initial side effects of trientine is not well defined, but these patients should be monitored as well.

Frequency of the Follow Up (Table 1)

After initiating treatment, the frequency of follow-up should be individualized. The frequency of monitoring depends on the severity of the disease and of the evolution of the liver and neurological disease. In our opinion, symptomatic patients should be assessed weekly intervals for the first 15 days, every 2 weeks until the end of the second month, at 2 months intervals until the sixth month and at 3 months intervals until the end of the first year. Presymptomatic patients should also be evaluated regularly during this initial phase.

Clinical Follow Up

It involves a global assessment that may include the following:

- Clinical assessment of hepatic status or decompensation: hepatomegaly, splenomegaly, spider nevi, jaundice, edema, ascites, bleeding, weight gain, encephalopathy, etc.
- Clinical assessment of neurological features. The utilization of standardized rating scales is very important to quantify changes in the neurological presentation. Various scales (GAS, UWDRS, etc.) have been proposed that can be administrated at the patients' bedside. The Unified Wilson's Disease Rating Scale (UWDRS), specific for Wilson disease, has been developed for clinical studies and is widely used by WD expert teams. It evaluates consciousness (part 1), functional scale based on Barthel scale and activities of daily living (part 2), and neurological examination: speech; gait; and dystonic, ataxic, tremor, and parkinsonian syndromes (part 3). The assessment scale consists of 35 items, the higher score corresponding to a severe handicap [12]. Its interrater reliability is excellent [13]. Deterioration is defined as an increase of at least four points in UWDRS part 3 or any increase in UWDRS part 2 [14]. The GAS is another scale for WD that includes neurological deficits, the liver, and cognitive/behavioral and osseomuscular items [15]. One of these two scales can be used in WD follow-up.
- Clinical assessment of psychiatric features: anxiety, depression, emotional lability, mania, behavioral abnormalities, and personality disorders [16, 17].

TABLE 1 Biological Follow-Up of Patients With WD

	Acute Phase							Maintenance Phase		
	W1	W2	W4	W6	M2	M4	M6	M9	M12	Then Twice a Year[a]
Liver enzymes, bilirubin, albumin, and prothrombin time										
Blood count										
Serum lipase and/or amylase[b]										
Antinuclear antibody[c]										
Urinary protein[c]										
Serum exchangeable copper										
Enzymatic determination of ceruloplasmin[d]										
Serum copper[d]										
24-Hour urinary copper										

[a]Follow-up is more frequent if worsening of symptoms, side effects of medications, treatment modifications, or suspicion of noncompliance.
[b]Patients under zinc salts.
[c]Patients under chelators.
[d]For calculation of nonceruloplasmin-bound copper or free copper (NCC), if exchangeable copper is not available.

- Clinical assessment of general physical health: fatigue; weight loss; ageusia; and sensitivity reactions with fever, rash, and adenopathy that are early adverse effects of treatment with D-pen; dyspepsia, nausea, and abdominal discomfort are often observed at the beginning of treatment with D-pen or zinc salts.

Biological Monitoring (Except Cupric Assessment)

Laboratory testing should include regular monitoring of the following:

- Hepatic synthetic function: liver enzymes, bilirubin, albumin and prothrombin time; in cases of low prothrombin time (<50%), determination of factor V is usually added.
- Complete blood count: platelets and leucocytes count may drop as a hypersensitive reaction to D-pen. This monitoring has to be very careful for patients with liver cirrhosis who may already have leucopenia and thrombocytopenia secondary to hypersplenism.

In patients with acute hepatic presentation, the modified Nazar Score whose components include serum bilirubin, serum aspartate aminotransferase, and prolongation of prothrombin time is a prognostic index used to determine the liver evolution [18]. A revision of this score including the white blood cell count has been validated in children [19] and in adults [20].

Concerning the detection of adverse effects:

- Urinary protein monitoring is recommended due to the risk of D-pen-induced nephropathy.
- Serum lipase and/or amylase could be checked in patients treated with zinc salts because elevations may occur, without clinical or radiological evidence of pancreatitis.
- Hemoglobin levels that drop during the maintenance phase might be indicative of overdecoppering and warrant investigation of the copper and the iron status as chelators act on both metals, but other causes of anemia must be excluded as well.

Copper Monitoring

Copper monitoring is based on two copper determinations: urinary copper excretion and serum nonceruloplasmin-bound copper (NCC) or "free copper," but the usefulness of this last assay remains debated.

Urinary copper excretion may be monitored with all treatments. During D-pen therapy, immediately after starting treatment, copper excretion increases. It should be between 8 and 16 μmol/24 h (500 and 1000 μg/24 h) [21]. Sometimes, it can be very elevated, especially during initiation of treatment, exceeding 16 μmol/24 h [22]. A rapid increase of the urinary copper should alert the clinician. Indeed, urinary excretion reflects the plasma free copper concentration, and high levels may indicate dangerous copper toxicosis that could lead to paradoxical neurological deterioration. So, in neurological forms, increase of the D-pen dose must carefully take into account urinary copper excretion.

During trientine therapy, the increase of 24 h urinary copper excretion is less important than with D-pen. This is probably due to its weak absorption by the gastrointestinal tract and to the fact that trientine and D-pen may mobilize different pools of body copper [23]. It should be between 4.8 and 16 μmol/24 h (300 and 1000 μg/24 h) [21]. As for D-pen treatment, very high elevation of urinary copper must be avoided in patients with neurological symptoms.

During zinc therapy, 24 h urinary copper excretion decreases in those who had elevations of the urine copper prior to treatment. Zincs' mechanism of action is different from that of chelators; zinc salts indirectly interfere with the uptake of copper from the gastrointestinal tract by blocking the transfer to the circulation of the normally absorbed copper by enterocytes, thereby increasing fecal excretion of copper as enterocytes are shed during their normal cycle. Zinc treatment does not induce urinary excretion of copper. The initial target for urinary copper excretion is <2 μmol/24 h [24].

During all treatments, current guidelines estimate serum NCC or "free copper" to be a useful parameter to control therapy [1]. The NCC can be calculated by subtracting from the total serum copper concentration the amount of ceruloplasmin-bound copper. It should be maintained at or near 1.6 μmol/L (100 μg/L) [25]. But, this calculation, based on the measurement of total copper and ceruloplasmin, can give a negative result, which is devoid of meaning. Ceruloplasmin may be overestimated in conditions of inflammation or estrogen therapy or through the use of immunologic assays. Indeed, immunologic assays measure both ceruloplasmin and apoceruloplasmin. Apoceruloplasmin assay is not appropriate, as it does not contain copper and has no enzymatic activity [26]. So, the serum "free" copper could be calculated more precisely if ceruloplasmin is determined enzymatically through measurement of its oxidase activity [25], but this assay is not readily available in most clinical laboratories.

So, a direct determination of exchangeable copper (CuEXC) is proposed [27, 28]. The published studies demonstrated the usefulness of the relative exchangeable copper (REC, ratio between CuEXC and total serum copper) in the diagnostic setting (familial screening, differentiation of liver diseases, etc.), whereas CuEXC is an indicator of the severity of the extrahepatic involvement [29–32]. In our experience, CuEXC is always increased (>2 µmol/L) in neurological forms and then normalizes progressively in 4–6 months. Further clinical and experimental data are necessary to establish if the normalization of CuEXC correlates well with clinical improvement. Setting this point is very important as it would allow clinicians to monitor exchangeable copper regularly before the clinical worsening.

THE MAINTENANCE PHASE

The goal of treatment monitoring is to confirm clinical and biochemical improvement, ensure compliance with therapy, and identify late adverse side effects. Compliance is a problem for WD patients in 25%–33% of all subjects and especially in presymptomatic patients where 55% are not adherent in taking the prescribed medication [33, 34]. Recognizing problems with compliance can be difficult but is essential because discontinuation of medical therapy is associated with the risk of intractable hepatic and/or neurological decompensation. This assessment is based on clinical and biological data.

Side effects are frequent with D-pen; 31.6% of patients stopped D-pen therapy because of long-term side effects [35]. They include nephrotoxicity; lupus syndrome with appearance of antinuclear antibodies; Goodpasture's syndrome; myasthenia gravis; and skin lesions such as elastosis perforans serpiginosa (Fig. 1), progeric changes of the skin (Fig. 2), pemphigus, lichen planus, and aphthous stomatitis [36]. Other side effects are very rare: polymyositis, immunoglobulin A depression, and serous retinitis.

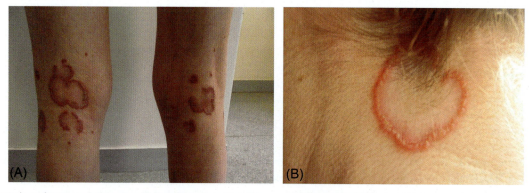

FIG. 1 Elastosis perforans serpiginosa in a 42-year-old woman treated by D-penicillamine since 26 years. (A) Knees and (B) neck.

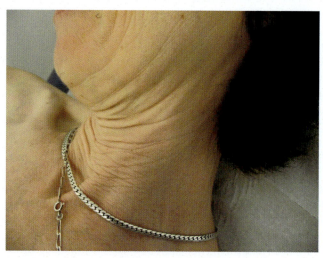

FIG. 2 Cutis laxa in a 51-year-old woman treated by D-penicillamine since 25 years.

Adverse effects are considerably rarer with trientine; they lead to the discontinuation of trientine treatment four times less frequently than with D-pen [35]. Adverse effects reported are reversible sideroblastic anemia, consequence of overtreatment with iron and/or copper deficiency and lupus-like syndrome, but de novo lupus syndrome without prior D-pen use is probably very rare.

The most common adverse effects during zinc treatment are gastric irritation (including erosions and ulcers) with dyspeptic symptoms [1, 37]. Elevations in serum lipase and/or amylase may occur without clinical or radiological evidence of pancreatitis [1]. Some authors reported insufficient control of liver disease with increased liver enzymes [9] or deterioration of liver disease during zinc therapy in 20% of patients with initial neurological presentations [38]. At last, copper deficiency as a consequence of overtreatment has to be identified and diagnosed before the occurrence of myelopathy [39].

Frequency of Follow Up During the Maintenance Phase (Table 2)

The frequency of follow-up may vary for an individual, but it should be performed at a minimum twice a year, even in asymptomatic or stable patients [22, 40]. More frequent monitoring is needed, for those experiencing worsening of symptoms or side effects of medications or after treatment modifications and in patients suspected of noncompliance with therapy. Monitoring must be particularly attentive in adolescents, in whom nonadherent behavior has been documented despite monitoring [21]. During pregnancy, clinical assessment and copper monitoring should be enhanced. The highest risks are fetal teratogenicity during the first trimester and insufficient copper supply to the fetus. So, dosages of drugs should be reduced and adapted to urinary copper excretion [1].

Clinical Follow Up

It associates hepatic, neurological, and psychiatric assessment as at the initial phase. Suggestive features of lupus, myasthenia gravis, polymyositis, and rare late adverse effects of D-pen should be investigated. Cutaneous changes should be carefully sought as in our experience, 40% of patients treated with D-pen develop progeric changes in the skin, cutis laxa, and elastosis perforans serpiginosa (preferentially placed on the neck and the posterior surface of the knees) (Figs. 1A, B and 2) [41].

TABLE 2 Monitoring of D-Penicillamine Therapy in Wilson Disease

	Effectiveness of D-pen Therapy	Main Adverse Effects	Suggestive of Weak Compliance	Suggestive of Overtreatment
Clinical assessment	Improvement or stable	Sensitivity reactions Ageusia Skin lesions as elastosis perforans serpiginosa, cutis laxa Myasthenia gravis Polymyositis Serous retinitis	Deterioration	
Blood/urine chemistry	Normalization of transaminases	Proteinuria	Increase of transaminases	
Hematology	Normalization of prothrombin time	Neutropenia and/or thrombocytopenia		
Copper monitoring				
Urinary copper	Initial phase: 8–16 μmol/24 h Maintenance phase: 3–8 μmol/24 h		>8 μmol/24 h	<2 μmol/24 h
NCC	5–15 μg/dL		>15 μg/dL	<5 μg/dL
CuEXC	0.60–1.15 μmol/L		>1.20 μmol/L	<0.40 μmol/L

Biological Monitoring (Except Cupric Assessment)

Laboratory tests performed at the initial phase are monitored over time. The appearance or the worsening of anemia and leucopenia suggests the possibility of overtreatment and copper deficiency that can be observed after numerous years of treatment. For patients treated with D-pen and trientine, lupus-like syndrome is sought with regular monitoring of antinuclear antibody (ANA). The reliability of ANA is currently a matter of discussion [42]. Late reactions with D-pen include also nephrotoxicity, usually signaled by proteinuria [1].

Copper Monitoring

Analysis of 24 h urinary copper excretion is essential for compliance monitoring.

Patients taking D-pen or trientine should have 24 h urinary copper excretion values between 3 and 8 µmol/24 h (190 and 500 µg/24 h) [1]. As there is important variability of copper excretion under treatment, the results must be interpreted with reference to the previous determinations. It's important to know that noncompliant patients often take D-pen just before control visits; in these cases, urinary copper excretion can be very high as in the initial phase. To document D-pen efficiency, some experts recommend collection of urine after stopping chelator for 48 h, to assess the urinary copper excretion while off treatment; a normalization of urinary copper excretion should be observed (0.8 µmol/24 h or 50 µg/24 h). Values of urinary copper excretion superior to 1.6 µmol/24 h or 100 µg/24 h may indicate nonadherence to treatment [43].

During zinc therapy, 24 h urinary excretion of copper is also monitored. Target to suggest satisfactory treatment is <2 µmol/24 h (130 µg/24 h) for Ala et al. [44] and <1.6 µmol/24 h (100 µg/24 h) in the guidelines of EASL [1]. Levels higher than 2 µmol/24 h suggest noncompliance with treatment or the lack of pharmacological effect. Urinary excretion of zinc may also be measured to check whether sufficient zinc is being taken and absorbed and to show patients' compliance; a 24 h urinary zinc level of <2 mg in those taking zinc salts three times daily containing 50 mg elemental zinc per dosage would indicate inadequate compliance [24], but this may vary depending on the zinc dosage and type of zinc salt.

Adequacy of treatment and compliance could also be monitored by measuring NCC while recognizing the limits of this assay. Negative and no interpretable results are reported in almost 20% of patients [45]. The serum NCC levels should be normalized [1, 22].

In our opinion, the more accurate biological methods to evaluate the efficacy of decoppering treatments are 24 h urinary copper excretion determination associated with serum exchangeable copper (CuEXC) determination. In our experience of 100 patients with a mean follow-up of 7 years, pathological increase of CuEXC was observed in 25% of patients. In 50% of them (12/25), this abnormal level of CuEXC along with high urinary copper excretion was associated with elevation of liver enzymes and a poor compliance to treatment.

Other Noninvasive Assessment of Wilson Disease

Repeat examination for Kayser-Fleischer Rings should be performed until their disappearance or if there is a question of patient compliance. Brain MRI, if abnormal at the onset of treatment, is controlled at 6 months and 1 year or according to the neurological state.

Noninvasive methods for liver assessment as elastographic methods could be useful to monitor liver disease evolution [46]. In pediatric patients with WD, liver stiffness is high at time of diagnosis and decreases during specific treatment [47]. Hepatocellular carcinoma is a rare complication of WD liver cirrhosis, even in compensated stages [48]. So, patients with cirrhosis should be examined with ultrasonography every 6 months to detect hepatocellular carcinoma; magnetic resonance imaging should be performed if there is a suspected liver lesion superior at 1 cm [36, 48]. Analysis of the tumor marker alpha-fetoprotein is not recommended routinely; its determination lacks adequate sensitivity and specificity for effective surveillance [49]. At the time of diagnosis of liver cirrhosis, screening for esophageal or gastric varices by esophagogastroduodenoscopy (EGD) should be performed. If there are no varices, revaluation can be avoided if liver stiffness <20 kPa with a platelet count >150,000. If liver stiffness increases or platelet count declines, these patients should undergo screening EGD [50]. Where varices are small, repeating EGD after 1–2 years is recommended. In patients with decompensated cirrhosis, annual EGDs are advised [51] for those not on therapy with nonspecific beta-blockers.

Red Flags

Weak Treatment Compliance (Tables 2–4)

Adequate compliance to therapy is often difficult in this lifelong disease that requires daily intake of pharmacological agents, usually two or three times daily. When clinical deterioration occurs, the possibility of no compliance or interruption of treatment should be considered. Worsening of liver function and/or neurological scales, newly developed neurological symptoms, reappearance or worsening of Kayser-Fleicher Rings should signal noncompliance. But, before clinical deterioration, clinicians should be alerted by an increase of urinary copper excretion associated with an increase of NCC or exchangeable copper. Psychological counseling and/or a reinforcement of therapeutic education program may be proposed [34].

Copper Deficiency (Tables 2–4)

It is important to look out for overtreatment. Adjustment of the maintenance therapy dose is very important to prevent neurological complications due to copper deficiency [10]. Copper deficiency can be observed after many years of treatment, essentially with zinc salts alone or in combination with trientine or D-pen [52]. This diagnosis should be evoked in front of anemia and leucopenia that precede the neurological symptoms characterized by ascending paresthesia and sensory ataxic gait due to a posterior cord syndrome. It's often associated with elevation of transaminases due to increased hepatic iron, accompanied by increased ferritin [21]. Urinary copper excretion is low and reduced as compared with patients' previous values. NCC and exchangeable copper are equally low. Dose of treatment should be lowered or temporary discontinued, before neurological complications occur. Neurological symptoms are similar to vitamin B 12 deficiency syndrome: axonal sensor or motor-sensor neuropathy and myelopathy with demyelination in dorsal columns visualized on spine magnetic resonance imaging (MRI). Symptoms can be reversed if found in time by lowering or stopping treatment as noted above [38].

Particular Attention During the Child-Adult Transition

Monitoring must be particularly attentive in children becoming adults. It's important to prepare the adolescent before transferring his medical follow-up in adult care. The success of this pathway requires an important work both on the side of the pediatrician, hepatologist, or neurologist, and on the side of the adolescent and his parents. The failure of this transition leads to the discontinuation of the treatment, the consequences of which can be very serious.

TABLE 3 Monitoring of Trientine Therapy in Wilson Disease

	Effectiveness of Trientine Therapy	Main Adverse Effects	Suggestive of Weak Compliance	Suggestive of Overtreatment
Clinical assessment	Improvement or stable	Lupus-like syndrome	Deterioration	Ascending paresthesias Appearance of sensory ataxic gait disorders
Blood chemistry	Normalization of transaminases		Increase of transaminases	
Hematology	Normalization of prothrombin time	Sideroblastic anemia		
Copper monitoring				
Urinary copper	Initial phase: 4.8–16 μmol/24 h Maintenance phase: 3–8 μmol/24 h		>8 μmol/24 h	<2 μmol/24 h
NCC	5–15 μg/dL		>15 μg/dL	<5 μg/dL
CuEXC	0.6–1.15 μmol/L		>1.20 μmol/L	<0.4 μmol/L

TABLE 4 Monitoring of Zinc Therapy in Wilson Disease

	Effectiveness of Zn Therapy	Adverse Effects	Suggestive of Weak Compliance	Suggestive of Overtreatment
Clinical assessment	Improvement or stable	Dyspepsia	Deterioration	Ascending paresthesias Appearance of sensory ataxic gait disorders
Blood chemistry	Normalization of transaminases	Increase of amylase and lipase	Increase of transaminases	
Hematology	Normalization of prothrombin time			Leucopenia and/or anemia
Copper monitoring				
Urinary copper	Initial phase: <2 µmol/24 h Maintenance phase: <1.6 µmol/24 h		>2 µmol/24 h	Important decrease <1 µmol/24 h
NCC	5–15 µg/dL		>15 µg/dL	<5 µg/dL
CuEXC	0.6–1.15 µmol/L		>1.20 µmol/L	<0.4 µmol/L

CONCLUSIONS

Follow-up of Wilson disease patients should be individualized according to the clinical state, and best care may involve evaluation by a multidisciplinary team that may include a pediatrician, hepatologist, neurologist, psychiatrist, psychologist, physiotherapist, speech therapist, dietician, and social worker. This clinical and biological pluridisciplinary approach present in French reference centers for Wilson disease is part of the French WD network [53]. Collaborations among professionals permit multidisciplinary care and improve the diagnosis and the outcomes of patients in terms of all their medical and social aspects. This allows also to improve the follow-up of children who become adults.

REFERENCES

[1] European Association for Study of Liver. EASL Clinical Practice Guidelines: Wilson's disease. J Hepatol 2012;56(3):671–85.
[2] Weiss KH, Stremmel W. Clinical considerations for an effective medical therapy in Wilson's disease. Ann N Y Acad Sci 2014;1315:81–5.
[3] Litwin T, Dzieżyc K, Karliński M, Chabik G, Czepiel W, Członkowska A. Early neurological worsening in patients with Wilson's disease. J Neurol Sci 2015;355(1–2):162–7.
[4] Walshe JM. Filterable and non-filterable serum copper. 1. The action of penicillamine. Clin Sci 1963;25:405–11.
[5] Stuerenburg HJ. CSF copper concentrations, blood-brain barrier function, and coeruloplasmin synthesis during the treatment of Wilson's disease. J Neural Transm (Vienna) 2000;107(3):321–9.
[6] Chen DB, Feng L, Lin XP, Zhang W, Li FR, Liang XL, et al. Penicillamine increases free copper and enhances oxidative stress in the brain of toxic milk mice. PLoS One 2012;7(5).
[7] Zhang JW, Liu JX, Hou HM, Chen DB, Feng L, Wu C, et al. Effects of tetrathiomolybdate and penicillamine on brain hydroxyl radical and free copper levels: a microdialysis study in vivo. Biochem Biophys Res Commun 2015;458(1):82–5.
[8] Poujois A, Mikol J, Woimant F. Wilson disease: brain pathology. Handb Clin Neurol 2017;142:77–89.
[9] Weiss KH, Gotthardt DN, Klemm D, Merle U, Ferenci-Foerster D, Schaefer M, et al. Zinc monotherapy is not as effective as chelating agents in treatment of Wilson disease. Gastroenterology 2011;140(4):1189–98.
[10] Poujois A, Devedjian JC, Moreau C, Devos D, Chaine P, Woimant F, et al. Bioavailable trace metals in neurological diseases. Curr Treat Options Neurol 2016;18(10):46.
[11] Merle U, Schaefer M, Ferenci P, Stremmel W. Clinical presentation, diagnosis and long-term outcome of Wilson's disease: a cohort study. Gut 2007;56(1):115–20.
[12] Członkowska A, Tarnacka B, Möller JC, Leinweber B, Bandmann O, Woimant F, et al. Unified Wilson's disease rating scale. A proposal for the neurological scoring of Wilson's disease patients. Neurol Neurochir Pol 2007;41(1):1–12.
[13] Leinweber B, Moller JC, Scherag A, Reuner U, Günther P, Lang CJ, et al. Evaluation of the unified Wilson's disease rating scale (UWDRS) in German patients with treated Wilson's disease. Mov Disord 2008;23(1):54–62.

[14] Członkowska A, Litwin T, Karliński M, Dziezyc K, Chabik G, Czerska M. D-Penicillamine versus zinc sulfate as first-line therapy for Wilson's disease. Eur J Neurol 2014;21(4):599–606.
[15] Aggarwal A, Aggarwal N, Nagral A, Jankharia G, Bhatt M. A novel global assessment scale for Wilson's disease (GAS for WD). Mov Disord 2009;24(4):509–18.
[16] Zimbrean PC, Schilsky ML. The spectrum of psychiatric symptoms in Wilson's disease: treatment and prognostic considerations. Am J Psychiatry 2015;172(11):1068–72.
[17] Demily C, Parant F, Cheillan D, Broussolle E, Pavec A, Guillaud O, et al. Screening of Wilson's disease in a psychiatric population: difficulties and pitfalls. A preliminary study. Ann Gen Psychiatry 2017;16:19.
[18] Nazer H, Ede RJ, Mowat AP, Williams R. Wilson's disease: clinical presentation and use of prognostic index. Gut 1986;27(11):1377–81.
[19] Dhawan A, Taylor RM, Cheeseman P, De Silva P, Katsiyiannakis L, Mieli-Vergani G. Wilson's disease in children: 37-year experience and revised King's score for liver transplantation. Liver Transpl 2005;11(4):441–8.
[20] Petrasek J, Jirsa M, Sperl J, Kozak L, Taimr P, Spicak J, et al. Revised King's College score for liver transplantation in adult patients with Wilson's disease. Liver Transpl 2007;13(1):55–61.
[21] Rodriguez-Castro KI, Hevia-Urrutia FJ, Sturniolo GC. Wilson's disease: a review of what we have learned. World J Hepatol 2015;7(29):2859–70.
[22] Roberts EA, Schilsky ML, American Association for Study of Liver Diseases (AASLD). Diagnosis and treatment of Wilson disease: an update. Hepatology 2008;47(6):2089–111.
[23] Sarkar B, Sass-Kortsak A, Clarke R, Laurie SH, Wei P. A comparative study of in vitro and in vivo interaction of D-penicillamine and triethylenetetramine with copper. Proc R Soc Med 1977;70(Suppl. 3):13–8.
[24] Brewer GJ. Wilson's disease: a Clinician's guide to recognition, diagnosis, and management. Norwell, MA: Kluwer Academic; 2001.
[25] Walshe JM. Wilson's disease: the importance of measuring serum caeruloplasmin non-immunologically. Ann Clin Biochem 2003;40(Pt 2):115–21.
[26] Merle U, Eisenbach C, Weiss KH, Tuma S, Stremmel W. Serum ceruloplasmin oxidase activity is a sensitive and highly specific diagnostic marker for Wilson's disease. J Hepatol 2009;51(5):925–30.
[27] Schmitt F, Podevin G, Poupon J, Roux J, Legras P, Trocello JM, et al. Evolution of exchangeable copper and relative exchangeable copper through the course of Wilson's disease in the Long Evans Cinnamon rat. PLoS One 2013;8(12).
[28] El Balkhi S, Poupon J, Trocello JM, Leyendecker A, Massicot F, Galliot-Guilley M, et al. Determination of ultrafiltrable and exchangeable copper in plasma: stability and reference values in healthy subjects. Anal Bioanal Chem 2009;394(5):1477–84.
[29] El Balkhi S, Trocello JM, Poupon J, Chappuis P, Massicot F, Girardot-Tinant N, et al. Relative exchangeable copper: a new highly sensitive and highly specific biomarker for Wilson's disease diagnosis. Clin Chim Acta 2011;412(23–24):2254–60.
[30] Trocello JM, El Balkhi S, Woimant F, Girardot-Tinant N, Chappuis P, Lloyd C, et al. Relative exchangeable copper: a promising tool for family screening in Wilson disease. Mov Disord 2014;29(4):558–62.
[31] Guillaud O, Brunet A-S, Mallet I, Dumortier J, Pelosse M, Heissat S, et al. Relative exchangeable copper: a valuable tool for the diagnosis of Wilson disease. Liver Int 2017;18.
[32] Poujois A, Trocello JM, Djebrani-Oussedik N, Poupon J, Collet C, Girardot-Tinant N, et al. Exchangeable copper: a reflection of the neurological severity in Wilson's disease. Eur J Neurol 2017;24(1):154–60.
[33] Masełbas W, Chabik G, Członkowska A. Persistence with treatment in patients with Wilson disease. Neurol Neurochir Pol 2010;44(3):260–3.
[34] Jacquelet E, Demain A, Girardot-Tinant N, Machado C, Pheulpin MC, Poujois A, et al. Complexité de l'observance médicamenteuse dans la maladie de Wilson: apport de l'étude Wilobs. Rev Neurol 2017;173S:S133–81.
[35] Weiss KH, Thurik F, Gotthardt DN, Schäfer M, Teufel U, Wiegand F, et al. Efficacy and safety of oral chelators in treatment of patients with Wilson disease. Clin Gastroenterol Hepatol 2013;11(8):1028–35. e1–2.
[36] Poujois A, Mikol J, Woimant F. Wilson disease: brain pathology. In: Członkowska A, Schilsky ML, editors. Wilson disease, volume 142 of the handbook of clinical neurology series. Amsterdam: Elsevier; 2017. p. 77–89.
[37] Wiernicka A, Jańczyk W, Dądalski M, Avsar Y, Schmidt H, Socha P. Gastrointestinal side effects in children with Wilson's disease treated with zinc sulphate. World J Gastroenterol 2013;19(27):4356–62.
[38] Linn FH, Houwen RH, van Hattum J, van der Kleij S, van Erpecum KJ. Long-term exclusive zinc monotherapy in symptomatic Wilson disease: experience in 17 patients. Hepatology 2009;50(5):1442–52.
[39] Dzieżyc K, Litwin T, Sobańska A, Członkowska A. Symptomatic copper deficiency in three Wilson's disease patients treated with zinc sulphate. Neurol Neurochir Pol 2014;48(3):214–8.
[40] Bandmann O, Weiss KH, Kaler SG. Wilson's disease and other neurological copper disorders. Lancet Neurol 2015;14(1):103–13.
[41] Woimant F, Trocello JM. Disorders of heavy metals. Handb Clin Neurol 2014;120:851–64.
[42] Seessle J, Gotthardt DN, Schäfer M, Gohdes A, Pfeiffenberger J, Ferenci P, et al. Concomitant immune-related events in Wilson disease: implications for monitoring chelator therapy. J Inherit Metab Dis 2016;39(1):125–30.
[43] Dzieżyc K, Litwin T, Chabik G, Członkowska A. Measurement of urinary copper excretion after 48-h D-penicillamine cessation as a compliance assessment in Wilson's disease. Funct Neurol 2015;30(4):264–8.
[44] Ala A, Walker AP, Ashkan K, Dooley JS, Schilsky ML. Wilson's disease. Lancet 2007;369(9559):397–408.
[45] Członkowska A, Litwin T. Wilson disease—currently used anticopper therapy. In: Członkowska A, Schilsky ML, editors. Wilson disease, volume 142 of the handbook of clinical neurology series. Amsterdam: Elsevier; 2017. p. 182–91.
[46] Sini M, Sorbello O, Civolani A, Liggi M, Demelia L. Non-invasive assessment of hepatic fibrosis in a series of patients with Wilson's disease. Dig Liver Dis 2012;44(6):487–91.

[47] Stefanescu AC, Pop TL, Stefanescu H, Miu N. Transient elastography of the liver in children with Wilson's disease: preliminary results. J Clin Ultrasound 2016;44(2):65–71.

[48] Pfeiffenberger J, Mogler C, Gotthardt DN, Schulze-Bergkamen H, Litwin T, Reuner U, et al. Hepatobiliary malignancies in Wilson disease. Liver Int 2015;35(5):1615–22.

[49] Bruix J, Sherman M, American Association for the Study of Liver Diseases. Management of hepatocellular carcinoma: an update. Hepatology 2011;53(3):1020–2.

[50] de Franchis R, Baveno VI Faculty. Expanding consensus in portal hypertension: report of the Baveno VI consensus workshop: stratifying risk and individualizing care for portal hypertension. J Hepatol 2015;63(3):743–52.

[51] de Franchis R. Updating consensus in portal hypertension: report of the Baveno III consensus workshop on definitions, methodology and therapeutic strategies in portal hypertension. J Hepatol 2000;33(5):846–52.

[52] Teodoro T, Neutel D, Lobo P, Geraldo AF, Conceição I, Rosa MM, et al. Recovery after copper-deficiency myeloneuropathy in Wilson's disease. J Neurol 2013;260(7):1917–8.

[53] Woimant F, Trocello JM, Chaine P, Broussolle E. Have centers of rare neurological diseases modified practices and the care of Wilson's disease? Rev Neurol (Paris) 2013;169(Suppl. 1):S18–22.

Part VII

Main Challenges in Diagnosis and Treatment

Chapter 23

Special Treatment Considerations for Wilson Disease

Michelle Angela Camarata[*,†,‡], Karl Heinz Weiss[§] and Michael L. Schilsky[¶]

[*]*Department of Surgery, Section of Transplant and Immunology, Yale School of Medicine, New Haven, CT, United States,* [†]*Department of Gastroenterology and Hepatology, Royal Surrey County Hospital, Guilford, United Kingdom,* [‡]*Department of Clinical and Experimental Medicine, University of Surrey, Guilford, United Kingdom,* [§]*Internal Medicine IV, University Hospital Heidelberg, Heidelberg, Germany,* [¶]*Division of Digestive Diseases and Transplant and Immunology, Department of Medicine and Surgery, Yale University School of Medicine, New Haven, CT, United States*

TREATMENT CHOICES AND THEIR INFLUENCE ON THE MONITORING OF THERAPY FOR OUTCOME AND SAFETY

Treatment of Wilson disease focuses on achieving a negative copper balance to reduce or detoxify copper in affected organs and prevent further accumulation of this metal [1] (EASL). As discussed previously by other authors, the main agents used in treatment include chelators, such as D-penicillamine or triethylene tetramine dihydrochloride (trientine). These bind circulating copper and are removed from the circulation by the kidney, making this class of treatment in part dependent on adequate renal clearance. Zinc therapy acts differently by reducing the absorption of copper by the digestive tract, creating a net negative for balance of copper over time. Both classes of agents, chelators, and zinc salts may even be combined for treatment when they are administered separately. Due to the lack of prospective clinical trials, the choice of treatment often depends on local experience, cost, and availability. Treatment response and decisions to alter management are therefore most often guided by clinical improvement, biochemical response, and side effects. However, there are also differences in expected results for testing for monitoring treatment efficacy and surveillance for side effects of each therapy. We will explore these in more detail.

Hepatic

Common to each therapy for Wilson disease is the aim to prevent the development of symptoms if none are present or to achieve stabilization and improvement in signs and symptoms of disease if any are present. It is easier to identify goals for hepatic disease where we have biochemical parameters to follow in addition to clinical signs of liver dysfunction such as jaundice or in those with cirrhosis of the liver and complications of portal hypertension (such as varices, ascites, or hepatic encephalopathy). Indeed, we have a scoring system for the critical ill liver patients with Wilson disease, the modified Nazar Score "New Wilson Index", to help determine if medical therapy is possible or if liver transplantation is advised for the individual patient. In addition to this score, for those with cirrhosis, we can follow the Child-Pugh-Turcotte status or MELD score to see if the patient is responding to therapy and assess their prognosis. However, for our patients without cirrhosis, these two scoring systems do not apply, and we need to distinguish true markers of arrested disease progression and improvement from testing needed for screening and surveillance of side effects for safety.

An area of increased importance in this population of patients without cirrhosis is the monitoring for fibrosis that occurs in response to hepatic injury. Fibrosis is a precursor to cirrhosis, and this parameter is valuable prognostically, potentially forecasting whether cirrhosis and later complications of portal hypertension are concerns for the patient or whether their fibrosis is regressing and the disease rendered quiescent. This is also of importance if other concomitant processes drive progression of liver disease, such as fatty liver disease, drug-induced liver injury, autoimmune disease, or other toxic injury such as alcohol. In these instances, liver disease may progress despite adequate treatment directed against excess copper.

The most accurate and reliable way of evaluating for hepatic fibrosis is liver biopsy with histological evaluation, providing information on the degree of inflammation and the stage of established fibrosis. While an initial biopsy may be performed to confirm the diagnosis of WD and the degree of hepatic fibrosis at presentation, follow-up liver biopsies are rarely performed, and little is known about fibrosis progression in patients with Wilson disease. Due to the invasive nature and potential risks associated with biopsy, noninvasive methods for assessing fibrosis are increasingly being used and can provide useful information to guide management decisions. Indirect markers are simple blood tests including liver enzymes ALT/AST and γ-GT; bilirubin; and parameters indicative of reduced liver synthetic function and portal hypertension such as INR, albumin, and platelet count. There are several direct markers of fibrosis that reflect collagen synthesis and breakdown. The most validated marker is hyaluronic acid (HA) [2], which correlates to fibrosis in ALD [3] and chronic viral hepatitis [4]. Others include laminin, YKL-40, procollagen type III, metalloproteinases MMP-1 and MMP-2, TIMPs, TGF-β1, and MP3 Index [5, 6]. Serological scores such as the APRI score [7] and the PGA score (P, prothrombin time; G, γ-glutamyltranspeptidase; and A, apolipoprotein AI) [8] have been developed using some of these parameters to predict significant fibrosis. The PGA was one of the first scores used for noninvasive detection of cirrhosis. There are several other scores that combine direct and indirect serum markers of fibrosis and clinical data. These have been validated in several disease populations in recent years. There is, however, a continuing need for increased accuracy of these instruments in estimating hepatic fibrosis and specifically further studies for their validation in the WD population (Table 1).

TABLE 1 Algorithms and Combined Scores to Predict Fibrosis

Score	Serum Parameters	Clinical Markers	Studied for Use in	Reference
APRI	AST to platelet ratio index		**Wilson disease** HCV, ALD	[7] [9] [10]
PGA	Prothrombin time, γ-GT and apolipoprotein		ALD	[8]
FibroTest (FibroSure in the United States)	Apolipoprotein-A A2-macroglobulin γ-GT Total bilirubin Haptoglobin	Patient age Gender	Chronic viral hepatitis	[11]
Hepascore	Bilirubin HA A2-macroglobulin γ-GT	Patient age Gender	Chronic viral hepatitis	[12]
FibroSpect	Serum HA TIMP-1 A2-macroglobulin		Chronic hepatitis C	[13]
FibroMeter	Platelet count Prothrombin index AST A2-macroglobulin HA Blood urea		Chronic viral hepatitis	[14]
Forns score	Platelet count γ-GT Cholesterol	Age	**Wilson disease** Chronic hepatitis C, nonviral chronic hepatitis	[15] [9]
European Liver Fibrosis (ELF) panel	HA TIMP-1 Amino-terminal propeptide type III collagen		ALD NAFLD	[16]
SAFE biopsy	Combination of APRI index with FibroTest		Chronic hepatitis C	[17]
Fib-4 score	AST Platelet count ALT	Age	**Wilson disease** NAFLD	[18] [9] [10]

Over recent years, there also have been advancements in imaging techniques to assess hepatic fibrosis. These are based on standard radiological tools such as ultrasound (US), computed tomography (CT), and magnetic resonance imaging (MRI). The most widely used imaging method is transient elastography (FibroScan and other devices). The FibroScan involves the use of a probe with an ultrasound transducer and a vibrator that emits low-frequency waves. The speed at which the waves travel through liver tissue relates to its stiffness, which in turn correlates with the degree of fibrosis [19]. Studies have found the accuracy of FibroScan for cirrhosis approaches 90%, with a sensitivity and specificity of 70%–80% for fibrosis [20, 21]. Accuracy is reduced in obesity and in patients with ascites. The acoustic radiation force impulses (ARFI) is a type of ultrasound elastography, which uses acoustic radiation force to generate images and assess liver tissue stiffness, which has been shown to correlate with fibrosis [22]. It has the advantage in that it can choose the area to be assessed and can selectively avoid structures such as large vessels or ribs. Steatosis also does not undermine its accuracy. Contrast-enhanced sonography uses intravenous injection of microbubbles that facilitate sonographic imaging [23]. The time needed for the microbubbles to pass through the liver (hepatic vein transit time) correlates with the degree of fibrosis [24, 25]. Other noninvasive imaging techniques to measure fibrosis include two-dimensional shear-wave elastography (2D-SWE) [26] and real-time sonography-based elastography [5]. Magnetic resonance elastography provides another measurement of liver fibrosis [27]. Its value lies in its ability to scan the whole of the liver, and it is effective despite ascites or obesity. High cost, the lack of center expertise, and availability of this technology limit its routine clinical use.

A few studies have evaluated noninvasive testing for fibrosis using clinical and histological correlation in patients with WD. A study by Sini et al. assessed liver stiffness in 35 WD patients using the FibroScan, APRI score, and FIB-4 and compared these tests with liver histology [10]. They found the FibroScan to be useful in predicting significant fibrosis from milder stages and might be helpful in managing chronic therapy. The accuracy of APRI and FIB-4 to detect significant and severe fibrosis was good; however, cases of mild or no fibrosis could not be differentiated with these tests. Karlas et al. in a study of 50 WD patients conducted assessments using FibroScan; ARFI; and serological scores APRI, FIB-4, and the Forns score. [9]. ARFI was successful in all patients, while there was a failure rate of 18% using the FibroScan. The main contributing factor to invalid measurements was a high BMI. The FibroScan was able to demonstrate a gradual increase in score with progressing liver disease and discriminated cirrhosis. ARFI was less useful at distinguishing between advanced liver disease and cases without liver disease, right liver lobe ARFI being more useful than left liver lobe ARFI. The study was limited in that the grade of hepatic disease was classified according to a clinical scoring system rather than using histology. Future clinical research will aim to increase the accuracy of noninvasive methods of assessing fibrosis and their use in WD as values and cutoffs for the stages of fibrosis may differ depending on the etiology of underlying liver disease [28, 29].

Neurologic

More difficult to ascertain is the predictability of improvement in neurological or psychiatric signs and symptoms in WD with treatment. Therefore, instruments that help define neurological disease status needed development. The UWDRS was developed to standardize the clinical assessment of neurological involvement in patients with WD. The use of this rating scale was validated independently, and the scale provides a useful tool for clinical studies [30, 31]. The Global Assessment Scale for Wilson Disease (GAS for WD) is another rating scale proposed by Aggarwal et al. in 2009. It is shorter than the UWDRS with two tiers, the first scoring global disability (Tier 1) and the second scoring neurological dysfunction (Tier 2) [32]. Other rating scales utilized previously include the modified Huntington's scale [33] or the Medici score [34]; however, these were not independently validated, the latter being only retrospectively applied to chart data sets. Unfortunately, few prospective treatment trials have been performed for Wilson disease, and in these few trials, validated rating scales were not uniformly utilized. A recent phase 2 trial of choline tetrathiomolybdate used the UWDRS to document changes over time in individual patients with treatment [35]. A prospective, monocentric study, cross-sectional study of 65 patients conducted by Volpert et al. (2017), has compared two of the previously described assessment scales the UWDRS and GAS for WD. The UWDRS neurological scale was found to be a more comprehensive assessment tool than the GAS for WD scale Tier 2, containing a larger number of items, thus allowing more detailed documentation of neurological disability [36]. The study also found that the "minimal UWDRS" score has a strong correlation with the UWDRS total score, the total UWDRS neurological subscore, and the GAS for WD Tier 2 score [36]. Although when used in isolation the "minimal UWDRS" does not provide a comprehensive neurological assessment, it may offer a convenient and time-efficient screening tool for use in clinical practice. Comparisons of the scoring systems used for evaluating hepatic and neurologic status in Wilson disease are shown in Table 2.

TABLE 2 Scores for Evaluating Hepatic and Neurologic Status in WD

Organ Evaluated	Rating Scale or Score	Strengths	Limitations	Reference
Liver	Child-Pugh-Turcotte	– Useful for assessing prognosis in cirrhotics	– Not useful in early disease – Includes some subjective entries	[37]
	MELD/MELD Na	– Prognostic score in cirrhotics, only objective entries – Useful for prioritizing transplants	– Not useful in early disease – Not specific for Wilson	[38]
	New Wilson Index	– Validated in pediatric and adults with WD		[39]
Brain	UWDRS	– Validated scoring system specific to WD	– Not yet extensively used in cohort studies	[30, 31]
	Unified Huntington's Disease Rating Scale	– Assessment of components common in neuropsychiatric WD—motor, cognition, behavior, independence, functional – Validated research tool	– Developed for use in Huntington's disease as a research tool – Use in WD would need validating	[33]
	Global Assessment Scale for WD (GAS-WD)	Assessment of components common in neuropsychiatric WD—motor, cognition, behavior, independence, functional	– Not commonly used	[32]

MRI has been used in the assessment of brain involvement in patients with WD. The most common brain MRI findings are high T1 signal intensity in the globus pallidus, putamen, and mesencephalon associated with hepatic dysfunction and high T2 signal lesions in the striatum in patients with neurology [40–44]. The characteristic brain MRI signs for WD in T2-weighted images are the midbrain "face of the giant panda" sign that is described as high signal in the tegmentum with preservation of signal intensity of the lateral portion of the pars reticulata of the substantia nigra and red nuclei with hypointensity of the superior colliculus and the pons "miniature panda" with hypointensity of the aqueductal opening to fourth ventricle [45, 46]. Use of MRI brain for assessing treatment response in WD suffers from a lack of uniform assessment criteria and validation [47–50]. A study by Sinha et al. of 50 patients who were treated with penicillamine and zinc underwent serial MRI studies at 9–52 month intervals. The anatomic distribution of abnormalities was noted, and scoring was performed based on the change in signal intensity of focal lesions and associated atrophy. The grading system provided a score of 0–30, with zero being normal and 30 indicating severe changes. The assessment of T2-weighted signal intensity changes was also performed subjectively. The degree of atrophy was quantified by measuring standardized dimensions of the brain and also subjectively based on the sulcal and ventricular enlargement. Improvement in overall MRI scores correlated well with clinical treatment response, although serial changes were variable [46]. Similarly, a study by Roh et al. in 1994 evaluated 16 patients with MRI follow-up scans post treatment with penicillamine. The study demonstrated improvement in abnormal high signal intensity lesions that correlated clinically with a reduction in neurological symptoms [50]. An MRI study by Kim et al. [51] imaged children between age 3 and 16 at diagnosis and on follow-up. The mean age of patients was 10.2 years. Only 15 patients had follow-up imaging, so conclusions may again be limited by the small study size. Results demonstrated that high signal intensity lesions in the basal ganglia on T1-weighted images reflect hepatic involvement of Wilson disease and lesions on T2-weighted images reflect cerebral involvement and correlated well with neurological symptoms. Reduction in lesion size and intensity on follow-up imaging was closely correlated to improvement in clinical findings. However, not all studies have shown consistent results. A further study in India by Prashanth et al.

evaluated clinical, laboratory, and MRI features in 29 patients with severe neurological WD [52]. Twenty-one patients underwent MRI imaging during their first evaluation, and serial MRI scans were carried out in 15 patients. To score changes, the anatomical distribution of abnormalities was noted, and scoring was performed based on the degree of change in signal intensity of focal lesions and any associated atrophy. The structures assessed for grading included the caudate, putamen, internal capsule, thalamus, midbrain, pons, medulla, cerebellum, white matter, and cortical changes. The grading system provided a score of 0–30, with zero indicating a normal scan and 30 indicating severe changes. Following treatment, 14 patients (group A) had worsening symptoms, and 15 patients (group B) demonstrated improvement. Regression of lesions on serial MRI was only seen among patients with clinical improvement. However, MRI severity scores did not significantly differ between the two groups. Diffuse white matter abnormalities were more extensive in the group with worsening symptoms post treatment, which may be a poor prognostic indicator. More studies are required to evaluate the use of MRI in assessing treatment response and prognosis.

Another advance, functional neuroimaging, looks more specifically at brain metabolism. A study by Tarnacka et al. looked at MR spectroscopy (MRS) in monitoring the treatment of WD patients [53]. The study evaluated 17 newly diagnosed WD cases for a year. N-acetylaspartate (NAA) is a marker of neuron activity, which is decreased during neurodegeneration such as in WD. In the study, pretreatment MRS showed low-level NAA/Cr. There appeared to be a correction of brain metabolism with treatment, and MRS demonstrated an increase in NAA/Cr in patients showing improvement, while in patients that deteriorated a reduced NAA/Cr was seen. This suggests that functional imaging such as MRS could be useful for monitoring treatment in WD patients. Other functional neuroimaging techniques include single-photon emission computed tomography (SPECT), and techniques that use radioisotopes such as positron emission tomography (PET) to analyze neurotransmission are also being explored. Further studies to determine their use in WD are required.

MONITORING COPPER PARAMETERS ON TREATMENT

In these paragraphs, we would like to highlight the differences and similarities between efficacy for chelation therapy and zinc therapy. Tests such as 24h urinary copper excretion and serum free copper can help demonstrate effective copper metabolism and can also help detect deterioration or the lack of compliance ([1, 54]. During treatment with the chelators trientine and D-penicillamine, the urinary copper excretion increases to >1000 μg/24 h (>16 μmol/24 h) after starting treatment and during chronic treatment (after a year) to 300–500 μg/24 h (4.8–8 μmol/24 h). Urinary copper levels below 300 μg/24 h (4.8 μmol/24 h) suggest overtreatment or noncompliance [55]. D-Penicillamine compliance can also be assessed using measurement of urinary copper excretion before and 48h after stopping D-penicillamine. Normalization of urine copper excretion (when measured after 48h off of the drug) might indicate a "decoppered state" and has been suggested to serve as a measure of efficacy and compliance [55].

Urinary copper excretion can also be used to demonstrate treatment effect in patients taking zinc although the values used are different from chelators due to the different mechanism of action. In patients receiving long-term treatment with zinc, urinary copper excretion should be <100 μg/24 h (<1.6 μmol/24 h). Levels higher than 100 μg/24 h (<1.6 μmol/24 h) may occur in noncompliance or the lack of treatment effect [55].

Serum nonceruloplasmin-bound copper (NCC) levels can also help in assessing treatment effect and may guide management decisions for both chelators and zinc. The NCC (μg/dL) = total serum copper (μg/dL) − 3.12 × serum ceruloplasmin (mg/dL). The NCC is thought to represent "free" or toxic copper that is thought to contribute to neurological deterioration on chelating agents in WD by rapidly mobilizing large stores of copper [56]. In patients receiving adequate treatment, levels should be between 5 and 15 μg/dL. An NCC level < 5 μg/dL is seen in overtreatment and > 15 μg/dL in noncompliance. However, when using commercial assays for determination of ceruloplasmin, performed mostly by immunologic methods, the calculation of NCC may be zero or even negative in a subgroup of patients. In these individuals, this parameter is not useful for monitoring treatment, and the use of other measures discussed above can be considered. While not fully validated as a marker of adequate treatment response, it is useful in the majority of patients where the NCC can be followed [1]. More accurate methods to monitor efficacy by standard copper and copper protein analysis remain to be developed. Determination of exchangeable copper and its derived relative exchangeable copper [57] may be useful for response-guided dosage regimens in the future (see Chapter 22).

The frequency of required monitoring is variable, but at least twice a year is recommended, with closer monitoring being necessary during the initial phase of treatment and during phases of disease progression or when dose adjustments or medication changes are made due to side effects (Fig. 1). At these time points, more frequent monitoring with cautious dose adjustments are often required until the maintenance phase is again achieved.

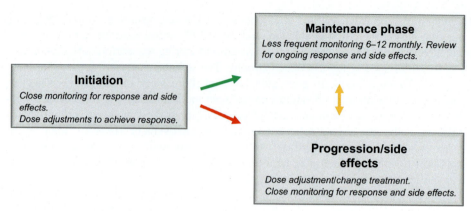

FIG. 1 Treatment monitoring during different phases of treatment.

WHAT CONSTITUTES TREATMENT FAILURE?

In the previous paragraphs, we have explored various ways of assessing disease progression through clinical and biochemical scores and imaging techniques some of which are readily available in clinical practice and some are being evaluated for use in future management. A combination of factors needs to be considered when evaluating for treatment failure. Treatment failure can be described clinically as deterioration or the lack of improvement in patients' symptoms, biochemically evidenced by deterioration or the lack of improvement of laboratory tests (liver tests, coagulation, and copper parameters) or due to intolerable side effects of medications or safety concerns. For example, in those patients with elevations in ALT, where values fail to improve to below two times the upper limit of normal despite an adequate treatment trial, this would be considered a potential treatment failure and cause for considering alternative therapy. In patients with cirrhosis, clinical worsening may present with persisting jaundice or decompensation with worsening symptoms of portal hypertension such as ascites, upper GI bleeding, hepatic encephalopathy, and hepatorenal syndrome. This may be captured by worsening in liver scores such as the MELD-Na score and Child-Pugh score predicting poor prognosis. These scores are not specific for WD and do not apply to patients without cirrhosis. The modified Nazar Score (New Wilson Index) is a tool developed for WD patients who have a significant hepatic disease burden and can help guide clinicians as to whether to continue medical treatment or alternatively, if liver transplant is necessary for survival. AASLD guidelines suggest a trial of treatment for 3 months in patients with a high modified Nazar Index Score, unless during this period, patients develop significant clinical or biochemical worsening [1].

Although the New Wilson Index can guide on treatment failure in severe hepatic disease, it is less useful at guiding treatment decisions in those with significant neurological and psychiatric disease. The UWDRS score has been developed to assess neurological disease severity in the Wilson disease population, and as previously discussed, imaging modalities for monitoring treatment response are being trialed. In addition to those whose primary presentation is with neuropsychiatric symptoms, paradoxical worsening of neuropsychiatric symptoms may also ensue following treatment initiation with chelators in up to 35% of patients [58–60]. In rare instances, patients can suffer grave disability. Transplant is currently not recommended purely for neuropsychiatric disease, and current therapies do not offer attractive alternatives. Adjusting the dose of chelators gradually reduces the probability of neurological worsening to <10% ([58, 59]. Currently, phase 2 trials in a new treatment bis-choline tetrathiomolybdate (WTX101) have demonstrated no cases of paradoxical drug-related neurological worsening and in the future may provide a promising alternative for those at risk [35].

As mentioned previously, scales such as the New Wilson Index, MELD score, and UWDRS can help guide response. Monitoring of laboratory tests is also vital for determining whether there has been biochemical improvement. In patients without significant clinical symptoms and without cirrhosis, these parameters may be the only indication of response to treatment. Clinicians are guided by monitoring of liver enzymes (AST, ALT, and γ-GT), complete blood count, coagulation, and copper parameters to assess for normalization and treatment response (reduction and stabilization of nonceruloplasmin copper and urine copper).

In a study by Weiss et al. [61], patients in tertiary care centers were retrospectively analyzed for adherence to therapy, survival, treatment failure, and adverse events on different Wilson disease treatments (chelators, zinc, or combination therapy). In this paper, hepatic treatment failure was defined as an increase in activity of two of three liver enzymes (aspartate aminotransferase (AST), alanine aminotransferase (ALT), and γ-glutamyltransferase (γ-GT)) to more than twofold the upper limit of normal or two times greater than baseline level at the time of discontinuation. This is needed to be associated with an increase in urinary copper excretion greater than the upper limit of normal, reflecting a positive copper balance, in three consecutive measurements taken at 3-month intervals at least. In patients receiving zinc therapy, urinary zinc levels had to be elevated by >35 μmol/day [61]. There is no clear evidence that elevated liver tests correlate with persistent histological inflammatory change in Wilson disease. However, in autoimmune hepatitis, a study by Luth et al. (2008) demonstrated that serum parameters were significantly associated with histological activity in 82 patients with autoimmune hepatitis. Serum taken at the time of biopsy was analyzed for ALT, AST, γ-GT, and immunoglobulin G (IgG) levels and compared with histology. All serum parameters were significantly associated with histological activity; in particular, ALT and IgG predicted high inflammatory activity with 99% sensitivity. Histological remission correlated with normalization of both serum parameters although about half of patients with normal values still demonstrated residual histological activity. The study concluded that normalized parameters identified patients at low risk of fibrosis progression [62].

Although the expectation is that liver tests will normalize on adequate treatment in Wilson disease patients, Iorio et al. performed a multicenter retrospective study to evaluate transaminase levels after penicillamine and zinc treatment [63] and found that a subset of patients treated with penicillamine (36%) and zinc (50%) had persistently high liver transaminases. The study suspected poor compliance in only 10% of cases. No causative factor for the persistent hypertransaminasemia was found. Interestingly, the persistently raised ALT level was not associated with worsening liver disease. ALT levels normalized within a median of 17 months (range 2–96 months) in patients treated with penicillamine, and in patients given zinc, ALT normalized within a median period of 6 months (range, 1–36 months). Due to the retrospective nature of the study, it is difficult to draw firm conclusions, especially as the data were not corrected for dosage of anticopper drugs. However, if we accept that a persistently low level (<twofold upper limit of lab normal) of raised ALT in the absence of other lab abnormalities represents a benign phenomenon, then extensive diagnostic work up and changes in therapy may not be required in this setting.

Treatment decisions are guided not only by the perceived efficacy of the drug—looking at clinical improvement and biochemical response—but also likewise by adverse reactions limiting the safety of the drug and impacting compliance. Similar data are required for treatment efficacy as for side effects and safety but with a different view. Compliance with treatment is key as stopping treatment can trigger significant disease progression. Thus, monitoring for adverse drug reactions limiting the compliance is crucial. D-Penicillamine is associated with side effects in 15%–30% of patients, which could lead to treatment discontinuation [54]. These include early hypersensitivity reactions, side effects related to nephrotoxicity, dermatological effects including skin changes and ulceration, and immune-mediated reactions. Therapy with trientine is associated with fewer side effects compared with D-penicillamine [1, 60]. No acute hypersensitivity reactions have been reported [1], and the most common side effects are sideroblastic anemia, lupus-like reactions, hemorrhagic gastritis, the loss of taste, and skin changes [55]. The most common side effect of treatment with zinc is dyspepsia due to gastric irritation with resulting erosions and ulcers [54, 64]. Other side effects that occur rarely are a biochemical pancreatitis resulting in elevated pancreatic enzymes (lipase/amylase) without clinical symptoms [54].

The severity of the side effect and impact on well-being and quality of life will dictate what changes to treatment are made. For example, mild dyspepsia with zinc therapy may be managed with antacids or proton pump inhibitors, while severe gastric irritation resulting in ulceration and bleeding would warrant a change in treatment.

During treatment for WD, symptomatic copper deficiency can result. The symptoms of the deficiency include hematological side effects such as anemia and neutropenia and neurological symptoms including sensory or motor-sensory neuropathies and myelopathies, which normally reverse with correction of the deficiency [65]. Monitoring for overtreatment is important on all treatments for Wilson disease (Table 3).

ALTERNATIVE REASONS FOR CHANGING THERAPY

Aside from treatment failure and side effects discussed above, there are two other main reasons for more frequent changes in a patient's treatment for their Wilson disease. These include economic reasons, such as cost of medication or nonavailability of a formulation in a specific region of the world. The price of chelation therapy, D-penicillamine, and trientine, in the United States and many other countries has skyrocketed and has caused many patients, who for years were maintained on these medications, to consider alternative treatments. Whether this was due to denial of insurance coverage or significant out-of-pocket expense that the patient could no longer afford, this represents a potential disruption in treatment [66]. Though some of the manufacturers have patient assistance programs to assist with copays, not all patients are eligible. In other countries where there are national formularies, only D-penicillamine is included on the national list, and availability of trientine may be limited and allowed only on a case-by-case basis

TABLE 3 Side Effects of Treatments in Wilson Disease

Drug	Mechanism	Side Effects
D-Penicillamine	Copper chelator, promotes urinary excretion	**Early** Hypersensitivity reactions, fever, neutropenia, lymphadenopathy, cutaneous erosions, thrombocytopenia, and proteinuria Risk neurological deterioration 10%–20% **Late (>3 weeks)** Nephrotoxicity: proteinuria, hematuria, and Goodpasture's syndrome Dermatological: elastosis perforans serpiginosa, progeric skin changes, pemphigoid, lichen planus, aphthous ulceration, and stomatitis Other: lupus-like syndromes, serous retinitis, myasthenia-like syndrome, polymyositis, the loss of taste, and aplastic anemia
Trientine	Copper chelator, promotes urinary excretion	Sideroblastic anemia Lupus-like reactions and skin changes Hemorrhagic gastritis Loss of taste Risk of neurological deterioration in 10%–15% in initial phase
Zinc	Metallothionein inducer. Reduces intestinal absorption of copper	Dyspepsia: including resulting erosions and ulcers Biochemical pancreatitis: Elevated lipase/amylase without clinical symptoms Symptomatic copper deficiency: Hematological—anemia and neutropenia. Neurological—sensory or motor-sensory neuropathies and myelopathies
Tetrathiomolybdate	Complexor of copper with protein, blocks absorption	Reversible anemia and elevation of LFTs. Dose-dependent

(see Chapter 17). In these instances, patients may not receive what their treating physician or other caregiver believes is the best therapy for them.

One other interesting reason for change is patient choice independent of cost of therapy. Some patients seek therapy based on what is perceived as "natural" or "organic," such as treatment with zinc since this is an essential trace element, while others fear developing side effects from D-penicillamine. Often, these patients will bargain with their providers to prescribe what the patient wishes to be treated with. It will then be up to the treating physician to provide their input as to the risks and benefits of these choices and monitoring needed to assess treatment efficacy.

EARLIEST AGE FOR INITIATING TREATMENT

One of the most difficult questions to answer is whether to initiate treatment for neonates and very young children under the age of 3–4 years. There is no large volume of data to support initiation of treatment in the asymptomatic young patient at this age or to follow most current recommendations to wait until the age of 3–5 years. This is likely to become an increasingly asked question given the ability to diagnose Wilson disease by genetic testing for ATP7B mutations even before there is any expectation of development of clinical signs or symptoms of disease. Indeed, in some instances, there is even a push for in utero diagnosis, especially now that there are ways to do this noninvasively by looking for fetal cells in the maternal circulation.

Arguments for a very early initiation of treatment are that one can prevent any development of hepatic copper toxicity by reducing the concentration of copper that is already elevated in newborn liver. Against early treatment is the concern for limiting systemic copper that will be needed for growth and development, in particular for neurological development. In the related disorder of copper transport, Menkes' disease, copper fails to be absorbed from the gut properly and also transport from cells to other tissues in the circulation is abnormal. Affected infants with Menkes' disease develop neurodegeneration and other clinical signs that to some degree may be ameliorated with copper administration. Though copper transport between cells in Wilson disease is normal, if copper itself is too low in the circulation at this critical age for neurological development, there is concern that neurological development may be inhibited.

In a report by Socio et al. [67] where they identified two asymptomatic patients out of their cohort of pediatric patients by family screening at ages 2 and 3 years, early treatment with D-penicillamine was recommended. However, data from both these patients showed a low nonceruloplasmin copper at baseline, and it is possible that the patient may have done just as well or better if treatment were delayed and dietary counseling given.

Therefore, if treatment is initiated in the very young, it is imperative that they not only be monitored for clinical signs of copper deficiency but also preferably be monitored biochemically as well so that the earliest signs of copper deficiency will be detected. Early signs for copper deficiency would include a sideroblastic anemia, neutropenia, or thrombocytopenia. In the extremely young (below 6 months), ceruloplasmin levels may not yet be at steady state, and therefore, the serum copper will be very low. Following the serum copper alone serially during this time would be potentially misleading as the expected rise in ceruloplasmin, albeit to less than normal values, would be accompanied by a rise in serum copper without the same connotation as if the copper rose in the nonceruloplasmin fraction (where it would be available for transport into extrahepatic sites where it could be toxic).

Another problem is if treatment is started in the very young, what therapy should be initiated and how should it be administered? There are advocates for using zinc salts that can be dissolved in water and given in bottles to infants. The appropriate dosage is not known, but can potentially be extrapolated from current recommendations for pediatric dosing of 25 mg three times daily. In one study by Abuduxikuer and Wang [68], the minimum dosage given to children that were 3 years of age was 30 mg of elemental zinc and zinc gluconate administered twice daily. In this report, dosages were adjusted based on liver tests, and most values remained normal or normalized within the first 12–24 months of therapy. However, of their larger cohort, there were 2/26 that remained with some abnormality of transaminase after this time period.

TREATMENT IN PREGNANCY

Treatment of WD during pregnancy is very important for the safety of the mother. However, because the risk to the fetus for any teratogenicity is highest in the first trimester of life, exposure to either D-penicillamine or trientine should be minimized by dose reduction as early as possible during the pregnancy or even in anticipation of the pregnancy. Additionally, the lower dose might prevent fetal copper deficiency. With dose reduction, typically by about 50% of the maintenance dosage, should come increased monitoring to assure safety. Therefore, while normal recommended intervals of monitoring treatment are about every 6 months for stable patients on maintenance therapy, in patients who are on lower dosages, there should be monitoring of liver tests every 3 months or even less. For patients on zinc therapy for their WD, there is no need to change course of care during treatment. If the patient or physician is concerned about keeping a patient on chelation therapy during pregnancy, then switching them to zinc ahead of pregnancy is an option for treatment. Alternatively, the current anticopper regimen might be maintained, as a recent series showed successful pregnancies under all available medical therapies [69].

REFERENCES

[1] Roberts E, Schilsky ML, American Association for Study of Liver Diseases (AASLD). Diagnosis and treatment of Wilson's disease an update. Hepatology 2008;47:2089–111.

[2] McGary CT, Raja RH, Weigel PH. Endocytosis of hyaluronic acid by rat liver endothelial cells. Evidence for receptor recycling. Biochem J 1989;257:875–84.

[3] Gibson PR, Fraser JR, et al. Hemodynamic and liver function predictors of serum hyaluronan in alcoholic liver disease. Hepatology 1992;15:1054–9.

[4] McHutchinson JG, Blatt LM, et al. Measurement of serum hyaluronic acid in patients with chronic hepatitis C and its relationship to liver histology. J Gastroenterol Hepatol 2000;15:945–51.

[5] Friedrich-Rust M, Ong MF, et al. Real-time elastography for non-invasive assessment of liver fibrosis in chronic viral hepatitis. AJR Am J Roentgenol 2007;188:758–64.

[6] Almpanis Z, Demonakou M, Tiniakos D. Evaluation of liver fibrosis: "something old, something new..." Ann Gastroenterol 2016;29:445–53.

[7] Loaeza-del-Castillo A, Paz-Pineda F, Oviedo-Cárdenas E, Sánchez-Avila F, Vargas-Voráčková F. AST to platelet ratio index (APRI) for the non-invasive evaluation of liver fibrosis. Ann Hepatol 2008;7:350–7.

[8] Poynard T, Aubert A, et al. A simple biological index for detection of alcoholic liver disease in drinkers. Gastroenterology 1991;100:1397–402.

[9] Karlas T, Hempel M, et al. Non-invasive evaluation of hepatic manifestation in Wilson disease with transient elastography, ARFI and different fibrosis scores. Scand J Gastroenterol 2012;47(11):1353–61.

[10] Sini M, Sorbello O, Civolani A, Liggi M, Demelia L. Non-invasive assessment of hepatic fibrosis in a series of patients with Wilson's disease. Dig Liver Dis 2012;44(6):487–91.

[11] Imbert-Bismut F, Ratziu V, et al. Biochemical markers of liver fibrosis in patients with hepatitis C virus infection: a prospective study. Lancet 2001;357:1069–75.

[12] Adams LA, Bulsara M, et al. Hepascore: an accurate validated predictor of liver fibrosis in chronic hepatitis C infection. Clin Chem 2005;51:1867–73.
[13] Patel K, Gordon SC, et al. Evaluation of a panel of non-invasive serum markers to differentiate mild from moderate to advanced liver fibrosis in chronic hepatitis C patients. J Hepatol 2004;41:935–42.
[14] Cales P, Oberti F, et al. A novel panel of blood markers to assess the degree of liver fibrosis. Hepatology 2005;42:1373–81.
[15] Forns X, Ampurdanès S. Identification of chronic hepatitis C patients without hepatic fibrosis by a simple predictive model. Hepatology 2002;36:986–92.
[16] Rosenberg WMC, Voelker M, et al. Serum markers detect the presence of liver fibrosis: a cohort study. Gastroenterology 2004;127:1704–13.
[17] Sebastiani G, Halfon P, et al. SAFE biopsy: a validated method for large-scale staging of liver fibrosis in chronic hepatitis C. Hepatology 2009;49:1821–7.
[18] Harrison SA, Oliver D, Arnold HL, Gogia S, Neuschwander-Tetri BA. Development and validation of a simple NAFLD clinical scoring system for identifying patients without advanced disease. Gut 2008;57:1441–7.
[19] Sandrin I, Fourquet B, et al. Transient elastography: a new noninvasive method for assessment of hepatic fibrosis. Ultrasound Med Biol 2003;29:1705–13.
[20] Friedrich-Rust M, Ong MF, et al. Performance of transient elastography for the staging of liver fibrosis: a meta-analysis. Gastroenterology 2008;134:960–74.
[21] Tsochatzis EA, Gurusamy KS, et al. Elastography for the diagnosis of severity of fibrosis in chronic liver disease: a meta-analysis of diagnostic accuracy. J Hepatol 2011;54:650–9.
[22] Lupsor M, Badea R, et al. Performance of a new elastographic method (ARFI technology) compared to unidimensional transient elastography in the non-invasive assessment of chronic hepatitis C. J Gastroinest Liver Dis 2009;18:303–10.
[23] Quaia E. Microbubble ultrasound contrast agents: an update. Eur Radiol 2007;17:1995–2008.
[24] Albrecht T, Blomley MJ, et al. Non-invasive diagnosis of hepatic cirrhosis by transit-time analysis of an ultrasound contrast agent. Lancet 1999;353:1579–83.
[25] Blomley MJ, Lim AK, et al. Liver microbubble transit time compared with histology and Child-Pugh score in diffuse liver disease: a cross sectional study. Gut 2003;52(8):1188–93.
[26] Castera L, Chan HL, et al. EASL-ALEH Clinical Practice Guidelines: non-invasive tests for evaluation of liver disease severity and prognosis. J Hepatol 2015;63:237–64.
[27] Huwart I, Sempoux C, et al. Magnetic resonance elastography for the noninvasive staging of liver fibrosis. Gastroenterology 2008;35:32–40.
[28] de Ledinghen V, Vergniol J. Transient elastography for the diagnosis of liver fibrosis. Expert Rev Med Devices 2010;7:811–23.
[29] Sebastiani G, Castera L, et al. The impact of liver disease aetiology and the stages of hepatic fibrosis on the performance of non-invasive fibrosis biomarkers: an international study of 2411 cases. Aliment Pharmacol Ther 2011;34:1202–16.
[30] Czlonkowska A, Tarnacka B, et al. Unified Wilson's disease rating scale—a proposal for the neurological scoring of Wilson's disease patients. Neurol Neurochir Pol 2007;41:1–12.
[31] Leinweber B, Möller JC, et al. Evaluation of the unified Wilson's disease rating scale (UWDRS) in German patients with treated Wilson's disease. Mov Disord 2008;23:54–62.
[32] Aggarwal A, Aggarwal N, Nagral A, Jankharia G, Bhatt M. A novel global assessment scale for Wilson's disease (GAS for WD). Mov Disord 2009;24:509–18.
[33] Huntington Study Group. The unified Huntington's disease rating scale: reliability and consistency. Mov Dis 1996;11:136–42.
[34] Medici V, Mirante VG, et al. Liver transplantation for Wilson's disease: the burden of neurological and psychiatric disorders. Liver Transpl 2005;11(9):1056–63.
[35] Weiss KH, Askari FK, et al. Bis-choline tetrathiomolybdate in patients with Wilson's disease: an open-label, multicentre, phase 2 study. Lancet 2017;1253(17):30293–5.
[36] Volpert HM, Pfeiffenberger J, et al. Comparative assessment of clinical rating scales in Wilson's disease. BMC Neurol 2017;17(1):140.
[37] Child CG, Turcotte JG. Surgery and portal hypertension. In: Child CG, editor. The liver and portal hypertension. Philadelphia: Saunders; 1964.
[38] Kamath PS, Kim WR, Advanced Liver Disease Study Group. The model of end-stage liver disease (MELD). Hepatology 2007;45(3):797–805.
[39] Dhawan A, Taylor RM, Cheeseman P, De Silva P, Katsiyiannakis L, Mieli-Vergani G. Wilson's disease in children: 37 year experience and revised King's score for liver transplantation. Liver Transplant 2005;11:441–8.
[40] Kozic D, Svetel M, Petrović B, Dragasević N, Semnic R, Kostić VS. MR imaging of the brain in patients with hepatic form of Wilson disease. Eur J Neurol 2003;10:587–92.
[41] Magalhaes AC, Caramelli P, et al. Wilson disease: MRI with clinical correlation. Neuroradiology 1994;36:97–100.
[42] Nazer H, Brismar J, et al. Magnetic resonance imaging of the brain in Wilson disease. Neuroradiology 1993;35:130–3.
[43] Saatci I, Topcu M, et al. Cranial MR findings in Wilson disease. Acta Radiol 1997;38:250–8.
[44] Thoumas KA, Aquilonius SM, Bergström K, Westermark K. Magnetic resonance imaging of the brain in Wilson disease. Neuroradiology 1993;35:134–41.
[45] Hitoshi S, Iwata M, Yoshikawa K. Mid-brain pathology of Wilson's disease: MRI analysis of three cases. J Neurol Neurosurg Psych 1991;54:624–6.
[46] Sinha S, Taly AB, et al. Sequential MRI changes in Wilson's disease with de-coppering therapy: a study of 50 patients. Br J Radiol 2007;80:744–9.
[47] Da Costa M, Spitz M, Bacheschi LA, Leite CC, Lucato LT, Barbosa ER. Wilson's disease: two treatment modalities. Correlations to pretreatment and post treatment brain MRI. Neuroradiology 2009;51:627–33.

[48] Engelbrecht V, Schlaug G, Hefter H, Kahn T, Mödder U. MRI of the brain in Wilson disease: T2 signal loss under therapy. J Comput Assist Tomogr 1995;19:635–8.
[49] Huang CC, Chu NS. Wilson disease resolution of MRI lesions following long-term oral zinc therapy. Acta Neurol Scand 1996;93:215–8.
[50] Roh JK, Lee TG, Wie BA, Lee SB, Park SH, Chang KH. Initial and follow-up brain MRI findings and correlation with the clinical course in Wilson disease. Neurology 1994;44:1064–8.
[51] Kim TJ, Kim IO, et al. MR imaging of the brain in Wilson disease of childhood: findings before and after treatment with clinical correlation. Am J Neuroradiol 2006;27:1373–8.
[52] Prashanth LK, Taly AB, et al. Prognostic factors in patients presenting with severe neurological forms of Wilson's disease. Q J Med 2005;98:557–63.
[53] Tarnacka B, Szeszkowski W, Golebiowski M, Czlonkowska A. MR spectroscopy in monitoring the treatment of Wilson's disease patients. Mov Disord 2008;23:1560–6.
[54] European Association for Study of Liver. EASL clinical practice guidelines: Wilson's disease. J Hepatol 2012;56:671–85.
[55] Litwin AC. Wilson disease—currently used anticopper therapy. In: Schilsky ML, Czlonkowska A, editors. Wilson disease. Amsterdam: Elsevier; 2017. p. 181–91.
[56] Brewer GJ, Askari F, et al. Treatment of Wilson disease with tetrathiomolybdate: V. Control of free copper by tetrathiomolybdate and a comparison with trientine. Transl Res 2009;154:70–7.
[57] El Balkhi S, Trocello JM. Relative exchangeable copper: a new highly sensitive and highly specific biomarker for Wilson disease diagnosis. Clin Chim Acta 2011;412:2254–60.
[58] Czlonkowska A, Litwin T, Karliński M, Dziezyc K, Chabik G, Czerska M. D-Penicillamine versus zinc sulfate as first line therapy in Wilson's disease. Eur J Neurol 2014;21:599–606.
[59] Litwin T, Dzieżyc K, Karliński M, Chabik G, Czepiel W, Członkowska A. Early neurological worsening in patients with Wilson's disease. J Neurol Sci 2015;355:162–7.
[60] Weiss KH, Thurik F, et al. Efficacy and safety of oral chelators in treatment of patients with Wilson disease. Clin Gastroenterol Hepatol 2013;11:1028–35.
[61] Weiss KH, Gotthardt DN, et al. Zinc monotherapy is not as effective as chelating agents in treatment of Wilson disease. Gastroenterology 2011;140:1189–98.
[62] Luth S, Herkel J, et al. Serologic markers compared with liver biopsy for monitoring disease activity in autoimmune hepatitis. J Clin Gastroenterol 2008;42(8):926–30.
[63] Iorio R, D'Ambrosi M, et al. Serum transaminases in children with Wilson's disease. J Pediatr Gastroenterol Nutr 2004;30:331–6.
[64] Wiernicka A, Jańczyk W, et al. Gastrointestinal side effects in children with Wilson's disease treated with zinc sulphate. World J Gastroenterol 2013;19:4356–62.
[65] Willis MS, Monaghan SA, et al. Zinc induced copper deficiency. Am J Clin Pathol 2005;123:125–31.
[66] Schilsky ML, Roberts EA, Hahn S, Askari F. Costly choices for treating Wilson's disease. Hepatology 2015;61(4):1106–8.
[67] de Andrade Sócio S, Ferreira AR, Fagundes EDT, Roquete MLV, Pimenta JR, de Faria Campos L, Penna FJ. Doença de Wilson em crianças e adolescentes: diagnóstico e tratamento (Wilson's disease in children and adolescents: diagnosis and treatment). Rev Paul Pediatr 2010;28(2):134–40.
[68] Abuduxikuer K, Wang JS. Zinc mono-therapy in pre-symptomatic Chinese children with Wilson disease: a single center, retrospective study. PLoS One 2014;9(3).
[69] Pfeiffenberger J, Beinhardt S, et al. Pregnancy in Wilson's disease: management and outcome. Hepatology 2018. https://doi.org/10.1002/hep.29490 [Epub ahead of print].

Index

Note: Page numbers followed by *f* indicate figures, and *t* indicate tables.

A

Acetylcysteine, 76–77
Acid sphingomyelinase (Asm), 217
Acoustic radiation force impulses (ARFI), 237
Acute fatal cirrhosis, 6
Acute liver failure (ALF), 100–101, 173–174
Alanine aminotransferase (ALT), 134, 240–241
Alkaline phosphatase (ALP), 136
Alzheimer's disease (AD), 47
Ambiguous Intentions Hostility Questionnaire, 163
American Association for the Study of Liver Diseases (AASLD), 130–132
Ammonium tetrathiomolybdate, 7*f*, 9
Animal model, 57–58*t*
   copper storage disorders
      dogs with copper-associated hepatitis, 56–57
      North Ronaldsay sheep, 57
   for Wilson disease, 51
      Atp7b$^{-/-}$ mouse, 55–56
      LPP$^{-/-}$ rat, 53
      tx-j mouse model, 54–55
      tx mouse, 53–54
Anticholinergics, 211
Anticopper therapy, 213
Antinuclear antibody (ANA), 228
Antioxidant 1 copper chaperone (ATOX1), 18–20, 19*f*, 26, 33, 105
Antioxidant therapy, 73–75*t*, 76–77
Antipsychotics, 162–163
Anxiety disorder, 159–162
Apoceruloplasmin (apoCp), 115
Astrocytes, 147
AST to platelet ratio index (APRI) score, 235–237
Ataxia, 149
ATOX1. *See* Antioxidant 1 copper chaperone (ATOX1)
ATP7A, 18–20, 33
ATP7B, 18–20, 66–67, 218, 242
   clinical molecular diagnosis, 109–110
   cloning, 10
   function, 10, 51
      biochemical function, 23–25
      biosynthetic and homeostatic functions, intracellular trafficking, 25
      in brain, 29
      in intestine, 28–29
      in kidneys, 29
      kinase-mediated phosphorylation, 27
      liver physiology, role in, 27–28, 33
      in mammary gland, 29
   genotype-phenotype correlation, 107–109
   molecular genetic studies, 103
   molecular structure, 105
   mutation analysis, 10, 134
   protein structure, 106
   regional gene frequency, 107
   variants
      actuator domain and ATPBD, mutations in, 45–46
      D1222V mutation, 45–46
      E1064A mutation, 46
      G626A mutation, 45
      G85V mutation, 45
      G1213V mutation, 45–46
      H1069Q mutant, 46, 106–107, 107*f*
      L1083F mutation, 46
      L492S mutation, 45
      molecular features, 34–35
      N41S mutation, 36–44
      N-terminal domain, mutations in, 44–45
      p.E1064A mutation, 106
      p.E1064K mutations, 107*f*
      p.Gly875Arg variant, 107
      p.G943S and p.M769V mutations, 106
      physical properties, 35–44
      p.R778L mutation, 106–107, 107*f*
      R1151H mutation, 46
      R616W mutation, 45
      SNPs, 47
      S653Y mutation, 47
      TM domain, mutations in, 46–47
      Trp779Stop mutation, 165
      variants of uncertain significance (VUS), 110
ATP-binding domain (ATPBD), 45–46
Auxiliary liver grafts, 178

B

Baclofen, 211
Barthel scale, 224
Basal ganglia, 4
N-Benzyl-D-glucamine dithiocarbamate (BGD), 76
Biochemical markers of copper metabolism
   ceruloplasmin, 115–117
   exchangeable copper and REC, 119–122
   liver biopsy with copper determination, 118–119
   serum copper and NCC, 117–118
   urinary copper excretion, 118
Bis-choline tetrathiomolybdate (WTX101), 240
   efficacy of, 199–200
   safety and tolerability, 200
Botulinum toxin (BTX), 148, 211
Brain stem evoked potentials (BSEP), 165
British anti-Lewisite (BAL), 6, 7*f*, 8

C

Cerebellar ataxia, 213
Ceruloplasmin (Cp), 17–18, 20, 115–117, 225, 243
   biochemical screening, 87
   diagnosis, 7
      diagnostic value, 125–129, 128–129*t*
      false-positive/false-negative serum levels, 125–129, 128*t*
      normal serum values, in various age groups, 125–129, 127*t*
   physiological function, 125
   serum concentration, determination of, 101
Chelation therapy, 52, 76, 151, 162, 239, 241–243. *See also* D-penicillamine (D-PA)
Child-adult transition, 229
Child-Pugh-Turcotte score, 235, 240
Chinese herbal medicines, 218
Choline tetrathiomolybdate, 237
Chorea, 150
Cirrhosis, 4–5
Cognitive deficits, 163
Cognitive impairment, 159, 163, 213
Compliance, 229
Confocal microscopy, 152
Coombs-negative hemolytic anemia, 133
Copper (Cu), 17
   deficiency, 66–67, 229
   exchangeable copper, 119–122, 226, 228
   homeostasis in hepatocyte, 33, 34*f*
   monitoring
      initial phase, 225–226
      maintenance phase, 228

248 Index

Copper (Cu) *(Continued)*
  normal human copper metabolism
    absorption, 18–20
    dietary copper, 18
    human health, role in, 17–18
    transport and excretion, 20
  toxicity, 6
    in animal models (*see* Animal model, for Wilson disease)
      Fenton-chemistry-based copper toxicity, 67, 77
      Haber-Weiss-based copper toxicity, 67, 77
    urinary copper excretion, 101–102, 118
Copper transporter 1 (CTR1), 18–20, 19f, 105
CuEXC. *See* Exchangeable copper (CuEXC)
Curcumin, 218
Cyclothymic disorder, 161–162

D
Decoppering therapy, 209
Deep brain stimulation (DBS), 211
Depression, 159
Diagnosis, of Wilson disease, 97
  clinical manifestations, 99–100, 99f
    acute hepatitis, 100–101
    Kayser-Fleischer rings, 6, 100
    in kidney, 100
    liver disease, 100
    in musculoskeletal system, 100
    neurological/neuropsychiatric disease, 100
  differential diagnosis, 97–98
  family screening, 103
  hepatic copper concentration estimation, 102
  Leipzig score, 97, 98t
  liver histology, 102
  molecular genetic studies, 103
  neuroradiology, 102–103
  nonceruloplasmin-bound copper, 101
  scoring system (*see* Scoring system, WD diagnosis)
  serum ceruloplasmin, 7, 101
  tissue copper, 7
  urinary copper excretion, 101–102
Dietary copper, 18
Direct Sanger sequencing, 109
Divalent metal transporter 1 (DMT1), 18, 19f
Dopamine β-hydroxylase (DBH), 17
D-penicillamine (D-PA), 6, 7f, 8, 76, 118
  efficacy, 184–185
  pharmacology
    drug-drug interactions, 183
    mode of action, 183
  side effects, 184
  trifunctional organic compound, 183
  use of, 184
Drooling, 150
Dysarthria, 150, 213
Dysphagia, 150
Dystonia, 210
  botulinum toxin, 211
  oral medications
    anticholinergics, 211
    baclofen, 211
  surgical treatment for, 211

E
Electroencephalogram (EEG), 165
Epidemiology, of Wilson disease
  clinical presentation
    age at presentation, 89
    diagnostic delay, 92
    gender differences, 92
  incidence and prevalence
    Hardy-Weinberg equilibrium, 85
    in high prevalence areas, 86, 89
    in larger areas, 86–87, 86t, 89
    population based genetic methods, 87–89
    relative to number of births, 87, 87t
    in Sardinia, 85–86, 89
  phenotype distribution, 90–91
  survival rate, 92
Epilepsy, 151
Esophagogastroduodenoscopy (EGD), 228
European Association for the Study of the Liver (EASL) guidelines, 125
Event-related potentials (ERPs), 165
Exchangeable copper (CuEXC), 119–122, 226, 228
Extraordinary drug price increases, 191–192
Extrapyramidal syndrome, 4

F
Facial Emotion Recognition Test, 163
Familial screening, 103, 119–121, 121f
Fenton-chemistry-based copper toxicity, 67, 77
FibroScan, 237
Fibrosis, 235–236, 236t
FIB-4 score, 237
Fulminant Wilson disease, 173–174

G
Gait and posture disturbances, 150
γ-glutamyltransferase (GGT), 240–241
G626A mutation, 45
Genetics, of Wilson disease, 165
  atypical inheritance patterns, 105
  autosomal dominant inheritance pattern, 105
  clinical molecular diagnosis, 109–110
  genotype-phenotype correlation, 107–109
  population screening, 110
  uniparental disomy, 105
  variants (*see* ATP7B, variants)
Gene transfer, 219
Genotype-phenotype correlation, 107–109, 134
Global Assessment Scale for Wilson Disease (GAS for WD), 145, 164, 237
Glutaredoxin 1 (Grx1), 25–26
Goodpasture's syndrome, 184, 226

H
Haber-Weiss-based copper toxicity, 67, 77
Hamilton depression scale, 164
Haplotype analysis, 103
Hardy-Weinberg equilibrium, 85
Hemolytic anemia, 97, 100–101
Hepatic copper determination, 118–119
Hepatic encephalopathy (HE), 147, 152–153
Hepatocyte/tissue transfer, 219
Hephaestin, 17–18
High-throughput next-generation sequencing, 110
Holoceruloplasmin (holoCp), 115
Hotspot detection, 134
H1069Q mutant, 46, 106–107, 107f
Huntington's scale, 237
Hydrophobicity index, 35–36, 36f, 37–43t, 44

I
Immunoglobulin G (IgG), 240–241
Immunosuppression, 179–180
Indian childhood cirrhosis, 144
Iron, 17–18

J
Jackson tx (tx-j) mouse model, 54–55

K
Kayser-Fleischer (K-F) rings, 5–6, 100, 103, 145, 152, 152f
  diagnosis, 133
Kinase-mediated phosphorylation, 27
Kyte-Doolittle hydropathy (KDH) index, 35–36, 36f, 37–43t

L
Leipzig scoring system, 97, 98t, 153
Linear block model, of ATP7B, 35–36, 36f
Lipoic acid, 76–77
Lithium, 166
Liver biopsy, 102, 118–119, 130–132, 132t
Liver disease, 5–6, 100
  and copper toxicity, in rodent models, 51
    Atp7b-/- mouse, 55–56
    biochemical impairments, 77
    LEC rat, 51–52
    LPP-/- rat, 53
    tx-j mouse model, 54–55
    tx mouse, 53–54
  hepatic encephalopathy, 152
Liver pathology
  acute liver failure, 139–140, 142–143
  ATPase protein, 139
  cellular alterations, 140
  chronic hepatitis, 139–140, 140f, 142–143
  cirrhosis, 142–143
  copper overload, 139, 144
  histochemical stain, 140
  histopathologic patterns, 139–140, 140f
  lipofuscin accumulations, 140
  Mallory-Denk bodies, 140, 143
  micro- and macrovesicular fat droplets, 140, 143f
  mitochondria, 143
  necrotic hepatocytes, 139
  nonhepatic tissues, 139
  orcein, 140
  paraffin-embedded specimens, 140
  rhodanine stain, 140, 142f

silver stain, 140
steatosis, 142
Liver transplantation, 9, 166
    acute liver failure (ALF), 173–174
    auxiliary liver grafts, 178
    immunosuppression, 179–180
    living-donor transplantation, 178
    neuropsychiatric manifestations, 176
    outcomes, 178
Living-donor liver transplantation, 178
Long-Evans Cinnamon (LEC) rat
    brain pathology and neurological presentation, 52
    clinical features, 51–52
    disadvantages, 52
    Hepatic mitochondrial changes, 52
    histological changes in liver, 52
    liver cancer, 52, 77
    metabolic changes, 52
    REC, 119
    therapeutic strategies, 52
        to ameliorate liver functions/support liver regeneration, 75–76t, 76–77
        antioxidant/radical scavenging therapies, 73–75t, 76–77
        hepatic copper load, reduction of, 72–73t, 76
Lupus syndrome, 226
Lysyl oxidase, 17–18

M

Magnetic resonance imaging (MRI), 145, 237–239
Medici score, 237
MELD-Na score, 235, 240
Menkes' disease, 10, 242
Metal-binding domains (MBDs), 25–26, 44–45, 106
Metallothionein (MT), 71, 77
Methanobactin (MB), 76, 218
Microarray method, 134
Mini-mental State Examination, 164
Modified Nazer score, 225, 235, 240
Molecular adsorbents recirculating system (MARS), 174
Molecular genetics, 9–10, 103
Molybdate, 198
Monitoring, WD
    initial phase
        biological monitoring (except cupric assessment), 225
        clinical follow up, 224–225
        copper monitoring, 225–226
        D-pen, 223
        free copper, 223
        frequency of follow-up, 224
        neurological manifestations, 223
    maintenance phase
        biological monitoring (except cupric assessment), 228
        clinical follow up, 227
        copper monitoring, 228
        frequency of follow-up, 227
        noninvasive assessment, 228
    red flags, 229
Mood disorder questionnaire (MDQ), 161–162, 164
Mood disorders, 4, 159–162
Movement disorder, 213–214
Movie for the Assessment of Social Cognition, 163
M645R mutation, 45
Multiplex ligation-dependent probe amplification (MLPA) test, 109
Mutation analysis, 10, 134

N

N-acetylaspartate (NAA), 239
NCC. See Nonceruloplasmin-bound copper (NCC)
Negative predictive value (NPV)
    liver copper, 132t
    serum ceruloplasmin, 125–129, 128–129t
    urinary copper, 131t
Neuroimaging, 165
Neuroleptic malignant syndrome, 166
Neurological Wilson disease
    classifications, 145, 146t
    features and frequency, 145, 146t
    pathophysiology of, 145–147
    supplementary tests, 145
    symptoms of, 145
        ataxia, 149
        chorea, 150
        drooling, 150
        dysarthria, 150
        dysphagia, 150
        dystonia, 148
        epilepsy, 151
        gait and posture disturbances, 150
        hepatic encephalopathy, 152–153
        neuropathies, 151
        parkinsonism, 149
        RLS, 151
        sleep disturbances, 151
        tremor, 147–148
Neuropsychiatric disease, 100
Neuropsychiatric inventory (NPI), 164
Neuroradiology, 102–103
Next-generation sequencing, 134
Niemann-Pick C (NPC1) protein, 27–28
Nonceruloplasmin-bound copper (NCC), 101, 117–118, 197, 225, 239
Normal human copper metabolism
    copper absorption, 18–20
    dietary copper, 18
    human health, copper role in, 17–18
    transport and excretion, 20
North Ronaldsay sheep, 57
NPV. See Negative predictive value (NPV)

O

Obsessive compulsive disorder (OCD), 162
Opalski cells, 147
Optical coherent tomography (OCT), 152
Oral chelator, 187, 191–193
Orphan drugs, 190
Oxidative stress, 67, 139–140
    redox-active copper, 190
    in WD patients and animal models, 71
        hepatitis and jaundice, stage of, 77
        markers, assessment in liver samples, 67, 68–71t, 71

P

Parkinsonism, 149, 209–210
D-Penicillamine challenge test (PCT), 130, 131t
Plant decapeptide OSIP108, 218
Plasmapheresis, 174–175
Population screening, 103, 110
Positive predictive value (PPV)
    liver copper, 132t
    serum ceruloplasmin, 125–129, 128–129t
    urinary copper, 131t
Positron emission tomography (PET), 153
Primidone, 212
Propranolol, 212
Protein biomarkers, 110
Pseudosclerosis, 5
Psychiatric symptoms, in WD
    apathy, 162
    belligerence, 162
    catatonia, 162
    cognitive deficits, 163
    cognitive impairment, 159
    diagnosis, psychiatric presentations
        copper and ceruloplasmin levels, 164
        EEG and evoked potential testing, 165
        neuroimaging, 165
        genetic studies, 165
    impaired social judgment/disinhibition, 162
    incongruent behavior, 162
    irritability, 162
    mood and anxiety disorders, 159–162
    neuropsychiatric variant, 159
    nonspecific psychiatric findings, 159
    OCD, 162
    prevalence of, 159–163, 160–161t
    psychotic disorders, 162
    sleep disturbances, 162
    treatment of, 166
Psychosis, 162
Psychotic disorders, 162
Psychotropics, 163

R

Radiocopper, 136
Relative exchangeable copper (REC), 119–122, 136
Response guided therapy, 217
Restless leg syndrome (RLS), 151

S

Sarcoplasmic-endoplasmic reticulum calcium (SERCA), 45
SASA. See Solvent-accessible surface areas (SASA)

Scoring system, WD diagnosis, 126t
　ceruloplasmin
　　copper transport mechanism, in human body, 125, 127f
　　diagnostic value, 125–129, 128–129t
　　false-positive/false-negative serum levels, 125–129, 128t
　　normal serum values, in various age groups, 125–129, 127t
　　physiological function of, 125
　Coombs-negative hemolysis, 133
　EASL guidelines, 125
　eye signs, 133
　genetic ATP7B mutation analysis, 134
　liver copper, 130–133
　neurological symptoms, 133–134
　proposed parameters
　　alkaline phosphatase, 136
　　ALP-to-TB ratio, 135–136
　　AST to ALT ratio, 134
　　REC and radiocopper, 136
　psychological symptoms, 134
　serum copper, 129
　urine collection
　　clinical conditions, 129–130, 130t
　　diagnostic value, 130, 131t
　　24h urinary copper measurements, 129–130
Serum copper determination, 117
Single nucleotide polymorphisms (SNPs), 47, 109
Single-photon emission computed tomography (SPECT), 153
Sleep disturbances, 151
Slit-lamp examination, 100
Social cognition, 163
Solvent-accessible surface areas (SASA), 35–36, 36f, 37–43t, 44
Somatosensory evoked potentials (SSEPs), 165
Structured Clinical Interview for DSM (SCID), 159, 163–164
Sunflower cataract (SC), 6, 152
Symptomatic treatment
　neurological manifestations
　　cerebellar ataxia, 213
　　dystonia, 210
　　parkinsonism, 209–210
　　tremor, 211–212
　psychiatric manifestations
　　cognitive impairment, 213
　　psychiatric manifestations, 213–214
Syprine, 191

T

Tetanoid chorea, 5
Tetrathiomolybdate (TTM), 76, 217
　bis-choline TTM (WTX-101), 198
　development and early clinical studies, 197–198
　mode of action, 199
　pharmacokinetics, 198
Therapeutic plasma exchange (TPE), 218
Toxic milk mouse (tx mouse), 53–54
Transcranial sonography (TCS), 153
Transient elastography, 237
Treatment, of Wilson disease, 7f
　ammonium tetrathiomolybdate, 9
　BAL, 6, 7f, 8
　changing therapy, alternative reasons for, 241–242
　Chinese herbal medicines, 218
　copper parameters on, 239
　curcumin, 218
　D-penicillamine (see D-penicillamine (D-PA))
　failure, 240–241
　gene transfer, 219
　hepatic, 235–237
　hepatocyte/tissue transfer, 219
　liver transplantation (see Liver transplantation)
　methanobactin, 218
　mutation specific therapy, 218–219
　negative copper balance, 235
　neurologic, 237–239
　plant decapeptide OSIP108, 218
　in pregnancy, 243
　response guided therapy, 217
　symptomatic treatment (see Symptomatic treatment)
　therapeutic plasma exchange, 218
　treatment initiation, earliest age for, 242–243
　trien (see Trientine (triethylenetetramine))
　zinc (see Zinc therapy)
Tremor, 147–148, 211–212
　primidone, 212
　propranolol, 212
Trientine (triethylenetetramine), 7f, 8, 205, 225
　adverse effects, 190
　chemistry, 187–188
　copper, 187–190
　emerging applications, 190
　pharmacology, 188–189
　social issues, 190–192
　therapeutics, 189–190
Tripartite complex, 198–200
TTM. See Tetrathiomolybdate (TTM)
Tyrosinase, 17–18

U

Unified Wilson's Disease Rating Scale (UWDRS), 145, 150, 165, 224, 237, 240
Urinary copper (UCu) excretion, 101–102, 118

V

Visual evoked potentials (VEP), 165
Vitamin E, 76–77

W

Wilson disease (WD)
　animal models (see Animal model, for Wilson disease)
　biochemical markers
　　ceruloplasmin, 115–117
　　exchangeable copper and REC, 119–122
　　liver biopsy with copper determination, 118–119
　　serum copper and NCC, 117–118
　　urinary copper excretion, 118
　definition of, 3–5
　developments, 3
　diagnosis (see Diagnosis, of Wilson disease)
　epidemiology (see Epidemiology, of Wilson disease)
　liver pathology (see Liver pathology)
　monitoring (see Monitoring, WD)
　neurologic symptoms (see Neurological Wilson disease)
　pseudosclerosis, 5
　psychiatric symptoms (see Psychiatric symptoms, in WD)
　treatment (see Treatment, of Wilson disease)
Wilson disease-acute liver failure (WD-ALF), 134, 135t, 136
Wilson Disease Mutation Database, 10, 47
Wilson-Jungner classic screening criteria, 110
Wilson, Kinnier, 3–5

Z

Zinc deficiency, 136
Zinc therapy, 7f, 8–9
　ATP7B mutations, 206
　chelators, 203–205
　combination therapy, 205
　copper intoxication, 203
　dosage, 203
　efficacy, 203–205
　Kayser-Fleischer rings, 203
　mechanism, 203
　overtreatment, 205
　pregnancy, 205
　side effects, 205

Printed in the United States
By Bookmasters